T0186830

HUMAN-ROBOT INTERACTIONS IN FUTURE MILITARY OPERATIONS

Human Factors in Defence

Series Editors:

Dr Don Harris, Managing Director of HFI Solutions Ltd, UK
Professor Neville Stanton, Chair in Human Factors of Transport at the
University of Southampton, UK
Dr Eduardo Salas, University of Central Florida, USA

Human factors is key to enabling today's armed forces to implement their vision to 'produce battle-winning people and equipment that are fit for the challenge of today, ready for the tasks of tomorrow and capable of building for the future' (source: UK MoD). Modern armed forces fulfil a wider variety of roles than ever before. In addition to defending sovereign territory and prosecuting armed conflicts, military personnel are engaged in homeland defence and in undertaking peacekeeping operations and delivering humanitarian aid right across the world. This requires top class personnel, trained to the highest standards in the use of first class equipment. The military has long recognised that good human factors is essential if these aims are to be achieved.

The defence sector is far and away the largest employer of human factors personnel across the globe and is the largest funder of basic and applied research. Much of this research is applicable to a wide audience, not just the military; this series aims to give readers access to some of this high quality work.

Ashgate's *Human Factors in Defence* series comprises of specially commissioned books from internationally recognised experts in the field. They provide in-depth, authoritative accounts of key human factors issues being addressed by the defence industry across the world.

HUMAN-ROBOT INTERACTIONS IN FUTURE MILITARY OPERATIONS

Human Factors in Defence

Series Editors:

Dr Don Harris, Managing Director of HFI Solutions Ltd, UK
Professor Neville Stanton, Chair in Human Factors of Transport at the
University of Southampton, UK
Dr Eduardo Salas, University of Central Florida, USA

Human factors is key to enabling today's armed forces to implement their vision to 'produce battle-winning people and equipment that are fit for the challenge of today, ready for the tasks of tomorrow and capable of building for the future' (source: UK MoD). Modern armed forces fulfil a wider variety of roles than ever before. In addition to defending sovereign territory and prosecuting armed conflicts, military personnel are engaged in homeland defence and in undertaking peacekeeping operations and delivering humanitarian aid right across the world. This requires top class personnel, trained to the highest standards in the use of first class equipment. The military has long recognised that good human factors is essential if these aims are to be achieved.

The defence sector is far and away the largest employer of human factors personnel across the globe and is the largest funder of basic and applied research. Much of this research is applicable to a wide audience, not just the military; this series aims to give readers access to some of this high quality work.

Ashgate's *Human Factors in Defence* series comprises of specially commissioned books from internationally recognised experts in the field. They provide in-depth, authoritative accounts of key human factors issues being addressed by the defence industry across the world.

Human-Robot Interactions in Future Military Operations

EDITED BY

MICHAEL BARNES
Army Research Laboratory (ARL-HRED), USA

&

FLORIAN JENTSCH
University of Central Florida, USA

ASHGATE

Published by
Ashgate Publishing Limited
Wey Court East
Union Road
Farnham
Surrey, GU9 7PT
England

Ashgate Publishing Company
Suite 420
101 Cherry Street
Burlington
VT 05401-4405
USA

www.ashgate.com

None of the opinions stated in this book represent official policy or endorsement by the United States Department of Defense and are solely the responsibility of the author.

British Library Cataloguing in Publication Data
Human-robot interactions in future military operations. --
  (Human factors in defence)
  1. Military robots. 2. Robotics--Military applications.
  3. Robotics--Human factors. 4. Psychology, Military.
  5. Human-machine systems--Psychological aspects.
  I. Series II. Barnes, Michael. III. Jentsch, Florian.
  355.4-dc22

  ISBN: 978-0-7546-7539-6 (hbk)
  ISBN: 978-0-7546-9801-2 (ebk)

Library of Congress Cataloging-in-Publication Data
Human-robot interactions in future military operations / edited by Michael Barnes and Florian Jentsch.
   p. cm. -- (Human factors in defence)
Includes bibliographical references and index.
ISBN 978-0-7546-7539-6 (hbk) -- ISBN 978-0-7546-9801-2
(ebook) 1. Military robots--United States. 2. Robotics--Human factors. 3. Human-computer interaction. 4. Robotics--Social aspects--United States. 5. Robotics--United States--Psychological aspects. I. Barnes, Michael (Michael Joseph), 1945- II. Jentsch, Florian.
UG479.H86 2009
355.4--dc22

                                                                    2009025444

Printed and bound in Great Britain by
MPG Books Group, UK

Contents

PART III UAV RESEARCH

PART IV UGV RESEARCH

PART V CROSS-PLATFORM RESEARCH

PART VI   FUTURE DIRECTIONS

List of Figures

List of Tables

About the Editors

Michael Barnes is a Research Psychologist with the U.S. Army Research Laboratory (ARL). For the past five years, he was the lead for the human robotic interaction (HRI) program that included both military and university research as part of an Army Technology Objective: Robotic Collaboration. His past experience includes tenure as a researcher with the U.S. Navy and as a human factors manager with the General Electric Corporation. Also, he has served on a number of international committees related to HRI. He has authored or co-authored over 70 articles on the human element of military systems. Located at Ft. Huachuca, AZ, his research interests include investigations of risk visualization, military intelligence processes, and unmanned aerial vehicles crew systems.

Florian Jentsch, Ph.D., is an Associate Professor at the University of Central Florida, Orlando, with joint appointments in the Department of Psychology and the Institute for Simulation & Training. He is also the Director of the Team Performance Laboratory. He received his Ph.D. in Human Factors Psychology in 1997, and he holds master's degrees in aeronautical science and aeronautical engineering. Dr. Jentsch is associate editor for *Human Factors and for Cognitive Technology*, and consulting editor for the *International Journal of Applied Aviation Studies*. His research interests are in team performance and training, pilot training and performance, human-robot interaction, and simulation methodology. Dr. Jentsch has co-authored over 200 publications and presentations; this is his second edited book.

List of Contributors

Laurel Allender
U.S. Army Research Laboratory

Michael J. Barnes
U.S. Army Research Laboratory

Jennifer L. Burke
University of South Florida

Gloria Calhoun
U.S. Air Force Research Laboratory

Roger A. Chadwick
New Mexico State University

Sheryl L. Chappell
*Systems Applications and
Technologies, Inc.*

Jessie Y. C. Chen
U.S. Army Research Laboratory

Erik S. Connors
*Systems Applications and
Technologies, Inc.*

Nancy J. Cooke
Arizona State University

Keryl Cosenzo
U.S. Army Research Laboratory

Ewart de Visser
George Mason University

Mark Draper
U.S. Air Force Research Laboratory

Linda R. Elliott
U.S. Army Research Laboratory

Mica R. Endsley
*Systems Applications and
Technologies, Inc.*

A. William Evans III
University of Central Florida

Douglas J. Gillan
North Carolina State University

Michael A. Goodrich
Brigham Young University

Ellen C. Haas
U.S. Army Research Laboratory

Chris Jansen
*Nederlandse Organisatie voor
Toegepast Natuurwetenschappelijk
Onderzoek (TNO)*

Florian G. Jentsch
*University of Central Florida
Team Performance Laboratory*

Brian Levinthal
Alion Science and Technology

Michael Lewis
University of Pittsburgh

Patricia McDermott
Alion Science and Technology

Claudia Meitinger
Universität der Bundeswehr München

Yaniv Minkov
Ben-Gurion University
Department of Industrial Engineering
and Management

Diane Kuhl Mitchell
U.S. Army Research Laboratory

Robin R. Murphy
University of South Florida

Tal Oron-Gilad
Ben-Gurion University
Department of Industrial Engineering
and Management

Scott Ososky
University of Central Florida

Raja Parasuraman
George Mason University

Skye L. Pazuchanics
New Mexico State University

Elizabeth S. Redden
U.S. Army Research Laboratory

Stephen Rice
Alion Science and Technology

Jennifer M. Riley
Systems Applications and
Technologies, Inc.

Charneta Samms
U.S. Army Research Laboratory

Merrill V. Sapp
New Mexico State University

Axel Schulte
Universität der Bundeswehr München

Laura D. Strater
Systems Applications and
Technologies, Inc.

Lori Foster Thompson
North Carolina State University

Jan B. F. van Erp
Nederlandse Organisatie voor
Toegepast Natuurwetenschappelijk
Onderzoek (TNO)

Jijun Wang
Quantum Leap Innovations, Inc.

Christopher D. Wickens
Alion Science and Technology

Endorsements

This book is remarkable because it covers a variety of topics including fundamentals of human-robot interaction and research in ground and aerial platforms as well as cross platform. It provides a deep understanding of cognitive and social perspectives of human-robot teaming as well as strengths and limits of the two components: human and machine. Reviews on adaptive automation, telepresence, and multimodal interface technologies offer solutions to many issues in human-robot interaction: improving situation awareness and human-robot ratio as well as reducing soldier workload. Experience gained and lessons learned from both simulation and field studies provide insights of interaction issues associated with uninhabited systems: teaming, coordination, communication, asset management, and system performance. This volume is a must read for all researchers and developers concerned with human-robot teaming in military domain.

Ming Hou, Defence Research & Development, Canada

Surprisingly, the more robots can do on their own, the more important becomes the ability to coordinate those activities with other people and groups. Barnes and Jentsch have assembled a comprehensive treatment of the latest data, concepts and interfaces needed to coordinate people and robots in demanding dynamic and uncertain situations.

David Woods, Ohio State University, USA

PART I
Introduction to HRI

Chapter 1

An Introduction to Human-Robot Interaction in Military Applications

A. William Evans III
University of Central Florida

> I'm completely operational and all my circuits are functioning normally.
>
> Hal in *2001: A Space Odyssey* (Kubrick & Clarke, 1968)

Already in the 1960s, when *2001: A Space Odyssey* was filmed, people imagined a world in which robots were able to seamlessly interact with humans. Since then, the dreams of young science fiction enthusiasts have grown into the realities of scientists, leading the robotics charge that we see today. While we have not quite achieved the level of interaction (or encountered the perils) imagined in Stanley Kubrick and Arthur C. Clarke's prediction of the year 2001, we have seen tremendous advances in the areas of robotics, artificial intelligence, and the interactions between humans and machines. Within that interaction between man and machine lies a bounty of questions, issues, and possibilities waiting to be explored.

Human-robot interaction (HRI) is an ever-growing research field with connections to both military and civilian applications. While there is a considerable overlap between military and civilian applications of unmanned vehicles, this volume focuses chiefly on military applications. Previous military uses of unmanned systems included, primarily, teleoperations of unmanned ground vehicles (UGVs) in search and rescue tasks and in ordnance disposal, as well as the use of unmanned aerial vehicles (UAVs) for surveillance and, more recently, strike operations. Advances in robot technology, however, will soon allow robots to perform advanced reconnaissance tasks, logistics supply, and battlefield casualty evacuation, among others. Hopes are that someday robotics will be able to supplement human counterparts in all military tasks, including a firefight. However, all of these advances, and others not yet conceived, are dependent upon the success of researchers within the HRI field to conduct sound and meaningful studies that help the HRI community learn more about the capabilities, uses, and potential misuses of such technology. This is the motivation for the current book. Included within this volume are examples of fundamental and domain-specific results and knowledge generated from the HRI community.

The chapters in this volume address such issues as basic HRI architecture, cognitive and social influences, and fundamental perceptual issues that are relevant to HRI. Beyond this, chapter contributors have addressed specific issues relating to UGVs, UAVs, and unmanned vehicle operator control units (OCUs). The overarching issues that concern all robotic systems in the military are addressed as well. In conjunction with these topics, several field experiments are outlined in detail, whereas other chapters discuss the theories relating to the future of HRI. In this brief foreword, a summary of the contributions is provided so as to give the reader an "advanced organizer" of what can be found in the remainder of the book.

Parts I and II: Introduction to and Foundations of HRI

This volume begins with an introduction to HRI, starting with an overview of current military research presented by one of the editors of this volume, Michael Barnes. Barnes' chapter is followed by a brief synopsis of current civilian applications of ground robotic assets, covering the lab and field research of Murphy and Burke. These authors have put particular emphasis on the human-robot ratio, with a specific focus on safety. Following this, a more detailed report of foundational research is presented, which starts with Gillan, Riley, and McDermott's description of the cognitive psychology involved with HRI. Specifically, Gillan and his co-authors have taken a look at which cognitive processes are critical in the control and use of robotic assets. The social factors affecting HRI are then discussed in Thompson and Gillan's work. Social influences such as acceptance and trust have been addressed within their work. Pazuchanics and Chadwick have continued the foundational theme by exploring the role of spatial ability and perception involved in human-robot teaming. One of the main foci of their chapter is the role of target recognition in HRI performance.

Expanding on some of the basic psychological foundations used in the previously mentioned chapters, Cosenzo, Parasuraman, and de Visser have discussed strategies for utilizing adaptive automation as a tool to increase performance in human-robot teams, by utilizing a dynamic "division of labor" between human and robotic teammates. Following this is Mitchell and Samms' chapter, which has taken a deeper look into soldier workload while utilizing robotic assets. In doing so, the authors have made use of the Army Research Laboratory's Improved Performance Research Integration Tool (IMPRINT).

Part III: UAV Issues and Research

The second section of the book deals with issues that are more closely related to the use of UAVs. UAVs were some of the first unmanned vehicles and thus are, in many instances, further along the developmental path than many of their ground-

and sea-based brethren. As such, UAV-specific research tends to provide a view of foundational issues that ultimately affect all unmanned vehicles.

The section begins with Schulte and Meitinger's chapter on the use of cognitive automation techniques and their potential uses in flight guidance systems for unmanned aircraft. The challenges of maintaining situational awareness while operating UAVs is discussed in detail by Riley, Strater, Chappell, Connors, and Endsley. Wickens, Levinthal, and Rice have investigated the issues of reliability or, more importantly, imperfect reliability, on workload and performance in supervisory control of UAVs. Continuing with the theme of understanding UAVs as teammates are Oron-Gilad and Minkov, who have taken a look at UAV usage from a bottom-up perspective. This section of the book is rounded out by a description of a research study, which was conducted by Calhoun and Draper, aimed at creating a better video display unit to help increase UAV operator's performance.

Part IV: UGV Issues and Research

Jansen and van Erp's chapter begins the third section, which is dedicated to UGV issues. While many global issues affecting UGVs have been foreshadowed in research with UAVs, there are issues specific to the more complex navigation and control of UGVs. One construct, for example, that is more critical for UGVs than UAVs is that of operator spatial and situational awareness. In this vein, Jansen and van Erp have presented the idea of telepresence and its usefulness for enhancing the situational awareness of human supervisors. The authors have explored the use of telepresence to gain some perspective on the needs of supervisors in human-robot teams when stepping back into the control loop, as operators. Following this, Haas and van Erp have discussed the use of multimodal displays to help create more efficient OCUs. These two chapters are followed by three research chapters, describing empirical studies in the area. Each of these chapters, like the first two in this section, is centered on the goal of improving operator awareness and control while utilizing UGVs. This begins with Chen's examination of operator workload in single-task and multi-task environments. Allender has followed up with a field investigation of the effects of temporal latency on the performance of robotic operators, and ways to combat those effects. Returning to the issue of display layouts, Redden and Elliott have presented their research on the scaling of OCUs specifically to be used by dismounted soldiers.

Part V: Cross-Platform Research

Research that attempts to span the gap between various unmanned vehicle platforms can become quickly very complex, simply due to the nature of comparing two dissimilar assets. Yet, it is what we can learn from one platform and apply to another that is of particular value to developers and researchers alike.

In an attempt to cross the bridge between UAVs and UGVs, Cooke and Chadwick have discussed lessons learned from research in different areas of HRI. Various issues such as piloting a plane versus piloting a UAV, and the limitations of UGV visual perception are addressed. Similarly, Goodrich has contributed an interesting chapter on the potential of a theoretical model for use in control of assets in human-robot teams, called a fan-out model. Fan-out refers to the number of assets an operator or operators can control at one time, and could have application in all manner of unmanned systems. Finally, Lewis and Wang have taken on the difficult task of creating metrics with which predictions can be made about task difficulty and coordination demand, as well as providing an algorithm to help in the control of large numbers of similar and dissimilar robotic assets.

Part VI: Summary and Future Directions

The chapters in this book have been assembled with the intention of providing an insight into current military HRI-related research. Based on this research, and coupled with the theories used by the chapter authors, readers are provided with an indication as to where unmanned systems use is headed in the military and the role that research will play in creating a more efficient and cohesive team union between man and machine. Evans and Jentsch have provided a summary of the research being done, and discuss how each of these individual projects is one piece of a much larger whole. In addition, examples of Jentsch, Evans, and Ososky's own research have been provided in Part V, discussing how past and concurrent research play a role in study design, theory, training, analysis, and conclusions. All in all, it is the hope of the editors that this book should provide a deep and diverse look into the world of HRI and unmanned systems in the military that can both inspire and guide future research within the field.

References

Allender, L. (2010). A Cognitive Systems Engineering Approach for Human-Robot Interaction: Lessons for an Examination of Temporal Latency. In M. J. Barnes & F. G. Jentsch (eds), *Human-Robot Interactions in Future Military Operations* (pp. 315–334). Farnham: Ashgate Publishing.

Barnes, M. J. (2010). Soldier-Robot Teams in Future Battlefields: An Overview. In M. J. Barnes & F. G. Jentsch (eds), *Human-Robot Interactions in Future Military Operations* (pp. 9–30). Farnham: Ashgate Publishing.

Calhoun, G. L. & Draper, M. H. (2010). Unmanned Aerial Vehicles: Enhancing Video Display Utility with Synthetic Vision Technology. In M. J. Barnes & F. G. Jentsch (eds), *Human-Robot Interactions in Future Military Operations* (pp. 229–248). Farnham: Ashgate Publishing.

Chen, J. Y. C. (2010). Robotics Operator Performance in a Multi-Tasking Environment. In M. J. Barnes & F. G. Jentsch (eds), *Human-Robot Interactions in Future Military Operations* (pp. 293–314). Farnham: Ashgate Publishing.

Cooke, N. J. & Chadwick, R. (2010). Lessons Learned from Human-Robotic Interactions on the Ground and in the Air. In M. J. Barnes & F. G. Jentsch (eds), *Human-Robot Interactions in Future Military Operations* (pp. 355–374). Farnham: Ashgate Publishing.

Cosenzo, K., Parasuraman, R., & de Visser, E. (2010). Automation Strategies for Facilitating Human Interaction with Military Unmanned Vehicles. In M. J. Barnes & F. G. Jentsch (eds), *Human-Robot Interactions in Future Military Operations* (pp. 103–124). Farnham: Ashgate Publishing.

Evans, A. W., III (2010). An Introduction to Human-Robot Interaction in Military Applications. In M. J. Barnes & F. G. Jentsch (eds), *Human-Robot Interactions in Future Military Operations* (pp. 3–8). Farnham: Ashgate Publishing.

Evans, A. W., III & Jentsch, F. G. (2010). The Future of HRI: Alternate Research Trajectories and Their Influence on the Future of Unmanned Systems . In M. J. Barnes & F. G. Jentsch (eds), *Human-Robot Interactions in Future Military Operations* (pp. 435–442). Farnham: Ashgate Publishing.

Gillan, D., Riley, J., & McDermott, P. (2010). The Cognitive Psychology of Human-Robot Interaction. In M. J. Barnes & F. G. Jentsch (eds), *Human-Robot Interactions in Future Military Operations* (pp. 53–66). Farnham: Ashgate Publishing.

Goodrich, M. (2010). On Maximizing Fan-Out: Towards Controlling Multiple Unmanned Vehicles. In M. J. Barnes & F. G. Jentsch (eds), *Human-Robot Interactions in Future Military Operations* (pp. 375–396). Farnham: Ashgate Publishing.

Haas, E. & van Erp, J. (2010). Multimodal Research for Human-Robot Interactions. In M. J. Barnes & F. G. Jentsch (eds), *Human-Robot Interactions in Future Military Operations* (pp. 271–292). Farnham: Ashgate Publishing.

Jansen, C. & van Erp, J. (2010). Telepresence Control of Unmanned Systems. In M. J. Barnes & F. G. Jentsch (eds), *Human-Robot Interactions in Future Military Operations* (pp. 251–270). Farnham: Ashgate Publishing.

Jentsch, F. G., Evans, A. W., III, & Ososky, S. (2010). Model World: Military HRI Research Conducted Using a Scale MOUT Facility. In M. J. Barnes & F. G. Jentsch (eds), *Human-Robot Interactions in Future Military Operations* (pp. 419–432). Farnham: Ashgate Publishing.

Kubrick, S. (writer/director) & Clarke, A. C. (writer) (1968). *2001: A Space Odyssey* [motion picture]. United States: MGM Studios.

Lewis, M. & Wang, J. (2010). Coordination and Automation for Controlling Robot Teams. In M. J. Barnes & F. G. Jentsch (eds), *Human-Robot Interactions in Future Military Operations* (pp. 397–418). Farnham: Ashgate Publishing.

Mitchell, D. & Samms, C. (2010). An Analytical Approach for Predicting Soldier Workload and Performance Using Human Performance Modeling. In M. J. Barnes & F. G. Jentsch (eds), *Human-Robot Interactions in Future Military Operations* (pp. 125–142). Farnham: Ashgate Publishing.

Murphy, R. & Burke, J. (2010). The Safe Human-Robot Ratio. In M. J. Barnes & F. G. Jentsch (eds), *Human-Robot Interactions in Future Military Operations* (pp. 31–50). Farnham: Ashgate Publishing.

Oran-Gilad, T. & Minkov, Y. (2010). Remotely Operate Vehicles (ROVs) from the Bottom-Up Operational Perspective. In M. J. Barnes & F. G. Jentsch (eds), *Human-Robot Interactions in Future Military Operations* (pp. 211–228). Farnham: Ashgate Publishing.

Pazuchanics, S., Chadwick, R., Sapp, M., & Gillan, D. (2010). Robots in Space and Time: The Role of Object, Motion, and Spatial Perception in the Control and Monitoring of Uninhabited Ground Vehicles. In M. J. Barnes & F. G. Jentsch (eds), *Human-Robot Interactions in Future Military Operations* (pp. 83–102). Farnham: Ashgate Publishing.

Redden, E., & Elliott, L. (2010). Robotic Control Systems for Dismounted Soldiers. In M. J. Barnes & F. G. Jentsch (eds), *Human-Robot Interactions in Future Military Operations* (pp. 335–352). Farnham: Ashgate Publishing.

Riley, J., Strater, L., Chappell, S., Connors, E., & Endsley, M. (2010). Situation Awareness in Human-Robot Interaction: Challenges and User Interface Requirements. In M. J. Barnes & F. G. Jentsch (eds), *Human-Robot Interactions in Future Military Operations* (pp. 171–192). Farnham: Ashgate Publishing.

Schulte, A. & Meitinger, C. (2010). Introducing Cognitive and Co-operative Automation into UAV Guidance Work Systems. In M. J. Barnes & F. G. Jentsch (eds), *Human-Robot Interactions in Future Military Operations* (pp. 145–170). Farnham: Ashgate Publishing.

Thompson, L. F. & Gillan, D. (2010). Social Factors in Human-Robot Interaction. In M. J. Barnes & F. G. Jentsch (eds), *Human-Robot Interactions in Future Military Operations* (pp. 67–82). Farnham: Ashgate Publishing.

Wickens, C., Levinthal, B., & Rice, S. (2010). Imperfect Reliability in Unmanned Air Vehicle Supervision and Control. In M. J. Barnes & F. G. Jentsch (eds), *Human-Robot Interactions in Future Military Operations* (pp. 193–210). Farnham: Ashgate Publishing.

Chapter 2

Soldier-Robot Teams in Future Battlefields: An Overview

Michael J. Barnes and A. William Evans III

U.S. Army Research Laboratory/University of Central Florida

Working with the Tank Automotive Research, Development and Engineering Center (TARDEC), the Human Research and Engineering Directorate (HRED), U.S. Army Research Laboratory (ARL) embarked on a five-year Army Technology Objective (ATO) research program that addressed human-robot interaction (HRI) and teaming for both aerial and ground robotic assets. The purpose of the program was to understand HRI issues in order to develop technologies and procedures that enhance HRI performance in future combat environments. Soldier-robot teams will be an important component of future battlespaces, creating a complex but potentially more survivable and effective combat force. The variety of robotic systems and the almost infinite number of possible Army missions create a dilemma for researchers who wish to predict HRI performance in future environments, requiring researchers to develop creative simulations of future combat systems as well to conduct field tests with actual systems (Barnes et al., 2006a). The purpose of this chapter is to summarize issues and solutions that have resulted from the research program. The researchers have published over 100 individual papers and more continue to be submitted. Furthermore, the larger field of HRI has grown immensely in the last five years. For obvious reasons, we have limited the chapter to a few crucial issues highlighting our own HRI results and suggested procedures to mitigate some of the identified problems. Other research is reviewed when it directly informs our HRI emphasis; but still the chapter covers a small portion of HRI.

Human-Robot Teaming

Human-robot teaming refers to the collaboration between human and robotic intelligent agents to perform common tasks. Woods et al. (2003) point out that both the human and the robot contribute unique qualities. The interaction is always asymmetric; the robot is never a stakeholder; only the human is responsible for the team's results. Because of the importance of the human stakeholder role, the human's unique contribution to the process is often underrepresented. Also, as robots become more automated, the teaming relationship may become more not less complex. Human partners will become responsible for understanding political context, changing tactics, meta-goals and also act as a back-up operator in cases where events do not emerge as planned (Chen et al., 2007).

Case studies of robotic rescue operations indicate some of the human-related problems (Murphy, 2004). Efficient rescue search consisted of at least two operators. Having a controller alone resulted in missed rescue opportunities because attention was focused on control functions. Also, even with a second operator, remote viewing was compromised because of the lack of the contextual cues that humans use when they are in the actual locale (Chen et al., 2007; Murphy, 2004). The two-person crew underestimates the human role in actual robotic missions. For example, in the World Trade Center (WTC) rescue operations, there were many other humans involved with setting priorities and coordinating activities to make the best use of the robotic assets (Murphy, 2004).

Similar logic suggests that the soldiers' role will increase as engineers find new uses for so-called autonomous and semi-autonomous systems. The operator control function will still be essential, but as a default mode for degraded operations. This will be difficult because humans will be called on to intervene under conditions that test the limits of the robot's capabilities (e.g., hidden trenches) and also because it will be difficult to maintain the operator's skill level and situation awareness if they intervene occasionally (Schipani, 2003).

In contrast, the role of the soldier as a stakeholder will increase in direct proportion to increases in autonomy. The larger the role of the autonomous robot in military operations, the more exacting the planning function, and the more commanders and staff will scrutinize and redirect robotic assets during the mission. Finally, because unmanned systems will become ubiquitous in future battlespaces, the human's intelligence and imagery interpretation roles will increase as robotic systems improve their imagery deliver systems (Barnes et al., 2006b). As Charles Kelley (1968) prophesized, automated systems will not replace humans; rather it will make their roles more cognitive and more goal centered.

Military Environments

An important feature of future battlespaces will be a greater use of intelligent systems to supplement traditional manned forces. Each service and each ally will have its own set of aerial and ground uninhabited systems with increasing levels of machine intelligence. The National Research Council (2003) reports that Congress mandated that up to one-third of future military systems are unmanned by 2015. For example, proposed Army systems include two new classes of unmanned aerial vehicles (UAV), small man packable unmanned ground vehicles (SUGV), multi-ton armed robotic vehicles (ARV), smaller platforms for logistics and extraction as well as intelligent stationary sentinels (Shoemaker, 2005). Because autonomous capability is increasing, control of the larger unmanned systems will not be the soldier's major focus; rather it will be maintaining situation awareness (SA) and reporting potential targets, the very things that have proved so difficult in actual robotic missions (Murphy, 2004; Hoeft et al., 2005).

Most of these systems are being designed to support mounted and dismounted soldiers at the lowest levels. These systems will increase the reach of future soldiers, multiply their firepower, and, most important, increase survivability (Barnes et al., 2006b). But in many ways unmanned is a misnomer: the soldier will acquire more responsibility and not less with the proliferation of these systems. Robotic tasks will be in addition to and not instead of their normal military tasks. For mounted operations, soldiers will be expected to conduct normal operations, including their crew tasks such as local security, gunnery, and command functions, while concurrently conducting remote targeting with robotic systems (Mitchell & Henthorn, 2005; Mitchell & Chen, 2006). Dismounted soldiers are expected to maintain situation awareness, conduct ground combat missions, perform their assigned military tasks, and use robotic assets as the military situation unfolds (Redden & Elliott, this volume). Robotic specialists will constitute a small percentage of future Army forces and be used mainly to coordinate the use of the diverse unmanned vehicles (UVs) (Mitchell & Samms, this volume).

HRI Multitasking

We have addressed a number of issues.

1. Soldiers will not be robotic specialists, but rather be expected to conduct multiple military tasks.
2. Their robotic role will not primarily be control, but rather targeting and intelligence gathering.
3. Future robotic military missions will be complex and difficult to predict.

Our initial HRI models based on operational requirements and inputs from miltary experts indicate that additional robotic roles will impact the soldier's ability to multitask efficiently (Mitchell & Chen, 2005; Mitchell & Henthorn, 2004; Mitchell & Samms, this volume). Also, field studies indicate that operators will need to teleoperate the robot under adverse conditions in addition to their other robotic duties (Chen et al., 2007). Thus an important thrust of the research was to understand the impact of the soldiers' multitasking environments and where possible to suggest mitigation strategies to overcome their adverse impacts.

Multitasking has been studied extensively in the laboratory, usually focusing on a limited number of artificial tasks. The findings are fairly straightforward: doing many things at once usually results in reduction in performance depending on the operator's priorities, task difficulties, interference effects, and amount of training. There is a switching cost and a limited subset of tasks that can be performed concurrently without performance decrements (Wickens & Hollands, 2000).

Of particular interest is recent Army-supported research by Wickens and his colleagues. They investigated both the theoretical and practical significance of multitasking based on realistic UAV simulations (Dixon & Wickens, 2003; Wickens et al., 2006, 2003). The number of tasks used for the simulation was fairly representative of real-world HRI conditions minus the combat stress and operational tempo.

Based somewhat loosely on Shadow-200 UAV training simulators at Ft. Huachuca, Arizona, the operators monitored one or two UAVs, performed a tracking task, identified targets of opportunity, specified targets, navigated, and monitored flight instruments. The results were compared to three theories of human cognitive capacity: Welford's single channel theory (SC), Kahneman's single resource theory (SR) and Wickens' version of multiple resource theory (MR). SC is the most elegant theory, positing that the processing time is a function of the number of tasks plus a switching cost, whereas SR allows humans to allocate resources among tasks, and MR suggests that human are even more flexible, that they can process information in parallel if the tasks involve different processing channels. For tasks performed in the same modality (visual), SC was a good fit because the processing time was the sum of the various tasks plus a switching cost. However, SC delay predictions were not upheld as a function of when tasks occurred during the processing of another task, suggesting some sharing of processing resources. More important, there were substantial savings when the monitoring task was reconfigured from a visual alert to an auditory alert. This supports the efficacy of MR predictions, indicating that the operator SC limitations can be mitigated if the operator's tasking is divided among modalities (Wickens et al., 2006).

Multitasking in Future Mounted Systems

The first scenario that we examined was the soldier's multitasking environment for simulated targeting missions in a future mounted vehicle (Chen et al., 2008). Previously, HRI workload analysis was conducted using ARL's Improved Performance Research Integration Tool (IMPRINT) software for the three crew members for a mounted combat system. The results, based on interviews with 25 experienced tank crew members from Ft. Knox, indicated potential overload for crew members if remote targeting using robotic assets was added to their taskings (Mitchell & Henthorn, 2006). The initial simulation investigated type and number of assets that the operators could control (span of control). Using scenarios developed by military subject matter experts (SMEs), the robotic operator's task was to use lasers to acquire potential targets while controlling either a semi-autonomous UV, a teleoperated UV, or a UAV. This was contrasted with the mixed condition with the human controlling all three assets. The UGV and UAV assets were assumed to be semi-autonomous; the operator designated waypoints on an automated map display to the final objective and monitored the progress of the vehicles, but the principal task was targeting. The operator controlled the UGV for the teleoperation conditions. The simulator was equipped with a steering wheel, and gas and brake pedals for control of the teleoperated vehicle.

Not surprisingly, teleoperations resulted in the poorer performance when compared to the other two conditions. However, the combined condition resulted in some intriguing findings. Figure 2.1 indicates that the mixed assets condition showed pronounced targeting degradation for the semi-autonomous UGV and teleoperated UGV when the operator was controlling three assets. In contrast, the UAV performance showed no such degradation, suggesting that operators' preferred strategy was to depend on the UAV when all three assets were available. This was true even though there was little difference between semi-autonomous UGVs and UAV targeting performance when the operator controlled them as individual entities (Figure 2.1). There were also individual differences. Those that tested high in spatial ability were able to conduct missions more rapidly using the UAV exocentric targeting views, whereas there was no advantage to having high spatial ability for UGV egocentric viewing conditions.

This was supported by other research investigating differences between ground robotic views and aerial views. Luck et al. (2006), using a gaming paradigm, found that even in conditions whererin participants performed their escape tasks better with UGV views, they still preferred UAV views. This suggests a generalized preference for the aerial views in comparison to the out of the window views obtained from ground robots.

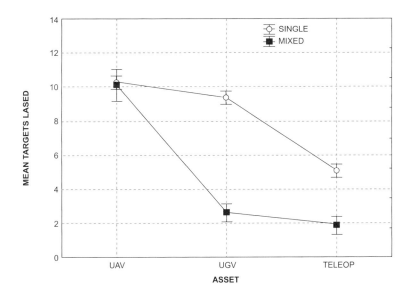

Figure 2.1   Interaction of asset condition and number of assets controlled

Multitasking and Remote Targeting with a Single UGV for Mounted Missions

Based on the modeling discussed previously, Chen and Joyner (2006), Chen and Terrence (2008a, 2008b), and Chen (this volume) conducted three additional simulation experiments. The goal of this research was to examine if gunners in a future armored vehicle were able to maintain local security (i.e., perform their primary gunners' tasks) while using the UGV as a forward asset for remote targeting (secondary task) (Chen & Joyner, 2006; Mitchell & Chen, 2006). The primary task required the operator to use the gunner's display to find targets around own vehicle (baseline).The concurrent tasks conditions required remote targeting using a separate robotic display in addition to the baseline gunner display in one of three secondary task conditions: 1) monitoring a UGV via the video feed; 2) managing a semi-autonomous UGV; 3) teleoperating a UGV. Participants also performed a tertiary communication task. The results indicated that that gunner's target detection performance degraded significantly when he or she had to concurrently monitor, manage, or teleoperate a UGV compared to the baseline condition (gunnery task only). The operator's workload and performance on both the primary and secondary tasks were degraded as a function of difficulty of the robotic tasks.

The results support the Mitchell and Chen's (2006) modeling predictions. Gunners were *not* able to conduct their primary task effectively if a secondary robotic task was added to the task inventory. Performance on the robotic task (secondary task) was also affected by multitasking, as was the communication task. In contrast to performance on the primary task, the teleoperated task resulted in better target detection than the semi-autonomous task for remote targeting using the robotic display. This is not as counterintuitive as it seems. Teleoperations is particularly difficult, causing operators to focus their attention on the robotic task during teleoperations; the increased attentional effort both improved performance for robotic targeting and had a deleterious effect on targeting for the primary gunner's task. Other experiments indicate that using an automated targeting recognition aid (AiTR) helped operators overcome their attentional deficits. More importantly, these experiments showed that individual differences in spatial ability and perceived control confidence had profound effects on performance (Chen, this volume; Chen & Terrence, 2008a, 2008b).

Soldier-Robot Teaming

Scientists from the Team Performance Laboratory (TPL) at the University of Central Florida (UCF) have focused their research on gaining a better understanding of collaboration among multiple operators controlling multiple robotic assets. They conducted a series of experiments using a 1/35 scale city of actual Iraqi urban terrain. The scaled terrain set-up allowed the researchers to change conditions when manipulating various urban-related parameters. The initial experiment investigated teaming relationship among UGVs and soldiers during an urban reconnaissance, surveillance, and target acquisition (RSTA) mission (Jentsch et al., 2005; Evans et al., 2005; Hoeft et al., 2005). The participants' task was to identify equipment and personnel while operating a UGV in the urban setting (Figure 2.2). Participants planned an RSTA route but the UGV was actually controlled by the experimenter, making the main task a target identification task similar to Chen and her colleagues' semi-autonomous conditions (Chen & Joyner, 2006; Chen & Terrence, 2008a, 2008b). Situation awareness (SA) was greatly improved for a two-person team compared to a single operator in all conditions. Surprisingly, for the most difficult conditions (barriers causing participants to replan their route), teams were almost three times as effective as single operators, suggesting that having a two-person crew had advantages beyond an extra set of eyes. Also, having an additional robot did not offer any advantage even for the two-person crew; in some cases it only added to their difficulties. From a systems design perspective, it reinforces the importance of designing for teams of soldiers coordinating multiple robots rather than focusing on single robotic control. More recent studies have investigated the collaboration of soldier UAV and UGV teams conducting collaborative tasks in an urban environment, such as finding a target under a bridge. These tasks were designed to maximize the advantages of collaboration between UAVs with aerial views coordinating with UGVs with a street-level view. Jentsch et al. (2007) found that performance was extremely poor when collaboration was conducted with chat communications and map views, but improved by nearly twofold when the operators of the two systems were able to share their imagery. Surprisingly, allowing an operator to control the other operator's sensor actually degraded performance when compared to chat plus imagery conditions. Research is continuing to investigate added communication options and additional collaborative environments in order to improve UV teaming arrangements.

Mitigation Strategies

The above research elucidates some of the unique Army problems involved in multitasking, asset coordination, and SA for future HRI missions from mounted (vehicular) systems. The same problems have been verified for dismounted operations (Oron-Gilad & Minkov, this volume; Redden & Elliott, this volume). The following sections investigate various mitigation technologies to improve HRI performance while maintaining the soldier's SA.

Figure 2.2   UGV camera view from ground level in UCF studies

Multimodal Interfaces

The multiple resource (MR) model of multitasking predicts processing savings using multimodality interfaces either as a redundant cue or as separate sources of information. Reviews by Sarter (2007) and Jones and Sarter (2008) enumerate many benefits of multimodal interfaces but also point out limitations, especially with cross-modality cueing. Operators build up expectations for a particular message transmitted in a specific modality presented in the same location; thus switching between modalities has performance costs. Also, modalities can mask other modalities in close temporal proximity. Certain modalities seem to pair better than others; tactile warnings related to icing conditions in aircraft were more efficient than visual warnings when both were paired with auditory warnings. Sarter concluded that multimodal interfaces had unexpected synergies in some cases and unintended negative consequences in others. Multimodal interfaces must be constructed for specific tasks and evaluated carefully for interference effects.

Haas and her colleagues have examined the possibility of using multimodal interfaces for HRI mitigation of multitasking limitations. Their research focused on multimodal cueing for remote targeting with a robotic system (Chen et al., 2007; Haas et al., 2007, 2008; Haas & van Erp, this volume). They found that 3-D audio spatial cues in conjunction with visual cues enhanced performance over baseline visual map cues; but the researchers noted that the participants would need a head tracker to disambiguate front and back audio signals. Verbal 2-D directional cues worked as well as the 3-D spatial audio cues, but the directional 2-D cues would have to be instantiated within a voice interactive system to be practical. In further studies, they found that tactile cues and a combination of tactile and auditory both showed promise as redundant cues for remote robotic targeting. Other researchers found no difference between tactile and auditory cues for infantry tasks in a fairly noisy environment, but showed a synergistic effect when the two modalities were combined; however, in a second experiment the auditory cue's effectiveness was diminished when back–front auditory cue differentiation was required of the dismounted soldiers. In general, empirical results indicated that tactile formats were superior to auditory ones for most of the soldier tasks tested (Gilson et al., 2007).

An obvious disadvantage of tactile cues is the possibility of their not being noticed when mounted vehicles are transversing rugged terrain. A number of researchers investigated attenuation effects for tactile signals in mobile environments. Haas et al. (2007) found that HRI-related tactile information was reliably transmitted under a variety of vehicular vibratory conditions during Humvee testing. Other Army researchers investigated three tactile display configurations using a motion simulator (Krausman & White, 2008). Results indicated effects due to signal strength, cross-country movements, and signal direction. However, the researchers concluded that the higher signal strength tactors gave consistently reliable performance. Of particular import was their finding that plunger-type tactors gave more reliable signals than the pancake-type tactors used by Haas et al., suggesting that crew performance might improve further as tactor technology improves (Gilson et al., 2007; Krausman & White, 2008).

Chen and Terrence (2008a, 2008b) found that tactile cueing reduced the dual tasking performance costs that the gunner suffered while conducting robotic targeting and local security tasks concurrently (see the Chen and Joyner paradigm above). However, as Sarter (2007) has cautioned, cross-cueing can be more complex than it first appears. Chen and Terrence found that tactile cues by themselves improved gunners' performance for their primary task as much as tactile cues combined with visual cues. Gunners were not only able to switch efficiently to their primary task display, but also to orient themselves within 360 degrees to find targets around the manned vehicle using tactile cues from an aided target recognition (AiTR) device. However, they also evaluated individual differences in spatial ability. They found that the participants who were low in spatial ability showed a marked preference for visual cues in addition to the tactile cues; however, participants with higher spatial ability could orient themselves using the tactile cues alone. This suggests that there may be important individual differences in the ability of operators to use modality cues and that multimodal displays may be useful in mitigating spatial orientation deficiencies in the operator population (Chen & Terrence, 2008b).

Multimodal solutions are proving to be a useful design option for future HRI interfaces. Voice systems and tactile interfaces are particularly promising for target alerting. Tactile systems have a design advantage in that they engage human sensory systems currently underused in operational environments (Haas et al., 2008; Jones & Sarter, 2008). Tactile interfaces have been effectively used for dismounted missions, such as robot navigation and alerting, as well as for target recognition during mounted missions. Even more intriguing is the recent success of Army researchers in using intuitive tactile lexicons to replace combat hand signals and sniper commands in realistic infantry field trials with minimum training (Gilson, et al., 2007; see also Jones & Sarter, 2005). However, multimodal systems require specific design considerations for possible interaction between modalities, the type of tasks serviced, and individual differences in the operator population (Chen, this volume; Gilson et al., 2007; Haas & van Erp, this volume; Jones & Sarter, 2008; Sarter, 2007).

Automation

Automation is an obvious solution to the increased workload of future military systems. However, in practice, automation is often impractical and it presents its own problems. Humans tend to over-rely and under-rely on automated systems, depending on the tasking environment. There are multiple reasons for this: decision biases, complacency, false alarms, social pressures, and failure to monitor automated tasks during high mental loadings (Barnes et al., 2006b; Lee & See, 2004; Parasuraman et al., 2000). Even a simple switch in decision order can change the automation bias from under-reliance to over-reliance (Dzindolet et al., 2001). Many of the biases are related to demands on cognitive resources; for example, automation complacency is more likely to occur in high workload multitasking environments (Parasuraman et al., 1993). Operators will not expend their limited resources on monitoring automated systems when their other tasks impose high demands. The automation paradox, wherein operators miss obvious errors in otherwise highly reliable automation, is a good example. Operators overtrust automation because their cognitive resources are allocated elsewhere; thus, the more reliable they perceive automation to be, the less likely they are to monitor it (Barnes et al., 2006a; Parasuraman et al., 2007).

Also, problems occur for unreliable automation. Wickens et al. (2005) based results from a meta-analysis found that if the automation reliability is below a certain level (70 percent), operator performance degrades, presumably because the operator overtrusts the automation in order to reduce overall multitasking workload. Further, the type of automated errors influences the operator's performance. For example, Chen and Terrence (2008b) simulated automated target recognition (AiTR) aid and varied type of unreliability—aids that missed targets vs. aids that reported false targets (Meyer, 2001). They found that automation errors interacted with personality factors—participants with high perceived control did poorly with false-alarm prone aids by ignoring correct alerts; whereas less confident participants did relatively poorly with miss-prone aids, presumably because they overtrusted the aid.

A number of possible solutions have been suggested to "put the operator back in the loop" without abrogating the advantages of automation (Endsley & Kris, 1995). Adaptive systems have been investigated for decades; their purpose is to minimize task difficulty without disengaging the operator. Adaptive systems vary the amount of automation depending on the operator's mental state or the state of the operational environment in order to achieve equilibrium between engagement and task demands (Kelley, 1968; Parasuraman et al., 1992). Army investigators studied adaptive automation to ameliorate multitasking effects involving UAV and UGV operations from a simulated mounted vehicle console (Cosenzo et al., this volume; Parasuraman et al., 2007). In their most telling experiment, Parasuraman et al. (2007) showed that adaptive automation could improve performance as well as enhance SA for an adaptive automation condition compared to static automation and manual conditions. The tasks included identifying UAV targets, monitoring UGV progress, answering SA probes, and conducting communications. For the adaptive condition, the trigger to invoke the AiTR device was the operator's failure to note changes in unit locations on the situation map. For static automation, the AiTR was on for half of the trial's duration and for the manual condition the AiTR was not invoked. The crucial difference between the types of automation was that the adaptive system was tailored to the individual's error rate, whereas static automation was invoked based on task difficulty (determined by averaged past performance). The results clearly showed an advantage to adapting to the individual operator's performance online. Based on this and other studies, Parasuraman et al. (2007) suggested the following guidelines to implement adaptive systems:

1. Information displays should adapt to the changing military environment.
2. Software should be developed to allow the operator to allocate automation under specified conditions before the mission (as in the Rotorcraft Pilot's Associate program).
3. At least initially, adaptive systems should be evaluated that do not take decision authority away from the operator. This can be accomplished in two ways: (a) an advisory asking permission to invoke automation; or (b) an advisory that automation will be invoked unless overridden.
4. For safety or crew protection situations, specific tactical or safety responses can be invoked without crew permission.

There are still important questions to be answered before adaptive systems can be implemented for robotic applications. We will need to compare the efficacy of different triggering devices, including neurophysiological indices. Also, we need to know both optimal engagement times and switching rates among multitaskings (Barnes et al., 2006b). Finally, we need to know how candidate adaptive systems work in complex tasking environments reflecting actual operations (Parasuraman et al., 2007). Adaptive systems are promising because they allow designers to combine the advantages of manual and automatic control without requiring absolute dependence on one or the other.

There are other approaches that capture the advantages of adaptive systems. Adjustable automation allows the operator to set the type of automation (Goodrich, this volume; Sellner et al., 2008); however, if the options are complex, adjusting may be another task in an already difficult multitasking environment. Adaptable systems involve preprogrammed behaviors that the operator can invoke during a robotic mission. The metaphor is a playbook used during a football game, with the operator being able to invoke different "plays" depending on the emerging situation (Barnes et al., 2006b). There is no reason to consider these solutions mutually exclusive, future systems (both manned and unmanned) will have complex requirements; the more options designers, have the more likely soldier-robot teams are to adapt effectively to the diverse nature of future combat missions.

Interface Solutions

Information presentation formats are a critical aspect of any human-operated system and are the first line of defense for reducing workload and improving SA (Wickens & Holland, 2000). HRI researchers have developed advanced interface concepts for controlling ground and aerial unmanned systems that improve SA, insert preprogrammed behavior, view multiple robots interacting, and create dialogs with robotic entities (Chen et al., 2007; Sellner et al., 2008). The Army program focused on three issues: 1 scalability; 2 stereo displays; 3 information presentation formats. Scalability is an important issue because in a network centric battlefield information must be pushed and pulled to a variety of unmanned systems that collaborate during an actual mission. Each system will have a requirement for common information as well as information specific to that system. The solution is to ensure that the formats are optimal for collaboration, while ensuring that robotic operators are able to operate their own systems. For example, information formats for two display sizes and different presentation options were assessed for their effectiveness in controlling moderate-sized, semi-autonomous UGVs during a field experiment at Ft. Knox. These tablet-size formats were scaled-down versions of information from the mounted vehicles with more extensive display systems. The software automatically rescaled the format, deleting buttons and information unnecessary or of low priority for the smaller display. The user evaluations verified the information content and discovered interface problems with the two scaled-down versions, resulting in design guidelines for tablet-size displays during mounted/dismounted missions (Cosenzo & Stafford, 2007).

For the dismounted soldier (infantry), scalability is particularly important because the soldier will most likely be controlling robots when he or she is most vulnerable. The ideal interface would be a type of helmet-mounted display (HMD) permitting soldiers to keep their heads up during combat. Small hand-held displays such as a personal digital assistant (PDA) have some disadvantages but are light and are more easily transported than a tablet-size display. Redden et al. (2008) investigated soldiers controlling a small UGV (SUGV) Talon robot (Figure 2.3) on an obstacle course at Ft. Benning, Georgia. The soldiers teleoperated the Talon with a 3.5-inch, 6.5-inch, 10-inch display, or goggle-mounted display (GMD) with a single eyepiece. The operators were able to use the Talon to transverse the course, detect targets, and avoid obstacles satisfactorily with all the display options. However, the GMD was the least successful option, causing eye strain, generally lower performance, and binocular rivalry. The important point is that, for these tasks, soldiers with minimal training were able to teleoperate an SUGV with PDA-size display, suggesting that the smaller displays may be a practical lightweight solution that could be incorporated as part of the soldier's gear, perhaps as a flexible display/communication system.

Researchers in Israel working with Israeli Defense Forces (IDF) personnel in the UAV domain found results that were both conflicting and supportive of the above results (Minkov et al., 2007; Oron-Gilad & Minkov, this volume). The IDF participants preferred hand-held displays for dismounted missions; they had positive assessments for a hand-held display shaped like a bicycle handle that, when the eyepiece was brought up to eye level, had video projection capabilities similar to an HMD. For navigation tasks, display size did not affect performance. However, for more complex tasks such image interpretation in an urban environment; performance was significantly worse for a PDA-size display than for the 7-inch or 14-inch options. The results suggest that, for SUGV operations, the smaller display or a small video window on a tablet may be sufficient for robotic control; however, for more complex missions, larger display areas and more resolution will be required. The GMD results are problematical; although GMDs have advantages because of their heads-up feature, these displays should be evaluated more thoroughly for possible health and performance issues (Redden & Elliott, this volume).

Figure 2.3   Talon robot next to a soldier participant at Ft. Benning

Army researchers examined the utility of true stereo displays for robotic operations at Ft. Leonard Wood, Missouri. Initial analysis identified two stereoscopic research areas: detection of obstacles, especially negative terrain (trenches, etc.); and robotic arm manipulation for grasping suspicious objects (Chen et al., 2007). Working with the maneuver support center battle lab, the Army researchers investigated the use of stereo displays for a number of engineering applications (Bodenhammer 2007; Pettijohn et al., 2007). For teleoperated route clearance tasks, the results were mixed; manipulation-related tasks were more easily accomplished with stereo displays. However, there were no benefits over 2-D displays for other tasks. Another experiment was conducted investigating the arm manipulation and detection capabilities of the mine clearance vehicles using stereo displays; again, the results were mixed, some tasks showing a benefit for stereo and others showing no benefit. The engineers concluded that there were still problems with implementing the stereo equipment on the mine clearance vehicles. Part of the problem was the immaturity of stereo displays for robotic systems. Some of the initial studies used a multiplexing system that sent half the display signals to each eye, resulting in a substantial reduction in display resolution compared to the 2-D systems (Bodenhamer, 2007). The Pettijohn et al. study used a polarized display system to project images to each eye without a reduction in resolution. Unfortunately, the stereo cameras were not aligned with the display at all depths of field. A more recent test used the polarized displays with cameras specially built to focus on all depths of field, resulting in greatly improved performance and user acceptance (Pettijohn, 2008). Also, other researchers are investigating both arm manipulations and driving characteristics using 3-D for robotic systems in other operational venues. Stereovision is a promising research area for robotic control because of its ability to present more veridical views of the terrain and to allow the operator to manipulate robotic arms using natural three-dimensional cues. However, traditionally, stereo displays have caused many problems, including nausea, foreshortening, etc. These negative results make it important to sort out not only the advantages of true stereo displays but also to determine what stereo technologies best serve the robotic operator and under what field conditions these benefits accrue (Barnes, 2003; Chen et al., 2007).

Another important aspect of interface design is the use of visualization techniques to portray operational parameters in complex situations such as coordinating multiple UAVs during combat missions (Hunn, 2006). U.S. Air Force designers developed visualization concepts to reduce pilot errors in situations where pilots must make rapid life-threatening decisions while monitoring multiple displays at high speeds. Synthetic environments portraying flight paths, flight data, threat envelopes, etc. were used successfully to improve SA for complex combat environments for a variety of aerial platforms (Calhoun & Draper, this volume). The purpose of creating these synthetic environments is to impart SA at a glance by reducing interpretation time to a minimum.

Army display designers encounter similar issues for robotic control from mounted vehicles. Simulating convoy scenarios, Cosenzo et al. (2009) investigated the advantages of graphical icons portraying manned and unmanned vehicular status in a mixed convoy. The improved display concepts resulted in the operators being able to note convoy deviations more rapidly and accurately compared to baseline displays. The advantages of creating a synthetic environment with iconic representations are well documented in the literature (Calhoun & Draper, this volume; Barnes, 2003). However, the guidelines are more intuitive than canonical, leaving a great deal of room for designer creativity. Because of the task-specific nature of format design, the Army is continuing to investigate candidate designs in their intended environments in order to determine the benefits and costs associated with each option (Cosenzo et al., 2009)

Conclusions

As alluded to in the introduction, the nature of HRI is changing as more intelligence is infused into both ground and aerial unmanned systems. Future robots will be able to make many tactical decisions on their own and even operate in traffic with human-controlled vehicles. The efficacy of swarms of robots reacting to each other without human intervention is also being studied by a number of researchers (see Lewis & Wang, this volume). These concepts are still years from being operational but the trend is almost inexorable; essentially, the human role will become supervisory. However, the ability of the robot to relay combat information and the emphasis on real-time intelligence also means that the operator will be in the loop for decades as a receiver and arbitrator of information, as well as a means for changing the goal of robotic missions as the combat situation unfolds. As Woods et al. (2004) pointed out, there are many philosophical issues to be resolved. In particular, the issue of autonomous systems in combat is worrisome. In anti-insurgent operations, it is difficult for humans to know who the enemies are and who the allies are; allegiances can change overnight. Robotic entities will not be able to make such distinctions reliably for the foreseeable future. It will be crucial to have software "brakes" to ensure that robotic decisions that make tactical sense do not have unintended consequences resulting in either unsafe or unethical acts. Ronald Arkin (2007) has developed a general approach and software architecture for instantiating ethical behaviors into robotic systems. The approach is admirable but still very much in its infancy. It probably does not need to be pointed out that after millions of years of evolutionary development humans still have a difficult time making ethical decisions. Any ethical "brakes" that are part of the robots' behavioral repertoire will be useful as long as they minimize harm. In the final analysis, it will be the human supervisor or the designer who will be held resonsible if they fail to intervene or fail to design the necessary safeguards. This is not a trivial issue; robots that are perceived to be dangerous (especially to the innocent) or unsafe will have very limited use.

The present program supported the Army's initial foray into future battlefields with arrays of unmanned systems. We examined important HRI problem areas, including: span of control, mixed assets, multitasking, overload, SA, aerial and ground perspectives, and soldier teams with complementary assets. We also investigated mitigation strategies to offset some of the problem areas we discovered: selection based on spatial abilities, training on UV visual perspective differences, tactile AiTR cueing, shared imagery, multimodal interfaces, adaptive systems, scalable interfaces, stereovision, and graphical user interface (GUI) designs. To summarize:

- Multitasking will be a crucial issue for integrating UVs into the future force.
- Participants with higher spatial ability performed better in multitasking robotic tasks.
- Two-person human-robot teams performed better than single operators for reconnaissance, surveillance, and target acquisition (RSTA)-type urban missions even when robotic control was not an issue.
- Shared imagery (but not shared control) improved the ability of UAV and UGV operators to conduct missions collaboratively.
- Even when performance was nearly equivalent, operators preferred aerial versus ground views.
- Visual training on different perspectives helped operators adjust to UGV operations rapidly.
- A mixture of visual cues and tactile cues indicating 360-degree target locations helped mitigate poor multitasking performance for those with low spatial abilities.
- Tactile, auditory, and visual cues can be used in a complementary mix for remote robotic targeting, especially during high workload mission segments.
- Multmodal systems must be designed to minimize cross-cueing interference.
- Tactile cues have advantages over spatial auditory cues because of back–front ambiguities and the many uses of auditory information in miltary environments.
- Fully automated systems can have deleterious effects because of decision biases, complacency effects, poor operator calibration, and even social pressure.
- Adaptive systems that modulate the degree of automation depending on the operator state or the state of the world offer a promising alternative to static automation in robotic environments.
- Simulation research indicated that adaptive automation (compared to static automation) can improve both performance and operator SA while conducting UAV/UGV multitasking missions.
- Small PDA 3.5-inch displays proved feasible for small robot control during a Ft. Benning field test.

- Larger displays (7 inch) were required for UAV imagery analysis in urban environments during Israeli simulations.
- Initially, stereovision showed mixed results for teleoperations and arm manipulation during field tests; however, recent results using a modified polarized stereo display showed both greater operator acceptance and superior performance for stereo conditions.
- Simulations investigating graphical interfaces for mixed manned/unmanned convoy operations indicated improved performance with HRI iconic information overlaid on proposed future Army formats.

References

Arkin, R. C. (2007). Governing lethal behavior: Embedded ethics in hybrid deliberative reactive robot architecture (TR GIT-GVU-07-11). Atlanta, GA: Georgia Institute of Technology.

Barnes, M. (2003). The human dimension of battlefield visualization: Research and design issues (ARL-TR-2855). Aberdeen Proving Ground, MD: U.S. Army Research Laboratory.

Barnes, M., Cosenzo, K., Jentsch F., Chen J. Y. C., & McDermott, P. (2006a). The use of virtual media for military applications. *Proceedings of the NATO HFM-136 Workshop*, West Point, NY.

Barnes, M., Parasuraman, R., & Cosenzo, K. (2006b). Adaptive automation for military robotic systems. NATO Technical Report (RTO-TR-HFM-078). *Uninhabited Military Vehicles: Human Factors Issues in Augmenting the Force*, 420–40.

Bodenhammer, A. S. (2007). Assessment of stereoscopic display systems for assisting in route clearance manipulation planning tasks (ARL-TR-4195). Aberdeen Proving Ground, MD: U.S. Army Research Laboratory.

Calhoun, G. L. & Draper, M. H. (2010). Unmanned aerial vehicles: Enhancing video display utility with synthetic vision technology. In M. J. Barnes & F. J. Jentsch (eds), *Human-Robot Interactions in Future Military Operations*. Farnham: Ashgate Publishing.

Chen, J. Y. C. (2010). Robot operator performance in multitasking environments. In M. J. Barnes & F. J. Jentsch (eds), *Human-Robot Interactions in Future Military Operations*. Farnham: Ashgate Publishing.

Chen, J. Y. C., Durlach, P. J., Sloan, J. A., & Bowens, L. D. (2008). Human robot interaction in the context of simulated route reconnaissance missions. *Military Psychology*, 20(3), 135–49.

Chen, J. Y. C., Haas, E. C., & Barnes, M. J. (2007). Human performance issues and user interface design for teleoperated robots. *IEEE Transactions on Systems, Man, and Cybernetics. Part C: Applications and Reviews*, 37(6), 1231–45.

Chen, J. Y. C. & Joyner, C. T. (2006). Effectiveness of gunnery and robotic control performance in a simulated multi-tasking environment. *Proceedings of the 25th Army Science Conference*. Orlando, FL, 27-30 November.

Chen, J. Y. C. & Terrence, P. I. (2008a). Effects of tactile cueing on concurrent performance of military and robotics tasks in a simulated multi-tasking environment. *Ergonomics*, 51(8), 1137–52.

Chen, J. Y. C. & Terrence, P. I. (2008b). Effects of imperfect automation on concurrent performance of gunner's and robotics operators's tasks in a simulated mounted environment (ARL Tech. Rep. ARL-TR-4455). Aberdeen Proving Ground, MD: U.S. Army Research Laboratory.

Cosenzo, K. & Stafford, S. (2007). Usability assessment of displays for dismounted soldier applications (ARL-TR-4326). Aberdeen Proving Ground, MD: U.S. Army Research Laboratory.

Cosenzo K., Parasuraman, R., & de Visser, E. (2010). Automation strategies for facilitating human interaction with military unmanned vehicles. In M. J. Barnes & F. J. Jentsch (eds), *Human-Robot Interactions in Future Military Operations*. Farnham: Ashgate Publishing.

Cosenzo, K., Capstick, E., Pomranky, R. A., Johnson, T., & Dungrani, S. (2009). Soldier machine interface for vehicle formations: Interface design and an approach evaluation and experimentation (ARL-TR-4678). Aberdeen, MD: U.S. Army Research Laboratory.

Dixon, S. & Wickens, C. D. (2003). Imperfect automation in unmanned aerial vehicle flight control. (AHFD-03-1/MAAD-03-1). Savoy, IL: University of Illinois Research Lab.

Dzindolet, M. T., Pierce, L. G., Beck, H. P., Dawe, L. A., & Anderson, B. W. (2001). Predicting misuse and disuse of combat identification systems. *Military Psychology*, 13(3), 147–64.

Endsley, M. R. & Kiris, E. O. (1995). The out-of-the-loop performance problem and level of control in automation. *Human Factors*, 37(2), Special issue: Telecommunications, pp. 381–94.

Evans, A. W., III, Hoeft, R., Rehfeld, S., Feldman, M., Curtis, M., Fincannon et al. (2005). Advancing robotics research through the use of a scale MOUT facility. *Proceedings of the 49th Annual Human Factors and Ergonomic Society Conference*. Orlando, FL.

Gilson, R. D., Redden, E. S., & Elliott, L. (2007). Remote tactile displays for future soldiers (TR-ARL-0152). Aberdeen Proving Ground, MD: U.S. Army Research Laboratory.

Goodrich, M. (2010). On maximizing fan-out: towards controlling multiple unmanned vehicles. In M. J. Barnes & F. J. Jentsch (eds), *Human-Robot Interactions in Future Military Operations*. Farnham: Ashgate Publishing.

Haas, E. C. & van Erp J. (2010). Multimodal research for human-robotic interactions. In M. J. Barnes & F. J. Jentsch (eds), *Human-Robot Interactions in Future Military Operations*. Farnham: Ashgate Publishing.

Haas, E. C., Stachowiak, C., White, T., Pillalamarri, K., & Feng, T. (2007). The effect of vehicle motion on the integration of audio and tactile displays in robotic system TCU displays. Unpublished manuscript. Aberdeen Proving Ground, MD: U.S. Army Research Laboratory.

Haas, E. C., Stachowiak, C., White, T., Pillalamarri, K., & Feng, T. (2008). Multimodal signals to improve safety and situational awareness in a multitask human-robotic interface. Unpublished manuscript. Aberdeen Proving Ground, MD: U.S. Army Research Laboratory.

Hoeft, R. M., Jentsch, F. G., & Bowers, C. (2005). The effects of interdependence on team performance in human-robot teams. *Proceedings of the 11th International Conference on Human-Computer Interaction Conference*, Las Vegas, NV.

Hunn, B. P. (2006). Video imagery's role in network centric, multiple unmanned aerial vehicle (UAV) operations. In N. J. Cooke, H. Pringle, H. Pedersen, & O. Connor (eds), *Human Factors of Remotely Operated Vehicles* (pp. 179–91). Oxford: Elsevier.

Jentsch, F. G., Evans, A. W., III, Curtis, M., & Fiore, S. (2005). Building low-cost simulation environments to study human-agent interaction. *Proceedings of the Interservice/Industry Training, Simulation and Education Conference*, Orlando, FL.

Jenstch, F. G., Evans, A. W., III, Fincannon, T., Keebler, J. R., Curtis, M., Ososky, S., Rehfeld, S., Feldman, M., & Cotton, J. (2007). Believability of the "man-behind-the-curtain": Methodology used in TPL HRI studies for SUS Plus Up Project.

Jones, L. A. & Sarter, N. B. (2008). Tactile displays: Guidance for their design and application. *Human Factors*, 50(1), 90–111.

Kelley, C. R. (1968). *Manual and Automatic Control*. New York: Wiley.

Krausman, A. S. & White, T. L. (2008). Detection and localization of vibrotactile signals in moving vehicles (ARL-TR-4463). Aberdeen Proving Ground, MD: U.S. Army Research Laboratory.

Lee, J. D. & See, K. A. (2004). Trust in automation: Designing for appropriate reliance. *Human Factors*, 46(1), 50–80.

Luck, J., McDermott, P. L., Allender, L., & Fisher, A. (2006). Advantages of co-location for effective human to human communication of information provided by an unmanned vehicle. In *Proceedings of the Human Factors and Ergonomics Society 50th Annual Conference*, San Francisco, October.

Meyer, J. (2001). Effects of warning validity and proximity on responses to warnings. *Human Factors*, 43, 563–72.

Minkov, Y., Perry, S., & Oron-Gilad, T. (2007). The effect of display size on performance of operational tasks with UAVs. *Proceedings of the Human Factors and Ergonomics Society 51st Annual Meeting* (pp. 1091–5).

Mitchell, D. & Chen, J. Y. C. (2006). Impacting system design with human performance modeling and experiments: Another success story. *Proceedings of the Human Factors and Ergonomics Society 50th Annual Meeting*, San Francisco, CA.

Mitchell, D. & Henthorn, T. (2005). Soldier workload analysis of the mounted combat system (MCS) platoon's use of unmanned assets (ARL-TR-3476). Aberdeen Proving Ground, MD: U.S. Army Research Laboratory.

Mitchell, D. & Samms, C. (2010). An analytical approach for predicting soldier workload and performance using human performance modeling. In M. J. Barnes & F. J. Jentsch (eds), *Human-Robot Interactions in Future Military Operations*. Farnham: Ashgate Publishing.

Murphy, R. (2004). Human-robot interactions in rescue robots. *IEEE Transactions on Systems, Man and Cybernetics: Applications and Reviews*, 34(2), 1–15.

National Research Council (2003). Building unmanned group vehicles requires more funding, greater focus. Retrieved from http://www8.nationalacademies. org/onpinews/newsitem.aspx?RecordID=10592.

Parasuraman, R., Bahri, T., Deaton, J., & Barnes, M. (1992). Theory and design of adaptive automation in aviation systems (Progress Report NAWCADWAR-92033-60), Warminster, PA: Naval Air Warfare Center.

Parasuraman, R., Barnes, M., & Cosenzo, K. (2007). Adaptive automation for human-robot teaming in future command and control systems. *The International C2 Journal*, 1(2), 43–68.

Parasuraman, R, Molloy, R., & Singh, I. L. (1993). Performance consequences of automation-induced complacency. *International Journal of Aviation Psychology*, 3(1), 1–23.

Parasuraman, R., Sheridan, T. B., & Wickens, C. D. (2000). A model for types and levels of human interaction with automation. *IEEE Transactions on Systems, Man, and Cybernetics. Part A: Systems and Humans*, 30, 286–97.

Pettijohn, B., Bodenhamer, A., Schweitzer, K. M., & Comella, D. (2007). Stero-vision on the mine protected clearance vehicle (Buffalo) (ARL-TR-4189). Aberdeen Proving Ground MD: U.S. Army Research Laboratory.

Redden E. & Elliott, L. (2010). Robotic control systems for dismounted soldiers. In M. J. Barnes & F. J. Jentsch (eds), *Human-Robot Interactions in Future Military Operations*. Farnham: Ashgate Publishing.

Redden, E., Pettitt, R. A., Carstens, C. B., & Elliott, L. (2008). Scalability of robotic displays: Display size investigations (ARL-TR-4556). Aberdeen Proving Ground, MD: U.S. Army Research Laboratory.

Sarter, N. (2007). Multiple-resource theory as a basis for multimodal interface design: Success stories, qualifications, and research needs. In A. F. Kramer, D. A. Wiegmann, & A. Kirlik (eds), *Attention: From theory to practice* (pp. 187–95). New York: Oxford University Press.

Schipani, S. L. (2003). An evaluation of operator workload during partially-autonomous vehicle operations. *Proceedings of PerMIS 2003*. Available: http:// www.isd.mel.nist.gov/research_areas/research_engineering/Performance_ Metrics/PerMIS_2003/Proceedings/Schipani.pdf.

Sellner, B., Heger, F. W., Haitt, L. M., Simmons, R., & and Singh, S. (2008). Coordinated multi-agent teams and sliding autonomy for large-scale assembly. *Proceedings of the IEEE: Special Issues in Multi-Robot Systems*.

Shoemaker, C. (2005). Interfaces for ground and aerial military robots: A workshop summary. T. Oron-Gilad (rapporteur), Washington, DC: National Academy of Sciences, pp. 5–7.

Wickens, C. D. & Hollands, J. (2000). *Engineering Psychology and Human Performance*. Upper Saddle River, NJ: Prentice Hall.

Wickens, C. D., Dixon, S., & Ambinder, M. S. (2006). Human factors of remotely operated vehicles. In N. J. Cooke, H. L. Pringle, H. K. Pedersen, & O. Connor (eds), *Advances in Human Performance and Cognitive Engineering*. Oxford: Elsevier.

Wickens, C. D., Dixon, S., & Chang, D. (2003). Using interference models to predict performance in a multiple task UAV environment—2 UAVs (Technical Report HFD-03-9/MAAD-03-1). Savoy, IL: Institute of Aviation, Aviation Human Factors Division.

Woods, D. D., Tittle, J., Feil, M., & Roesler, A. (2004). Envisioning human-robot coordination for future operations. *IEEE SMC Part C*, 34(2), 210–18.

Chapter 3
The Safe Human-Robot Ratio

Robin R. Murphy and Jennifer L. Burke
University of South Florida

Abstract

This chapter discusses the generation of the appropriate human-robot ratio for sociotechnical systems where a robot is being used for a novel task. Overall safety is considered as a driver in selecting the human-robot ratio. The human-robot literature has ignored safety, making default assumption that low human-robot ratios are desirable per se, and often pursuing an arbitrary goal of *1:many* based on expected advances in vehicle autonomy. Instead, the real goal is the human-robot ratio that maximizes both the safety of team, bystanders, and robots (success of the enterprise) as well as logistical efficiency. The chapter presents a new technique for projecting the safe human-robot ratio: viewpoint-oriented CWA. Viewpoint CWA is an extension of cognitive work analysis (CWA); it concentrates on the perceptual foci of the team members as the key affordance in determining which roles can be safely merged or made autonomous. The technique is illustrated by a review of actual human-robot teaming for urban search and rescue, a remote presence application similar to military operations in urban terrains (MOUT), where tactical users are projecting themselves into the environment via the robot. These teaming experiences include one deployment of a heterogeneous unmanned aerial vehicle-unmanned surface vehicle (UAV-USV) team to Hurricane Wilma, three deployments of UAVs to incidents (Hurricane Katrina, Hurricane Wilma, Berkman Plaza II collapse), and deployments of unmanned ground vehicles (UGVs) to five incidents (World Trade Center, LaConchita mudslides, Midas Gold Mine Nevada, Crandall Canyon Gold Mine, Utah, Berkman Plaza II collapse), plus demonstrations of basic research in UAV-UGV and UAV-USV teaming, and formal studies of mixed teams. Using the findings from applying viewpoint-oriented CWA, the chapter proposes a formula establishing the baseline safe human-robot ratio, $N_h = N_v + N_p + 1$, from which any changes would need to be justified. The chapter concludes with a discussion on how the human-robot ratio might be safely reduced and concerns that must be addressed.

1 Introduction

The expectation of several Department of Defense initiatives is a single-operator, multiple robot (SOMR) team (Agah and Tanie, 1999) or a human-robot ratio of *1:many*. However, the current state of practice for aerial systems is *many:1*, as witnessed by Global Hawk and Predator crews, and field studies indicate that ground robots are more effective with a *2:1* ratio (Burke et al., 2004; Burke and Murphy, 2004). Increased vehicle autonomy is expected to be the key to the inversion of the current human-robot ratio, with task execution as the primary metric.

The focus on task execution (*can it be done?*) neglects the larger enterprise issue of safety (*can it be done safely?*). The inability to prove safe operations of UAVs has severely curtailed the use of UAVs in the U.S. national airspace. The Federal Aviation Administration (FAA) is concerned with how UAVs will avoid collisions with other aircraft (aka "sense and avoid"), will not injure people or property on the ground, and will not impact other users of the national airspace. On the other hand, military operations have a different safety standard, possibly in part due to the military control of the airspace and the lack of liability. In terms of ground vehicles, safety concerns have retarded the transfer of autonomous navigation capabilities available since the late 1990s to highway systems in the United States. In general, safety considerations in the civilian sector have outweighed the benefits of the unmanned systems. Safety considerations in the military sector may be only an incident away from becoming equally paramount.

There appear to be four primary categories of safety concerns with unmanned vehicles: *damage or loss of vehicle*, *harm to bystanders or property*, *harm to the operations team*, and *disruption of other aircraft or missions*. The risk to the vehicle is well documented, but the risks to the bystanders, team, or disruption of other missions have not captured as much, if any, attention. The risk of injury to bystanders or property might come about from a crash, literally an aircraft falling out of the sky onto someone or something, or from a ground, air, or sea collision. In the case of UAVs, high-altitude platforms typically have some glide ratio or descent time to correct a problem and micro UAVs (MAVs) are typically derived from lightweight platforms which are cleverly engineered to minimize the consequences of a crash. However, there is always some risk to a human. The third group with safety risks is the operations team itself. Tactical teams are by definition operating in dangerous conditions, having to physically travel through, and set up on, dangerous areas. In some cases, the environment may no longer be intact, instead covered with rubble, or may have unacceptable exposure to a sniper. Teams operating strategic assets, for example Global Hawk, may be located outside the theater of war. The risk of disrupting other missions is a low probability, given the centralized control chain and improving communications in military operations, but it remains a possibility. Tactical unmanned assets are intended to be used on demand. This on-demand courts a Murphy's Law scenario where the deployment of a tactical asset has an unforeseen consequence on another operation. In civilian rescue operations, there is significant concern as to overlapping manned and unmanned activity; for example, a MAV from a rescue team may be launched near a coastguard helicopter conducting a tactical rescue. The probability of a MAV hitting a manned helicopter and causing a crash is low. What is of higher probability, given that there is little or no way of communicating in real time during a disaster, is that a MAV could interfere with manned missions. A manned helicopter has no choice except to cease operations if the pilot sees unknown aircraft. Thus, a MAV operating near a tactical rescue could cause a life-saving rescue to be aborted and thereby lead to loss of life. While this is a civilian example, it is reasonable to expect that similar scenarios exist, or will emerge, in MOUT.

One aspect of insuring the highest level of safety in unmanned systems is establishing the correct human-ratio for a particular system and application. The human-robot ratio specifies the number of operators responsible for the robot. If that ratio is too low, the system becomes at risk because anomalies and developing situations may go unnoticed until too late, or there may not be sufficient resources to manage the response or mitigate consequences. If the ratio is too high, the overall enterprise is inefficient and tactical operators may be exposed to unnecessary risks. In addition, a high ratio may lead to distractions and back-seat driving (Burke et al., 2004a).

This chapter addresses the generation of a safe human-robot ratio for tactical military operations in urban terrain, based on our extensive experience deploying ground, aerial, and sea surface robots for search and rescue at N disasters. Section 2 compares urban search and rescue with MOUT, highlighting the similarities that suggest a transfer of results between domains. Section 3 introduces the baseline human-robot ratio and describes the evolution of that formula, starting with direct observations of humans using ground robots at the 2001 World Trade Center disaster to fielding single unmanned aerial (USV) and sea surface vehicles (USV) and in heterogeneous teams. The key technique in generating the $N_h = N_v + N_p + 1$ ratio is viewpoint-oriented cognitive work analysis, which is described with examples in Section 4. The $N_h = N_v + N_p + 1$ ratio is a baseline for discussing how to reduce the ratio safely. A typical assumption is that autonomy will reduce the need for a robot operator and thus that role can be combined with a person holding a payload or mission-oriented role. Section 5 describes issues in combining roles that suggest more research is needed. We conclude that ensuring overall safe operations, at the level that would pass a National Transportation Safety Board type of accident investigation, is not trivial and requires significant additional research in human-robot interaction.

2 Similarities of Urban Search and Rescue (US&R) and MOUT

Our research has explored the US&R domain, which is similar to MOUT, and thus our findings on the safe human-robot ratio should transfer to tactical military unmanned systems. US&R has been an active research area in the U.S. and Japan since 1995, following the Oklahoma City bombing and Kobe earthquake disasters that year, which motivated the transfer of research in small rovers for planetary exploration and special operations to the rescue domain. A more complete description of search and rescue robotics can be found in (Murphy et al., 2008c).

US&R and MOUT are similar in two key ways. First, they both deal with close urban operations, where the workspace may present obstacles less than a few multiples of the vehicle's length from the structure or associated obstacles. An unmanned system may be expected to operate near a complex structure, such as a bridge, or in the cluttered region on the sides of an urban canyon. Due to the cluttered nature of the workspace, there is a high risk of collision. The obstacles in the workspace cannot be accurately modeled a priori due to scale (accurate GIS information on building layout versus actual location of power lines, signage, etc.) or variations between design and construction of a building with additional ornamenting, or damage. Furthermore, there is the possibility of human intrusion into the workspace, or below it, as the vehicle is operating in populated regions. The complexity and unpredictability of the environment tends to drive higher human-robot ratios. Second, US&R and MOUT are domains where the robot is a real-time projection of a human(s) into a remote environment, which we refer to as remote presence domains. In remote presence domains, a human will remain involved with the robot in order to "see through its eyes" regardless of the level of autonomy. This is in contrast to taskable agent domains, such as planetary exploration, where a robot is expected to operate as an autonomous agent for a task and then return with valuable information. Remote presence domains suggest that the human-robot ratio has a lower limit of the number of observers using the robot as a proxy.

As already noted, MOUT does share some significant differences with US&R which could impact the human-robot ratio. MOUT has less emphasis on safety, and liability does not appear to be a major concern. Likewise, there is a greater personal risk to tactical operators, which creates an incentive to minimize the human-robot ratio. A less obvious difference is that soldiers may have more rigorous training with an unmanned system, whereas fire-rescue departments typically have 80 hours or less, which may influence the human-robot ratio.

3 Baseline Human-Robot Ratio

Our research proposes a baseline safe human-robot ratio of:

$$N_h = N_v + N_p + 1$$

where N_h is the number of humans, N_v is the number of vehicles, and N_p is the number of payloads. This should be considered as the starting point for a human-robot system, where the impact of autonomy, interfaces, attentional demands, sparse environments, etc. may permit a lower human-robot ratio. The development of the baseline human-robot ratio represents an evolution due to expanding experiences in deploying different types of unmanned systems, increased familiarity with the human factors literature, and the development of new techniques. This section briefly reviews the evolution that started by focusing on what ratio produces the highest task performance and then considers what ratio produces the safest performance. The following sections describe viewpoint-oriented cognitive work analysis (CWA) and discuss implications for attempting to reduce the ratio by having a person hold multiple roles.

Ratio from Performance Concerns

Initial direct observation using CWA of ground robots used in disasters starting with the 2001 World Trade Center collapse (Casper and Murphy, 2003) and in staged experimentation (Burke et al., 2004a; Burke and Murphy, 2004) suggested a 2:1 ratio was appropriate. While this may be interpreted as "one to drive, one to look," experimentation in Burke and Murphy (2004) showed that responders were nine times more likely to find simulated victims if they worked *cooperatively* in teams of two and found one more victim than they would otherwise (Burke and Murphy, 2007). The roles of robot operator and payload operator, which we refer to as Pilot and Mission Specialist respectively, appear fluid. The Mission Specialist role appears shared by both team members, or "one to drive, two to look" and "two heads are better than one." Details of the experimental results are provided below.

Three field studies of mobile robots for urban search and rescue in highly realistic training exercises refute the assumption of a 1:1 human-robot ratio as an effective configuration. While only one person can drive the robot at a time, both people are required to look, and performance is significantly improved by sharing the robot's perspective in a remote search environment. Therefore, a minimum 2:1 human-to-robot ratio is required. In addition, a job-relevant task performance outcome (victim location) provides compelling evidence that effective team processes are positively related to robot operator performance.

In an observational study of five robot operators conducted during a 16-hour high-fidelity US&R disaster response drill, communication analysis was used to examine operator situation awareness and team interaction (Burke et al., 2004). The human:robot ratio was consistently 2:1 or 3:1, and team processes, particularly communication between team members, contributed to the development of situation awareness. The findings indicate that operators spent significantly more time gathering information about the state of the robot and the state of the environment than they did navigating the robot. Operators had difficulty integrating the robot's view into their understanding of the search and rescue site. They compensated for this lack of situation awareness by communicating with team members at the site, attempting to gather information that would provide a more complete mental model of the site. They also worked with team members to develop search strategies.

In a subsequent operationally standardized field study conducted in 24-hour high-fidelity training exercise in a collapsed eight-story housing project, 28 operators teleoperated a robot to search for victims in a repeatable scenario (Burke and Murphy, 2005). High SA operators engaged in more goal-directed communication (reporting, search strategy, environment) with teammates, and better SA was not only associated with higher task performance ratings (r = .81), but also led to more successful performance in locating victims. Teams with high SA operators were nine times as likely to locate the victim.

Finally, in a series of large-scale field experiments conducted with 25 dyadic US&R teams at the NASA-Ames DART facility and Lakehurst NAVAIR in New Jersey, teams where both team members had access to the robot's view of the search environment (remote shared visual presence) performed significantly better in a repeated measures search task scenario (Burke and Murphy, 2004). Moreover, distributed teams (where the team members had no visual contact) essentially found one more victim than they would have without having the shared robot's view. This confirms the claim that robot-assisted technical search requires two heads thinking, not just two bodies physically working with the robot.

Ratio from Safety and Complexity Concerns

While work with ground robots suggested an $N_h = 2 \times N_v$ ratio, experiences with small unmanned aerial vehicles at Hurricane Katrina in 2006 (Murphy et al., 2006) suggested a 3:1 ratio, or $N_h = 2 \times N_v + 1$. In the case of UAVs, a third role was identified, that of Director, who was responsible for safety and general flight operations. The Director, in part, scanned the skies for other aircraft in order to prevent disrupting the manned mission, and scanned the work envelope to make sure that bystanders had not entered the area and that the Pilot and Mission Specialist were not in danger from trip hazards. The identification of the Director role was in part due to our use of cognitive work analysis, but also to an increasing emphasis on using *roles* (Long et al., 2007) to formally define human and robot responsibilities, individually and to the team.

Experiences with a USV-UAV team at Hurricane Wilma (2006) (Murphy et al., 2008b) and two staged basic research projects in cooperative UGV-UAV and USV-UAV heterogeneous teams also showed the practical efficacy of the Director role; one person is needed to maintain an overall awareness of the entire situation. This reinforced the $N_h = 2 \times N_v + 1$ equation.

In addition, the USV-UAV experiences also illustrated that there may be multiple Mission Specialists, leading to a reformulation of the ratio as $N_h = N_v + N_p + 1$. The AEOS-1 and Sea-RAI unmanned surface vehicles in the studies above required a minimum of three people to operate when inspecting littoral structures: a Pilot, a Mission Specialist trained at interpreting the underwater sensor, and a Mission Specialist to look above the waterline, directly overhead, or $N_v = 1$, $N_p = 2$. Observations have shown that it is hard for a single Mission Specialist to adequately inspect above and below waterline simultaneously. Thus, N_h is not merely a function of the number of vehicles, but the number of *perceptually distinct* payloads.

As we began to formally define the roles held by the human and the robot in order to better understand how to reduce the ratio, we created a variation of cognitive work analysis which captures the perceptual viewpoints held by each role. As a result of viewpoint-oriented cognitive work analysis, described in Section 4, we identified potential safety issues in attempting to reduce $N_h = N_v + N_p + 1$ to $N_h = N_p + 1$ strictly by automating the Pilot role. In particular, the significant differences in viewpoints suggest that a Mission Specialist would not be able to transition quickly enough to take over an autonomy failure in a UAV, though perhaps for a UGV or USV depending on the circumstances. This inability for a human to be a reliable backup for autonomous operation of a robot is an example of the human out of the loop (OOTL) control problem (Endsley and Kaber, 1999) from the human factors literature. This is discussed in more detail in Section 5 on how to reduce the ratio.

One concern is about this $N_h = N_v + N_p + 1$ formula is that it projects a higher number (three humans) to safely operate a ground vehicle, while in practice, a 2:1 ratio has emerged as effective for ground vehicles for US&R (Burke et al., 2004). In the case of ground vehicles for US&R, a 2:1 ratio is sufficiently safe because the US&R enterprise already has a safety officer who is explicitly charged with the safety of all rescuers. A separate safety officer who would include the safety of the robot is not needed in this case because the robot operates beyond the line of sight. However, if a safety officer is not present, then a 2:1 ratio may not be justified. In the case of ground vehicles for military operations, the 3:1 ratio may be more realistic: two soldiers directly working with the robot and sensors, the third serving as a lookout. The studies by Burke et al. (2004) and Burke and Murphy (2004) suggest that having a single soldier focused on the robot, while the other serves as lookout, will not be as productive as a 3:1 ratio. And the concerns raised by viewpoint-oriented CWA suggest that a single soldier will be at risk if expected to share attention between personal safety and the robot. This is discussed more in Section 5.

4 Viewpoint-Oriented Cognitive Work Analysis

Our approach to determining human-robot ratios for novel, tactical uses of robots has been to use the broader cognitive work analysis (CWA) (Rasmussen et al., 1994; Vicente, 1999) technique rather than the more focused cognitive task analysis (CTA) and its variants, such as goal-directed task analysis (GDTA), which dominate the HRI literature. CWA is more appropriate for novel, tactical uses for three reasons. First, CWA is explicitly designed to capture domains where new technology is being inserted into the workplace and the practices have not been codified, i.e., it is formative. In contrast, task analysis assumes that the task definition and mechanisms are mature, i.e. it is normative. Thus, CWA would be applied first to form the initial, broad understanding of the domain. Second, in novel applications, there is a high rate of innovation and worker adaptation. As a result, the tasks and roles typically evolve through practice in the field, and new uses or workarounds for the technology are discovered. If there are manuals or prior practices (such as how a human would do it), these are generally adapted on the fly, the changes are not codified or may not be even noticed, and may take some time to converge on a mature form. The dynamically changing nature of technology insertion is a major concern of CWA (Vicente, 1999) and CWA can be thought of as providing the ontology of the domain which permits CTA techniques to be applied as the specific tasks and procedures evolve. Third, CTA, especially its GDTA variant, is generally used to create user interfaces to the technology. While user interfaces influence the human-robot ratio, understanding the human-robot ratio is a broader endeavor. Figure 3.1 illustrates the relationship between CWA and CTA for novel sociotechnical systems.

We find CWA essential in establishing a human-robot ratio for *formative* tasks with technology that has never been deployed, or only a few times. But while CWA is excellent at capturing worker adaptation, it has been insufficient for projecting how the human-robot ratio can be *reduced* with the addition of autonomy. Therefore, we have developed a variation of CWA which takes into account the viewpoint of each role. We believe the viewpoint—what the team member looks at (the robot, the environment, the sensor data, etc.)—is the affordance for whether roles can be shared or whether the perception needed for autonomy is sufficient for that role. This section presents the viewpoint-oriented cognitive work analysis and provides two examples of how it leads to the projection of a safe human-robot ratio and the constraints on autonomy.

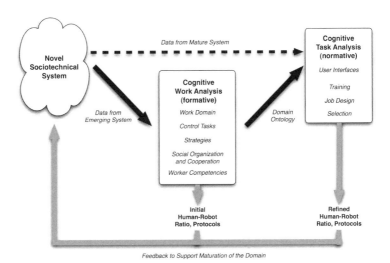

Figure 3.1   Relationship of CWA and CTA

Viewpoint-Oriented Cognitive Work Analysis

Following Vicente (1999), cognitive work analysis can be divided into five components or "mini-analyses." Each of these components contributes to an understanding of the safe human-robot ratio as described below. The data collected for each of the mini-analyses is largely from direct observation and participation in the US&R enterprise.

Work domain analysis   The work domain analysis provides a model of the domain which can then be used by goal-directed techniques. As noted in Vicente (1999, p. 168), a work domain analysis is helpful because "the greatest threat to safety in complex sociotechnical systems is events that are unfamiliar to workers and that have not been anticipated by designers." By establishing the safety needs and potential sources for surprise (though, by definition, the list of sources can never be complete), a work domain analysis contributes to the generation of an appropriate human-robot ratio.

Control task analysis   The control task analysis identifies the specific, reoccurring tasks in the domain and focuses on what needs to be done, not by whom or how Vicente (1999). The abstraction of tasks supports a discussion of whether a robot, software agent, or human is best suited for the task, which is fundamental to the human-robot ratio.

Strategies analysis   The strategies analysis captures how the tasks are done. How a task is done may include decision-making algorithms or heuristics, what actions require teams and sharing, and general transfer of control or responsibilities. The strategies used by mixed teams provide a basis for the human-robot ratio, as they reflect how people have adapted the sociotechnical system. Any changes to the human-robot ratio will have to consider the impact on these strategies and mitigate unforeseen consequences.

Social organization and cooperation analysis   The social organization and cooperation analysis captures the social and technical factors that exist in the domain that influence how work is accomplished. These factors include communication mechanisms and paths, management or behavioral constraints, and location of team members. The social organization and cooperation analysis is especially helpful for projecting an acceptable human-robot ratio. For example, team members in the US&R domain work in groups of two (the "buddy system"), so a 2:1 human-robot ratio is natural and there is little pressure to have a 1:1 or *1: N* ratio (Murphy, 2004). The social organization of MOUT may be different, and thus the analysis may identify a different ratio.

Worker competencies analysis   The worker competencies analysis addresses the human factors aspect of a work domain. Our work expresses worker competencies as *roles*. A role has one or more tasks associated with it (control task analysis) and will execute a strategy (strategies analysis). A team member may hold one or more roles concurrently or sequentially (social organization and cooperation analysis). Each role requires some core knowledge or skill. The roles influence the human-robot interaction for at least two reasons. One is that the demands of a role may be unique to a particular team member. For instance, the robot is the physically situated agent and that role cannot be transferred to another team member, whereas a role of observer may become automated by pattern recognition software. Another reason is that the number of roles will influence the workload that can be absorbed by a human team member; clearly a human may not be able to reliably handle a dozen distinct and different roles.

We have extended CWA to emphasize the consideration of the *perceptual viewpoints* associated with each team member or role. Our experience has been that role assignment is directly related to the necessary viewpoint or focus of perception. To date, we have identified three perceptual viewpoints in mixed teams: egocentric (view through the robot's sensory systems); exocentric (view of the external relationship of the robot or team to the environment); and mixed (swapping or shared ego- and exocentric views) (Murphy et al., 2008a, 2008b). For example, in US&R there is a need for a safety overwatch—someone to make sure that bystanders have not entered the workspace, that the operators are not about to step off rubble and injure themselves, or that the robot is not at risk from some danger not seen by the operator. This safety overwatch viewpoint is exocentric and quite different from the egocentric viewpoint of the mission specialist looking through a sensor. As will be discussed later in the chapter, viewpoint-oriented cognitive work analysis provides a foundation for considering how roles may be merged. If one role relies on an egocentric viewpoint and another relies on an exocentric viewpoint, the gap between the viewpoints may be too extreme to safely bridge.

Example: Homogeneous UAV Team at Hurricane Katrina

The advantage of viewpoint-oriented cognitive work analysis for role assignment can be seen by in the first known use of MAVs for structural inspection in the aftermath of Hurricane Katrina. Using lessons learned from flights during the actual response to Katrina in Mississippi, the Center for Robot-Assisted Search and Rescue (CRASAR) returned to survey damage to multistory commercial structures along the Gulf Coast two months after the storm. The four-person team flew five days for a total of 30 flights, giving ample opportunity for best practices to emerge.

In this application, a 3:1 human-robot ratio emerged based on the viewpoint-oriented cognitive work analysis. The work domain analysis indicated that safety was an important consideration and that line-of-sight (LOS) operations were the only acceptable deployment scheme. The control task analysis showed that the primary task of flying to obtain structural data consisted of short flights (in the order of 3–8 minutes) at low altitudes (under 400 feet), often within two body lengths of a structure, tree, or power line. This meant that a significant portion of the flight was either take-off and landing or maneuvering near obstacles in extremely unpredictable wind shear conditions. These conditions are challenging for a human pilot, and even more so for autonomous control; therefore the robot had to be teleoperated. The strategies and worker competencies analysis documented the emergence of a three-person team to handle the responsibilities associated with safe flights. By examining the viewpoints needed to accomplish these roles and considering the technical factors from the social organization and cooperation analysis, the 3:1 ratio was deemed to be the safe ratio for the current state of technology.

Three distinct roles for humans emerged, Pilot, Mission Specialist, and Flight Director, and after the second day of flying, one human was assigned to execute one role exclusively. The three humans are shown in Figure 3.2. Each of these roles relied on a different perspective of the operation: egocentric (or through the robot's "eyes"), exocentric (external view), or mixed (swapping between egocentric and exocentric). Each role also utilized a different degree of computer mediation of perception: total (human sees only mediated output); none (human sees world directly); and on-demand (human can choose between mediation or none). The unique perspective and mediation of each role is shown in Figure 3.3.

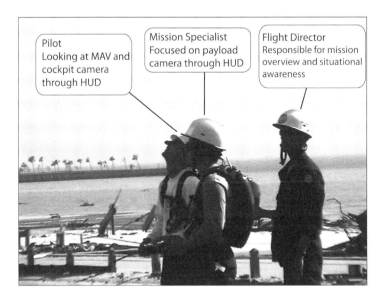

Figure 3.2   Human team members and role assignment

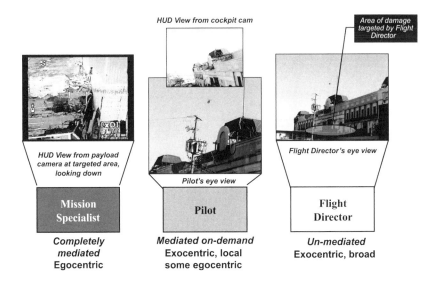

Figure 3.3   Roles: perspective and mediation of each role

Pilot   The Pilot was responsible for teleoperating the vehicle within LOS, assessing general airworthiness of the vehicle prior to and during flight, and general maintenance. This role demands a high degree of training and concentration, more than a typical radio-controlled hobbyist, because of the operations in close proximity to structures. Weather conditions, especially sunlight, added to fatigue. In general, operating at midday provided the best flying conditions; however, this is not possible for the response phase of a disaster. The Pilot had the authority to refuse to fly or to terminate a flight for any reason. The Pilot worked with the Flight Director to determine the appropriate Landing Zone and envelope of operations. As shown in Figure 3.3, the Pilot's perspective of operations was usually exocentric and unmediated (i.e., directly watching the MAV), but during the post-Hurricane Katrina activity it could shift on demand to an egocentric, mediated view from the MAV's forward-facing camera in the heads-up display (HUD).

Mission Specialist   The Mission Specialist was responsible for collecting the data. The Mission Specialist would look at video output from the MAV high-resolution camera and direct the Pilot to a hovering station, thereby framing the shot; then, while the MAV hovered, he would adjust the pan-tilt and zoom to get the best photograph. The Mission Specialist would declare when enough data had been collected and the flight had met its objectives. The skill set for this role was the ability 1) to determine damage to photograph and 2) to operate the pan-tilt gimbal and the zoom camera. Due to the complexity of the gimbal, the camera operation skill set took about four hours of training to become proficient. Ideally the Mission Specialist would be a structural specialist. In the Katrina recovery case, the team structural specialist (who was serving as the Flight Director) instructed the Mission Specialist on what to take pictures of during rehearsal. The Mission Specialist's perspective was egocentric with the robot and mediated by the HUD of the camera imagery.

Flight Director   The Flight Director was responsible for the overall safety of the team members, both human and MAV, and mission situation awareness, and had the authority to cease the flight at any point. The Flight Director performed the initial site safety assessment to make sure the flight area was unoccupied and to watch for intrusions from humans entering the workspace and other aircraft approaching the airspace. The Flight Director also proscribed the placement and movement of the Pilot and Mission Specialist during the safety review, and stood near throwable personal floatation devices when the team was working near water. The Pilot had a tendency to move to get a better viewing angle but could not look down without losing sight of the MAV and concentration. Whenever the Pilot moved, the Mission Specialist would move as well to stay within verbal conversation range, but also did not look away from his task. Therefore, the Flight Director would move to physically protect or guide the team when they moved.

Example: Heterogeneous USV-UAV Team at Hurricane Wilma

Hurricane Wilma in 2005 provided the first use of a USV-UAV team for urban search and rescue. The robots were used to evaluate damage to the building, seawall, and docks of the Marco Island Yacht Club and to the super- and substructure of the CR571 bridge connecting Marco Island with the mainland of Florida. Based on the experiences at Wilma, we consider the baseline human-robot ratio for an MAV-USV team to be a five-person team: MAV Pilot, MAV Mission Specialist, USV Pilot, USV Mission Specialist, Flight Director. The USV Mission Specialist operated the below-the-waterline sensor. This would be a 5:2 ratio, with three distinct human roles (Pilot, Mission Specialist, and Flight Director), but multiple instantiations of the Pilot and Mission Specialists roles. However, viewpoint-oriented CWA suggests a sixth person would be needed for complete bridge inspection—the USV Mission Specialist role was really two roles, one focused on interpreting below-the-waterline imagery and the other focused on the above-the-waterline sensing for damage. In the field trials, there was no visual inspection above the waterline.

5 Viewpoint-Oriented CWA and Combining Roles

$N_h = N_v + N_p + 1$ poses a fairly high human-robot ratio and requires justification for what roles can be safely combined or shared. Viewpoint-oriented CWA can help determine candidates for combination, as well as explain tendencies to combine or share roles independently of safety. UGVs provide an example of the tendency to combine or share roles independently of safety. Since many UGVs use video for both navigation and for the mission, there is the assumption that the Pilot and Mission Specialist roles can be combined. This is not supported by field studies and does not generalize when the UGV carries a more diverse sensor payload. MAVs provide an example of the importance of considering role combination or sharing in terms of the out-of-the-loop (OOTL) control problem, as OOTL control studies show that humans cannot switch roles quickly. The push for controlling multiple UAVs with a single human appears predicated on an air traffic controller metaphor; viewpoint analysis suggests this metaphor is flawed and will not lead to safe operations. The point of the discussion is not to insist on high human-robot ratios or to argue that autonomy cannot reduce these ratios; rather the point is that the challenges in reducing the human-robot ratio while maintaining safety are being underestimated.

Misleading Assumptions from Shared Sensors

Viewpoint-oriented CWA may explain why the current 2:1 human-robot ratio for UGVs for US&R is often ignored and a 1:1 ratio is used in simulated events such as RoboCup Rescue. This 1:1 assumption appears to be an artifact of the Pilot and Mission Specialist using the single exproprioceptive sensor, the video camera, to perform their missions, rather than a careful analysis of the roles and perceptual activities needed to accomplish those roles. Viewpoint-oriented CWA helps uncover these hidden assumptions, which can then be explicitly justified or retracted.

The 1:1 ratio in RoboCup Rescue might be explained with viewpoints as follows. In general, the Pilot and Mission Specialist both use a single, common, egocentric video camera. This means that the Pilot and the Mission Specialist both typically share the same viewpoint, despite different functions derived from that viewpoint. Since they are sharing identical viewpoints, it is easy to assume that the roles can be shared, despite the performance issues identified in Burke et al. (2004) and Burke and Murphy (2004). Note that $N_h = N_v + N_p + 1$ makes these assumptions explicit.

Viewpoint-oriented CWA shows that although the viewpoints are shared, the functions are not and therefore safe role sharing is not guaranteed. If the use of additional mission sensors such as thermal imaging become prevalent in RoboCup, the role sharing would become harder (as observed in our field studies). According to $N_h = N_v + N_p + 1$, the initial 2:1 ratio would need to be expanded to 3:1 to handle the added egocentric viewpoint from a different sensor, i.e., that there are two Mission Specialist roles. Indeed, our work with the Localize, Observe, report Victims, Report (LOVR) protocol has maintained the 2:1 ratio by serializing the Mission Specialist roles so that they can be conducted by one person. The protocol has the search of an area conducted first with one sensor (Mission Specialist 1's role), then with the second sensor (Mission Specialist 2's role).

Our experiences in the field with combining the Mission Specialist roles for each sensor via side-by-side displays of the sensors have not been positive, though this may be a function of technology and interface design. Side-by-side displays allow multiple streams of sensor-centric data to be displayed, but do not guarantee that the human is attending to the right stream at the right time. We have noted that video and thermal imaging have different fields of view, resolutions, and general information characteristics. Certainly work in better displays, particularly fusing the streams into a single display, will be important. We see fused displays as being helpful when the sensors are operating over a relatively small viewpoint, such as with ground vehicles where the sensors are facing the same direction. The possibility of fusion becomes less credible for combining more disparate sensors operating over larger volumes of space, such as the above and below the waterline sensors on a USV. While a combined display such as a 3D volumetric range reconstruction of the environment is possible, it may actually degrade the Mission Specialist's performance.

Autonomous or semi-autonomous perception will likely be of benefit in combining Mission Specialist roles, though perception has resisted automation and the Mission Specialist role may never be totally eliminated from a remote presence task where the intent is for a human to project themselves into an environment. As far back as 1998, we have advocated the use of software to extract affordances, or cues, of victims to permit concurrent sensor processing by a single Mission Specialist. In this case, the Mission Specialist used the video camera to manually search for a victim while autonomous software agents searched the video and thermal images for affordances (color, motion, heat). It would then alert the Mission Specialist to the presence of a cue. It is possible that with advanced software agents, the Mission Specialist role(s) can be reduced to reactive, on-demand interpretation. Then the Mission Specialist role could perhaps be executed serially with the Pilot role for a UGV.

The OOTL Problem in Combining Roles

The previous section has already used viewpoint-oriented CWA to project a minimum 3:1 safe human-robot ratio for single robots. Unfortunately, the viewpoint of each role suggests that these roles will be extremely difficult to safely combine due to the out-of-the-loop (OOTL) control problem. A good example of the OOTL problem is the use of MAVs for US&R.

As MAVs become more autonomous, the role of Pilot may become a shared role between the vehicle and the human Pilot. The human portion of this shared role is generally referred to as the Pilot-in-Command; that is, the human retains ultimate responsibility but the flight authority remains with the vehicle until an exception occurs. In an emergency, the Pilot-in-Command would be expected to assume control of the vehicle.

However, viewpoint-oriented CWA suggests that combining the Pilot-in-Command with either the Safety Officer or the Mission Specialist roles will lead to a *human out-of-the-loop (OOTL)* control problem. The human factors community notes that there are often significant delays in a human assuming control of an autonomous process (Endsley and Kaber, 1999). In general, the more autonomous the process, the harder it is for the human to react to a problem correctly in time because the human does not maintain the necessary situation awareness for control of the process. The egocentric, environment-based frame of reference held by the Mission Specialist is quite different from the exocentric or possibly mixed exocentric/egocentric on the horizon viewpoint needed by a Pilot. The differences in these two views seem difficult to overcome in a few seconds. The exocentric view of the Safety Officer is closer to the exocentric view needed for the Pilot; but the scope is quite different, as the Safety Officer is responsible for a large context and the Pilot is focused on a narrow element. In addition, during an emergency, the Pilot role will, by necessity, capture the human's attention. This means the Safety Officer role will be vacated, possibly at the moment when it is needed most for human team member safety (e.g., preventing the humans from becoming distracted and stepping into danger, etc.).

A common approach to sharing the Pilot role is to serialize the sharing of roles between robot and human; yet the OOTL problem with the human as the Pilot-in-Command suggests serialization will not be sufficiently safe with current levels of autonomy. In serialization scenarios, the flight would begin with the human in the Pilot role focused on manual or supervised take-off; take-off and landing take up a significant portion of close urban operations flight time, yet are notoriously difficult to automate (Murphy et al., 2008a). When the MAV reached the nominal operating altitude, the vehicle would be switched to autonomous operation. The human would then assume the role and viewpoint of Mission Specialist, directing the vehicle by directing the sensor, for example by clicking on an area that the Mission Specialists wanted to inspect in more detail. The robot would be expected to maintain stable flight and avoid collisions with buildings, trees, power lines, etc. When the mission was completed, the MAV would hover until the human felt comfortable reassuming the Pilot role. The human would then land or supervise the landing of the MAV.

While serialization may appear to manage viewpoints (exocentric, egocentric, then exocentric), it is not a safe solution because it neglects the Pilot-in-Command role. In practice, the human does not swap the Pilot role with the robot and then add the Mission Specialist role, but rather swaps the Pilot role for a shared Pilot role, where the robot is the *Local* Pilot and the human the Pilot-in-Command, and adds the "full-time" Mission Specialist role to the "part-time" Pilot-in-Command role. But, as seen with viewpoint-oriented CWA, the Mission Specialist and Pilot-in-Command roles have fundamentally different viewpoints (egocentric and exocentric, respectively) and these differences in viewpoints exacerbate the OOTL problem. Another approach to reducing the human-robot ratio is increased perceptual autonomy, which could in theory eliminate the Mission Specialist role altogether. However, as noted in the discussion of UGVs, autonomous perception remains a challenge and is not expected to materialize in the near future. At best, the workload associated with the Mission Specialist role could be reduced. Could the workload be reduced enough that the egocentric Mission Specialist could successfully absorb the exocentric Pilot-in-Command role? Again, the same issues arise as seen in role combining under serialization. Given the conservative nature of safety considerations, extensive experimentation would be needed to justify a single human handling multiple roles.

The Air Traffic Control Metaphor Fallacy

The previous discussion suggests that the safe ratio is currently *N:1* for a single robot or *N:M* where *N>M* for multiple vehicles, whereas the introduction noted that the implicit goal is *1:N*. In some cases, the justification for a single human operating multiple robots appears to be based on the *air traffic controller (ATC) metaphor*: UAVs will become sufficiently autonomous such that a single person can manage the fleet or swarm much as an air traffic controller manages flights within a region, and since an ATC can handle multiple aircraft safely, a single human will be able to handle multiple UAVs safely. The ATC metaphor is fallacious for projecting the safe human-robot ratio because it ignores the Pilot-in-Command role, which is unique to unmanned vehicles. If a manned air vehicle encounters difficulties, the ATC does not assume control of the aircraft; the Pilot on board remains in charge of each individual aircraft and responds to the local problem. In unmanned aircraft, the human is expected to be the Pilot-in-Command of each vehicle and to take over operations in case of a problem. In terms of perceptual viewpoints, the human is expected to go from a broad, exocentric viewpoint of "dots on a screen" to actually flying the vehicle in a degraded and possibly unknown state, which will likely require mixed exo- and egocentric viewpoints. Clearly, this is an extreme form of the OOTL control problem and will have to be addressed before *1:N* systems will be truly safe.

6 Summary

In summary, this chapter proposes a formula based on viewpoint-oriented cognitive work analysis for projecting the baseline, or worst-case, human-robot ratio needed for safe operation of one or more robots: The $N_h = N_v + N_p + 1$ formula states that the number of humans N_h is the sum of the number of vehicles N_v and the number of payloads on those vehicles N_p, plus an additional safety officer (1). It is independent of type of vehicle or payload.

The resulting number is generally high and it is expected that this formula will be used as the starting point for justifying why a different (lower) ratio is used. Given that this formula is essentially the human-robot ratio for teleoperation, autonomy or semi-autonomy is expected to be able to reduce the overall ratio for a specific team. However, the different visual perspective (egocentric, exocentric, mixed) and scope (narrow, broad) of the team roles identified to date (Pilot,

Pilot-in-Command, Mission Specialist, Flight Director, Safety Officer) suggest that combining some roles may not be safe and will require close scrutiny. In particular, the out-of-the-loop control problem appears to be a hurdle to shared autonomy. Expecting a soldier to drive/fly a robot, interpret sensor data, and remain sufficiently aware of the surroundings is unrealistic with significant advances in mixed-team processes, user interfaces, and computer support work.

Autonomous operations will have to be reliable and address human factors issues in order to earn approval from civilian safety agencies such as the Federal Aviation Administration (FAA), which controls the National Airspace and impacts Department of Defense operations in the United States. Safety concerns may impact military operations in general. These concerns go beyond the safety of military personnel, and extend to safety of larger missions and civilian personnel and property.

- $N_h = N_v + N_p + 1$ is the baseline for a human-robot ratio for safe operations, based on observational studies at disasters with UGV, USV, UAVs, and heterogeneous teams; and based on staged experimentation with urban search and rescue personnel.
- Urban search and rescue is very similar to military operations in urban terrain.
- Two human ground robot operators working cooperatively have significantly higher performance than a single operator.
- The ratio can be reduced based on the density of the work envelope, mission needs, similarities in viewpoints, and general risk factors.
- Autonomous navigation may not be sufficient to safely reduce the ratio due to the human out-of-the-loop control problem.

References

Agah, A. and Tanie, K. (1999). Taxonomy of research on human interactions with intelligent systems. In *Proceedings of 1999 IEEE International Conference on Systems, Man, and Cybernetics*, pp. 965–70.

Burke, J.L. and Murphy, R.R. (2004). Human-robot interaction in USAR technical search: Two heads are better than one. *IEEE RO-MAN 13th International Workshop on Robot and Human Interactive Communication*, Kurashiki, Okayama, Japan.

Burke, J.L. and Murphy, R.R. (2005). Situation awareness and task performance in robot-assisted technical search: Bujold goes to Bridgeport. Tampa, FL: Center for Robot-Assisted Search and Rescue (CRASAR-TR2004-23).

Burke, J.L. and Murphy, R.R. (2007). RSVP: An investigation of remote shared visual presence as common ground for human-robot teams. *ACM/IEEE Human-Robot Interaction*.

Burke, J.L., Murphy, R.R., Coovert, M.D., and Riddle, D.L. (2004). Moonlight in Miami: A field study of human-robot interaction in the context of an urban search and rescue disaster response training exercise. *Human-Computer Interaction*, 19, 85–116.

Casper, J. and Murphy, R.R. (2003). Human-robot interactions during the robot-assisted urban search and rescue response at the World Trade Center. *IEEE Transactions on Systems Man and Cybernetics. Part B: Cybernetics*, 33, pp. 367–85.

Endsley, M. and Kaber, D. (1999). Level of automation effects on performance: Situation awareness and workload in a dynamic control task. *Ergonomics*, 42(3), 462–92.

Long, M.T., Murphy, R.R., and Hicinbothom, J. (2007). Social roles for taskability in robot teams. *IEEE International Conference on Intelligent Robots and Systems*, pp. 2338–44,

Murphy, R.R. (2004). Human-robot interaction in rescue robotics. *IEEE Systems, Man and Cybernetics Part A*, special issue on Human-Robot Interaction, 34(2, May), 138–53.

Murphy, R.R., Griffin, C., Stover, S., and Pratt, K. (2006). Use of micro air vehicles at Hurricane Katrina. *IEEE Workshop on Safety Security Rescue Robots*, Gaithersburg, MD, August.

Murphy, R.R., Pratt, K., and Burke, J.L. (2008a). Crew roles and operational protocols for rotary-wing micro-UAVs in close urban environments. *ACM/IEEE Human-Robot Interaction*, Amsterdam, March.

Murphy, R.R., Steimle, E., Griffin, C., Cullins, C., Hall, M., and Pratt, K. (2008b). Cooperative use of unmanned sea surface and micro aerial vehicle at Hurricane Wilma. *Journal of Field Robotics*, 25(3), 164–80.

Murphy, R.R., Tadokoro, S., Nardi, D., Jacoff, A., Fiorini, P., and Erkmen, A. (2008c). Rescue robotics. In A. Zelinkski and B. Sciliano (eds), *Handbook of Robotics*. Berlin: Springer-Verlag.

Rasmussen, J., Pejtersen, A., and Goodstein, L. (1994). *Cognitive Systems Engineering*. New York: Wiley.

Vicente, K. (1999). *Cognitive Work Analysis: Toward Safe, Productive and Healthy Computer-based Work*. Mahwah, NJ: Lawrence Erlbaum Associates.

PART II
Foundations of HRI

Chapter 4

The Cognitive Psychology of Human-Robot Interaction

Douglas J. Gillan
North Carolina State University

Jennifer Riley
SA Technologies

Patricia McDermott
Alion Science

Technology extends human capabilities. For much of human history, technological development focused on enhancing physical abilities—for example, hammers increase our strength; transportation technologies extend our ability to move (Gillan & Schvaneveldt, 1999). In contrast, many technological innovations since the mid-twentieth century have focused on enhancing human cognitive abilities—for example, computing technologies extend our ability to process, store, and represent symbolic information (Gillan & Schvaneveldt, 1999).

When the Czech playwright Karel Čapek first coined the term robot (from the Czech *robota*, meaning work or forced labor) in his 1921 play *R.U.R.* (*Rossum's Universal Robots*), he foresaw technological development that would extend the ability to perform virtually any task (Čapek, 1921/2004). Interestingly, he foresaw the use of robots to perform both physical and intellectual tasks, since the Czech word *rozum* means "mind" or "intellect." The robots that Čapek described were human-like androids capable of supplanting humans. In the play, the robots seem to value physical work in that they spare the life of one human, based on their observation that he works with his hands.

As in Čapek's vision of robots, modern robots' capabilities span a wide variety of tasks. We constrain this chapter by focusing on a set of robotic tasks related to military contexts. The goal of this chapter is to examine selected cognitive processes that operators of robotic systems will need to apply as they control, supervise, and/or monitor robots performing these tasks. Consequently, we have also restricted our analysis of cognitive processes to those most closely associated with an operator's activities while a robot performs one of these tasks.

Robotic Tasks

Our functional classification focuses on the task goal and tries to avoid artificially splitting the activity into independent tasks when they are truly interrelated. Four functions, which soldiers currently perform and for which they are likely to use robots, include:

- navigation;
- remote sensing;
- telemanipulation;
- enemy engagement.

Navigation

Navigation can be considered as an individual function or as a subtask to a function, such as remote sensing. As its own function, navigation might be simply defined as movement from point A to point B, but, from the operator's point of view, navigation is more complex. The operator must understand where the robot is in relation to where it needs to be, detect and avoid obstacles, and make adjustments to the course as necessary. Often there is latency between the control inputs and robot action or between robot action and the update of the robot's position. Depending on the sophistication of the robot, the operator could be teleoperating the robot or monitoring its progress and providing input only when necessary (e.g., acknowledging a waypoint reached or approving a course deviation). However, even the most sophisticated autonomous robots may need to be teleoperated, for example, if the robot were stuck in the terrain or were to lose global positioning system (GPS) satellite communications and become lost. Operators may perform navigation (without remote sensing) when moving an unmanned vehicle to a named area of interest, returning an unmanned vehicle to base, flying an unmanned aerial system to a target, or when moving an ordnance robot within the vicinity of a suspected explosive device.

Remote Sensing

Remote sensing encompasses two interrelated tasks, scanning and navigation. Scanning involves target search, detection, location, and identification. For reconnaissance, the target could be an enemy tank or a suspicious truck. For search and rescue, the targets are wounded or trapped victims. Operators could also use robots to search for enemy weapons and materials, diagram the layout of a building, monitor traffic patterns in and out of a suspected terrorist headquarters, or measure hazardous material levels in the air or water. Navigation has its own demands, as described in the previous paragraph. These challenges are compounded by the fact that the operator may have to perform both scanning and navigation simultaneously. Not only does this involve multitasking, but the motion itself may alter perception.

Operators must filter the data and determine what information to pass to other stakeholders who are interested in interpreting the remotely sensed data, for example, to find a trapped person in an urban search and rescue task. Unmanned aerial vehicles (UAVs) allow operators to monitor ground activities from a bird's eye view while the UAV is perpetually in motion, usually on a predetermined track. Unmanned ground systems are capable of collecting streaming video and still snapshots while navigating from point to point.

Telemanipulation

Robots can be used to perform manipulations—disarming bombs, inspecting potential explosive devices, installing equipment in space, or performing repairs under water—with the operator located out of harm's way. Typically, the operator uses the controls to move a robotic arm into place and to perform fine control movements, such as connecting wires or carefully handling explosives, with an end effector. The operator and robot manipulate objects by pushing, extracting, inserting, and rotating. The operator determines what needs to be manipulated, either specifies how to do the manipulation or executes the manipulation, and ensures that the intended effect is accomplished. One unique challenge for the operator is to use mental imagery to understand where an object is in space and how it is oriented in relation to other objects and in relation to the robot.

Enemy Engagement

Robots can be used to engage the enemy, either in offensive or defensive operations. Although not currently prevalent, this function is more likely in the future, when more armed robotic systems are developed. The Predator Unmanned Aerial System can be used to suppress enemy targets. The U.S. Army's planned Armed Robotic Vehicle-Assault (ARV-A) is a semi-autonomous platform designed to accompany Future Combat System (FCS) platoons. It will mount anti-tank missiles, automatic cannon, and a high-rate-of-fire machine gun. The ARV-Reconnaissance version will carry a cannon for self-defense. Because these engagements will take place at a distance from human operators, remote sensing and targeting will be essential aspects of engaging the enemy.

Cognition and Human-Robot Interaction (HRI)

The tasks previously described—navigation, remote sensing, telemanipulation, and enemy engagement—all make use of robots' capability to extend the distance across which humans operate by allowing us to perceive information from distant environments and to perform manipulations in those environments. As a result of that capability, humans can operate in hostile environments such as outer space, under the ocean, amidst disaster rubble, and surrounded by nuclear or biological hazards. Robots can also increase our ability to detect, locate, identify, and track remote objects, including other humans. However, the restriction of the information bandwidth from those remote locations can disrupt the ability to develop and maintain situation awareness and to make sense of the situation (Riley et al., 2006). Thus, a critical component of the human-robot interface in support of the operator's cognition will be to help enhance situation awareness.

In addition to developing and maintaining situation awareness, the operator will need to use spatial information to understand the characteristics of the environment in which the robot performs tasks, including its position relative to landmarks and memory for paths that the robot has taken. As a consequence, we believe that an understanding of human spatial cognition, especially mental maps, will be key to the design of human-robot interfaces.

The section above describes each task separately. However, operators who control or monitor robots performing military functions on a battlefield will often find themselves interweaving tasks like scanning and navigation throughout these functions (e.g., Chadwick et al., 2004). Consequently, another challenge facing operators is to switch among these tasks (often very rapidly), and a corresponding challenge facing HRI interface designers will be to improve the efficiency of this task switching and multitasking.

Within the topic of cognition, we will address these key issues—situation awareness and sensemaking, spatial cognition, and task switching. Thus, we have selected higher-level macrocognitive functions characteristic of experienced operators (situation awareness and sensemaking), as well as lower-level, microcognitive processes employed to deal with HRI challenges of attention management during task switching and the memory and navigational processes involved with spatial cognition. There is no question that the set of cognitive processes that we have selected is far from exhaustive. Other chapters could be and should be written about HRI and naturalistic decision-making; planning, adaptation, and replanning; and team cognitive processes (on the macrocognitive side) and about HRI and working memory, reasoning, and motor control (on the microcognitive side).

Situational Awareness (SA)

Several aspects of the operator or system can influence the effectiveness of human interaction with robotic systems:

1. the user's skill (in the task context and in operating unmanned systems);
2. the knowledge that the operator can acquire for task completion;
3. the system design and capabilities, including interfaces and control mechanisms.

The overall operational performance in tasks involving control of unmanned vehicles (UVs) can be affected if there are limitations and/or breakdowns in any of these areas.

An important issue in the control of remotely operated vehicles is the development of operator situation awareness (SA), defined as "the *perception* of elements in the environment within a volume of time and space, the *comprehension* of their meaning, and the *projection* of their status in the near future" (Endsley, 1988, 1995). Endsley's definition and model of SA illustrate three levels of SA: perception, comprehension, and projection. Level 1 is perceiving basic elements of the environment—the status, attributes, and dynamics of relevant elements in the environment. This is the process of monitoring, detecting critical cues, and recognizing objects, events, people, environment aspects, etc. and their current states. Level 2 involves comprehending a situation by integrating the often disjointed perceptual level items. Comprehending a situation requires synthesizing information to understand the impact on goals and objectives; developing a mental picture of the relevant portion of the world that is of interest to a robot operator or payload specialist. Higher-level SA (Level 3) involves the ability to anticipate future states, actions, events, or occurrences based upon what has been perceived and comprehended. It involves near future projections of a situation, for example, anticipating the presence of mines in an upcoming field during a robotic minesweeping task. Limited development of SA by robotic system operators may be caused by impoverished sensory information, attentional resource limits, task and environmental stressors, and interface or system design faults (Endsley & Riley, 2004; Riley et al, 2006).

In general terms, SA involves being aware of what is happening around you so that you can understand how information, events, and your own actions affect both current and future goals. In robotics system control, operators have information to acquire in both the local and remote locations. For example, operators of remote systems often need knowledge of system status, position, and orientation, and how the system functions, capabilities, and modes of operation support meeting objectives. The successful robotics operator also understands task factors, like mission requirements, task states, and issues of priority and task criticality. Beyond this robotics operators might also work in team situations in which they need to acquire and maintain SA related to the team, including team actions, locations, and information needs. The amount of information to be acquired, understood, tracked, and managed increases in situations in which robot operators have to control multiple vehicles, and where robot autonomy is used in an attempt to help operators work with multiple vehicles simultaneously.

General Difficulties with High-Level SA

In general, we have observed that it is difficult to develop higher levels of SA, in part because of the lack of integrated, situation-relevant information provided through the interface. Interface designers need (1) to understand what robotics operators in various tasks or domains need to know; and (2) to determine how to present the information in an integrated fashion in support of SA and decision making. Higher-level SA (comprehension and projections) deficits have also been observed to be a result of workload associated with tasking and controlling unmanned systems. When observing robotics operators involved in urban search and rescue, the interfaces associated with the robotic systems provided either no interface support or too much low-level data that had to be cognitively integrated by the operators, in both cases hindering the operators' ability to acquire SA on the system. The result was lack of understanding of system location and status, and, ultimately, difficulties locating (simulated) victims and difficulties leading victim extraction teams to the victims.

Participants in experimental robotics tasks examined by Riley and Strater (2006) developed SA on low-level SA items, but experienced difficulties in comprehension and projection. The level of SA was assessed through direct queries on mission knowledge, robot and task status, and environmental elements. Participants were able to provide 70 to 80 percent correct responses to SA queries on perceptual items like robot component system status (battery level, communication link, etc.) or recent items like their last control action or last recognized landmark. These participants struggled, however, with translating perceptual knowledge into higher-level understanding. This comprehension deficit can be related to difficulties in sensemaking—that is, the "motivated, continuous effort to understand connections (which can be among people, places, and events) in order to anticipate their trajectories and act effectively" (Klein et al., 2006, p. 71). Difficulties in this process affect the outcome (i.e., SA)—the overall state of knowledge that is acquired with the robot. For example, participants failed (1) to integrate status data acquired through monitoring of individual systems into an understanding of the overall system status; (2) to comprehend the impact of the overall status on robot functioning; or (3) to make decisions on the basis of this understanding. Further experimental operators were not able to integrate the series of control actions that they completed (e.g., turned left, moved for X feet, turned right, etc.) into an understanding of the environment layout.

Spatial Cognition

Difficulties in Robot Localization

In field observations (and experimentation) a good deal of time is spent trying to orient oneself and reorient in the environment. For example, robot operators in urban search and rescue appear to expend considerable cognitive effort tracking control actions in order to figure out where the robot is and where it has been. Operators also do a substantial amount of backtracking to familiar landmarks or past locations. In a study to assess interface designs for supporting SA in robot-assisted search and rescue, Yanco and Drury (2004) observed trials in which participants spent about 30 percent of the trial time trying to acquire SA related to robot localization. Operators often expressed confusion about where the robot was located and whether it was navigating toward obstacles.

In terms of navigation, SA and sensemaking issues are also related to a lack of understanding regarding how orientation of the robot affects its mobility in an environment. Traversing slopes, inclines, or uneven terrains is sometimes difficult. Operators struggle to understand how the robot is situated on the terrain (pitch and roll of the vehicle), manage camera perspectives, and maintain appropriate speeds for traversing difficult terrain without error (rollovers and collisions). Riley and Strater (2006) observed that operators had low SA regarding orientation and direction of the system in an experimental task. Participants were only able to provide correct responses regarding current robot direction and orientation 20–30 percent of the time. Furthermore, when asked to provide rough sketches of the environment they had recently explored, participants were unable to produce representations of the space and layout.

When considering the issues in localization, navigation, and mobility in robot control, future research must assess the effects of the mobility of the operator. The issues observed above were in scenarios where the operator was fairly stationary (at a robot control station or only moving periodically to get closer to the robot for maintaining a strong communications link). A goal for future military operations will involve robot operators who are in a moving vehicle. Such operator mobility raises issues concerning the operators' understanding of where a robot is with respect to the operator, areas of interest, task locations, future operator position, etc. of mounted control.

The Importance of a Good Mental Map

Many military tasks will involve navigating robotic assets into specified areas. Whether the operator directly controls the robot or simply monitors it, development of a mental map of the space that the robot moves through may have a profound affect on the subsequent performance by the operator. In a task in which navigation was a central component, Chadwick et al. (2005) showed that UGV operators of robots who had good mental maps later navigated the space substantially better than operators with poorer mental maps. Key issues for HRI include the acquisition of mental maps, the nature of the representation of mental maps, and the later access of information from the mental map.

Acquisition of Mental Maps

Humans acquire knowledge of a space when they navigate through that space (Golledge, 1999). Likewise, robot operators can learn about the layout of the distant spaces through which they navigate their robots. However, this may have disadvantages for the acquisition of spatial knowledge. First, in many cases a robotic camera provides a limited view of the world around the robot—a view that has been compared to looking at the world through a soda straw (Chadwick et al., 2005). As a consequence of this limited view, the operator (1) may miss key elements of the landscape, such as landmarks, as the robot travels through it; (2) may have distorted views of features of the landscape; and (3) will have to integrate many different views to construct a mental map.

An additional disadvantage of acquiring a mental map through navigation is that the operator only learns about the space in the routes that the robot travels. Accordingly, if the previously taken route is not available or is missed, the prior experience will be of little use. In fact, the prior experience could be detrimental to navigation if the operator believes that the robot is on the previous route.

Another means by which people learn about space is by examining overviews of the spatial layout, frequently by looking at a map (Kitchin & Blades, 2002). In addition, a UGV operator might also have UAV views available as an overview. The advantage of maps and other overviews is that they can reveal spatial relations among many elements that cannot be viewed from even the best ground views, let alone a soda straw's eye view. However, a map view has to be integrated with the ground view in order to be turned into useful information for navigation, identification of target location, and other spatial tasks.

A third way that people learn about space is through directions to a location or other verbal descriptions of a location in a spatial environment. Survey descriptions of the spatial arrangements of objects in an environment allow people to answer questions about novel aspects of the environment, such as locations of objects as they face a different way from that in the original description (Franklin & Tversky, 1990). In fact, survey descriptions and route description appear to be as good as reading a map in helping people verify survey or route statements (Taylor & Tversky, 1992). These findings suggest that people can convert verbal descriptions of various types into mental maps, especially if the descriptions include key spatial directions such as above-below or left-right (Franklin & Tversky, 1990).

Representation of Spatial Knowledge

One of the major arguments in the cognitive representation literature in general, as well as in the spatial cognition literature, is whether all knowledge is represented propositionally. Evidence from mental map research strongly suggests that spatial knowledge can be represented propositionally—perhaps the most frequently cited example is that, when asked which city is further north, Seattle or Montreal, most people respond, incorrectly, that Montreal is north of Seattle (Stevens & Coupe, 1978). The propositional hypothesis explains this finding by suggesting that people know the proposition that Canada is north of the United States and access this information when answering the question about cities in the U.S. and Canada. Chadwick and Pazuchanics (2007) had robot operators navigate a space and search for specific target objects, and then asked them to place the target objects on a map of the space. Many of the errors in this task also seemed to be related to a propositional representation of the environment. For example, their participants tended to place targets that were near a building corner close to a corner in the map placement task, but frequently placed the targets on the wrong side of the corner or near the wrong corner of the building, or even near the corner of the wrong building. The participants appear to have encoded that the target was near a landmark or salient feature—a building corner—but did not have a complete representation of the space.

Do findings that support some degree of propositional representation preclude a spatial representation of environmental space? Certainly research on mental imagery that indicates that people can manipulate a mental image in much the same way that they can manipulate a directly perceived image supports the hypothesis that percepts and mental images of space are processed similarly (Kosslyn, 1973). Likewise, neurophysiological studies of mental imagery implicate the same brain areas that process directly perceived visual stimuli (Kosslyn, 1994). It may well be that the cognitive system represents space both propositionally and spatially; likewise, it may represent much concrete, spatially oriented verbal information like route directions both propositionally and spatially. The system may be biased one way or the other on the basis of the prominence of spatial layout information versus landmarks during learning, with a landmark orientation leading to greater propositional encoding. Thus, robot operators may make different kinds of errors in navigation or in post-task recall depending on their experience with a space.

Recall of Spatial Information

In addition to issues in acquisition and representation, characteristics of the recall task may also influence spatial performance. For example, a question like "Is the distance between A and B greater than that between C and D?" may require access of spatial relations (as would be indicated by a distance effect in which the greater the difference in distance, the shorter the response time), whereas questions that focus on relations that might be hierarchical, like "Is A north of B?"—where A and B are both subparts of larger spatial entities—might be responded to by accessing propositional information.

Finally, the relation between the method of acquisition of spatial information and the method of assessing that knowledge can also have an effect (Troberg and Gillan, 2007). People who acquired knowledge about a virtual space by reading a map did better in a map drawing task than did people who acquired knowledge about the space by navigating through it. In contrast, performance in a navigational task was somewhat better for those who had learned about the space by navigation than by those who had read a map. So, operator performance could be expected to be a complex function of the interaction of experience in a space and the task needs for information about that space.

Multitasking and Task Switching

Why would a robot operator switch tasks? One reason is that the external world may interrupt the operator. For example, the operator might be interrupted by an alert telling her to check a system indicator. Second, sometimes task demands or task characteristics motivate a switch between tasks—when a subtask is completed and it is convenient to move to a different task or when working on another task becomes more pressing (such as, an operator who is monitoring two robots may need to check to see if one of the robots is close to a minefield). Third, task switching may occur because of the operator's state. A task might become uninteresting and, to maintain an optimal level of overall attention, the operator may decide to switch. Likewise, after reaching a level of frustration in one task, an operator might switch to another, simpler task. How and why people switch tasks is an issue that will require much additional clever research in the context of domains of practice or in laboratory simulations of those contexts.

One source of multitasking and consequent switching between tasks is the inherent multiple task requirements of a single robot operator who has to navigate and perform functional work with a robot. Another source of multitasking is multiple vehicle control. The need to control multiple vehicles can have a detrimental effect on SA, especially if the UGVs are involved in different or uncoordinated tasks. Issues in SA and workload appear to be due to limits on attention. Under multiple vehicle control (and teleoperation in general), an operator could theoretically develop SA on one robot (or one location) to the exclusion of developing awareness on the other. Alternatively, an operator could possibly develop some level of awareness on both robots (or locations) simultaneously. Success in simultaneous awareness is dependent upon an operator's ability to divide attention effectively across the multiple systems and perhaps multiple locations for planning, decision making, change detection, and SA. With respect to SA and decision making, we have observed performance decrements caused by the need to control multiple systems in close temporal proximity. When navigating one vehicle through an unfamiliar environment and local area search for landmarks, participants were able to provide on average 80 percent correct responses to queries. When controlling two vehicles, average participant accuracy dropped to about 49 percent (Riley & Strater, 2006). Participants also reported higher mental workload under multiple vehicle control. Results showed that as workload increased, SA on the robotics task and environment decreased. Similar effects were observed when operators remotely control robots while also completing a local secondary task—switching to a local task (e.g., monitoring locally displayed system status information) resulted in decreased SA on the robotics task, as measured by a drop in percent correct responses to SA queries.

Although task switching can lead to decreased SA, the need to control multiple robots or to navigate a robot while searching for target objects in a search and rescue operation may be a necessary part of the task environment. In addition, users may decide to switch during periods of low activity in one task to another, more fruitful task. Research that used virtual robot control tasks, such as searching for a target by one robot and minesweeping by another, found that the optimal performance occurred when operators switched between the tasks an intermediate number of times in a session; poorer performance was observed in participants who switched very infrequently or very frequently (Chadwick et al., 2004). Likewise, a diary study of interaction with technology across the day showed that people engage in substantial task switching (Czwerwinski et al., 2004).

Note that both the Chadwick et al. (2004) and the Czerwinski et al. (2004) studies involved participants who could switch between tasks freely. This free switch method contrasts with the more common procedure in psychological research on attention and task switching in which participants switch when signaled to, often after only a few seconds working on an artificial task or even a subtask (e.g., switching between adding columns of numbers and searching text for target letters or switching between two different geometric pattern categorization tasks)—for an example of this paradigm, see Rubinstein et al. (2001). Studies using the controlled switch paradigm are designed to maximize the costs of task switching, and they do an excellent job of that. What they do not do is provide insights into task switching in real-world situations. Research that focuses exclusively on switching costs will produce theories that, when applied to real-world situations, meet with the fierce resistance of real people functioning in real task environments. Those real people switch tasks—from reading only part of a book while commuting on the subway (the real-world need of exiting the train intrudes on the more pleasurable task of reading a book) to a soldier switching between terrain analysis and communicating progress to superiors. Thus, one important characteristic of task switching may be the control over the point in the task that the switcher exerts. Control over switching may be one difference that explains why people engage in multitasking despite the switching costs—someone who has hit downtime in a task or has reached their threshold for boredom in a task may perceive a value in switching that exceeds the costs.

Summary

Design of an interface by which an operator or operators can interact with robots (for purposes of control, supervision, and/or monitoring) will require designers to understand and support basic and high-level cognitive functioning by the operator to support task performance on such key military tasks as navigation, remote sensing, telemanipulation, and enemy engagement. We hope that the following guidelines, which serve to summarize this chapter, can serve as the starting point for interface design concepts that accommodate cognitive abilities and constraints.

1. Robotic systems operators have to acquire both information that is proximal to them, such as system status, and information that is proximal to the robot, such as the position and orientation of the robot. In addition, the operator needs to understand task factors and, in many cases, process information from team members.
2. Sources of poor SA by robot operators include impoverished sensory information, limited attentional resources, stress induced by task demands or the environment, and flaws in the interface or system design:

    –   Interface design can enhance information provided to the operator, help funnel attention appropriately, and reduce task demands, thereby increasing SA.

3. Interface designers need to understand what robot operators in various task domains need to know:

    –   The interface should be designed to present the needed information in an integrated fashion to support SA.
    –   Interfaces tend to present too much low-level data that required the operators to integrate the information, thereby interfering with the development of SA.
    –   Interfaces that present low-level data may make it difficult for the operator to translate the information into higher-level understanding, thereby disrupting sensemaking.

4. The interface should support the development, retention, and recall of an accurate mental map of the spatial environment through which the robot has moved:

    –   Operators expend considerable cognitive effort extracting spatial information (where the robot is, the characteristics of the terrain and the robot's relation to the terrain, and where the robot has been).
    –   Operators with good mental maps of the spatial environment of the robot tend to perform better in navigation tasks.
    –   Operators have difficulty acquiring good mental maps because of the impoverished spatial information from the robot.
    –   Operators who acquire a mental map only via navigation through an environment tend to only learn the routes that the robot traveled.
    –   Operators who acquire a mental map via an overview of the spatial layout (e.g., from a physical map or a bird's-eye view) have to integrate that information with the ground view from the robot.
    –   Propositional information concerning the spatial environment of a robot (e.g., a target object is near a corner) will interact with visual and imagery-based information, which can lead to errors.

– How an operator recalls spatial information and the relation between the processes of acquisition and recall of spatial information can both affect operator performance in navigation tasks.

5. Multitasking and task switching are inherent in the multiple task requirements of a robot operator and, consequently, the interface should support task switching and multitasking.

   – Sources of task switching include external interruptions, task demands, the operator's state, and the need to control multiple vehicles.
   – Operators who have to control multiple robots simultaneously have been observed to show poor performance.
   – Forcing an operator to switch from control of the remote robot to a local task (monitoring a local display) can lead to poor performance.
   – When allowed to freely switch between tasks in a two-robot control situation, optimal performance was observed with an intermediate level of switching.
   – Free switching may allow operators to make optimal use of downtime in a multiple task situation and to switch at appropriate times in the tasks.

References

Čapek, K. (1921/2004) *R.U.R.* (translated by C. Jones). New York: Penguin.

Chadwick, R.A., Gillan, D.J., Simon, D., & Pazuchanics, S. (2004). Cognitive analysis methods for control of multiple robots: Robotics on $5 a day. *Proceedings of the Human Factors and Ergonomics Society 48th Annual Meeting.* 688–92.

Chadwick, R. A. & Pazuchanics, S. (2007). Spatial disorientation in remote ground vehicle operations: Target localization errors. *Proceedings of the Human Factors and Ergonomics Society 51st Annual Meeting*, 161–5.

Chadwick, R., Pazuchanics, S. L., & Gillan, D. J. (2005). What the robot's camera tells the operator's brain. In N. Cooke, H. Pringle, H. Pedersen, & O. Conner (eds). *Advances in Human Performance and Cognitive Engineering Research: Human Factors of Remotely Piloted Vehicles* (pp. 373–84). Amsterdam: Elsevier.

Czerwinski, M., Horvitz, E. & Wilhite, S. (2004). A diary study of task switching and interruptions. *Proceedings of ACM Human Factors in Computing Systems CHI 2004*, p. 175–82.

Endsley, M. R. (1988). Situation awareness global assessment technique (SAGAT). *Proceedings of the National Aerospace and Electronics Conference (NAECON)*, 789–95.

Endsley, M. R. (1995). Toward a theory of situation awareness in dynamic systems. *Human Factors*, 37(1), 32–64.

Endsley, M. R. & Riley, J. M. (2004). *Supporting Situation Awareness in Human Operator—UAV/UGV Collaboration*. Presentation to the Army AMRDEC Science and Technology Workshop, Moffet Field, AFB, CA.

Franklin, N. & Tversky, B. (1990). Searching imagined environments. *Journal of Experimental Psychology: General*, 119, 63–76.

Gillan, D. J. & Schvaneveldt, R. W. (1999). Applying cognitive psychology: Bridging the gulf between basic research and cognitive artifacts. In F. T. Durso, R. Nickerson, R. Schvaneveldt, S. Dumais, M. Chi, & S. Lindsay (eds), *The Handbook of Applied Cognition* (pp. 3–31). Chichester: Wiley.

Golledge, R. G. (1999). Human wayfinding and cognitive maps. In R. G. Golledge (ed.), *Wayfinding Behavior: Cognitive Mapping and other Spatial Behaviors* (pp. 5–46). Baltimore, MD: Johns Hopkins.

Kitchin, R. & Blades, M. (2002). *The Cognition of Geographic Space*. London: I. B. Tauris.

Klein, G., Moon, B, & Hoffman, R. R. (2006). Making sense of sensemaking 1: Alternative perspectives. *IEEE Intelligent Systems*, 21(4), 70–73.

Kosslyn, S. M. (1973). Scanning visual images: Some structural implications. *Perception and Psychophysics*, 14, 90–94.

Kosslyn, S. M. (1994). *Image and Brain*. Cambridge, MA: MIT Press.

Riley, J. M., Murphy, R., & Endsley, M. R. (2006). Situation awareness in control of unmanned ground vehicles. In N. Cooke, H. Pringle, H. Pedersen, & O. Conner (eds). *Advances in Human Performance and Cognitive Engineering Research: Human Factors of Remotely Piloted Vehicles* (pp. 359–71). Amsterdam: Elsevier.

Riley, J. M. & Strater, L. D. (2006). Effects of robot control mode on situation awareness and performane in a navigation task. *Proceedings of the Human Factors and Ergonomics Society 50th Annual Meeting*.

Rubinstein, J. S., Meyer, D. E. & Evans, J. E. (2001). Executive control of cognitive processes in task switching. *Journal of Experimental Psychology: Human Perception and Performance*, 27, 763–97.

Stevens, A. & Coupe, P. (1978). Distortions in judged spatial relations. *Cognitive Psychology*, 10, 422–37.

Taylor, H. A. & Tversky, B. (1992). Spatial mental models derived from survey and route descriptions. *Journal of Memory and Language*, 31, 261–92.

Troberg, E. & Gillan, D. J. (2007). Measuring spatial knowledge: Effects of the relation between acquisition and testing. In *Proceedings of the Human Factors and Ergonomics Society 51st Annual Meeting*. (pp. 368–71). Santa Monica, CA: Human Factors and Ergonomics Society.

Yanco, H. A. & Drury, J. L. (2004). Where am I? Acquiring situation awareness using a remote robot platform. In *Proceedings of the 2004 IEEE Conference on Systems, Man, and Cybernetics*.

Chapter 5

Social Factors in Human-Robot Interaction

Lori Foster Thompson and Douglas J. Gillan
North Carolina State University

Social Factors in Human-Robot Interaction

Much of this book centers on technical, usability issues pertaining to humans' interactions with robots. Although usability is a critical antecedent of technology acceptance, it is by no means the only determinant. Even an intuitive, easy-to-use robot will be disregarded if military personnel dislike it, distrust it, doubt it, or resent its ability to contribute effectively and efficiently to the mission at hand. Burke et al. (2006) outlined four major human-robot interaction (HRI) issues requiring attention from human factors professionals. Included on this list was the need to consider the factors driving acceptance of robots in work settings. The failure to accept a robotic assistant can result in situations where operators or soldiers sharing a field of operations with robots (1) do not place enough weight on the information or suggestions provided by the robot; (2) refuse to interact with the robot altogether; or (3) actively reject or react counter to the information from the robot (Madhavan et al., 2006).

The purpose of this chapter is to discuss the social factors in HRI, especially as they affect the acceptance of robots and their integration into military teams. Implicit in the approach to this chapter is a broad use of the term "interface." Our use of "interface" covers any nexus of interaction between a human and robot, including both the means by which operators control and monitor the robot and, most importantly, the points of contacts between soldiers in the field and robots. Thus, the interface may involve social as well as cognitive and perceptual interactions. With this in mind, we begin this chapter by considering the dynamics that may influence the acceptance of robots by individual and/or teams of soldiers. We then discuss a number of issues that emerge as robots are designed to interact with soldiers in *social* terms normally associated with human-to-human exchanges.

Individual and Team Dynamics of Acceptance

Research and theory devoted to understanding the variables that encourage or deter acceptance of emerging information technologies in the workplace can be applied to HRI in a military context. The United Theory of Acceptance and Use of Technology (UTAUT) highlights four key determinants of technology acceptance: effort expectancy, performance expectancy, social influence, and facilitating conditions (Venkatesh et al., 2003). Akin to usability, effort expectancy is the degree of ease associated with the use of a technology, in this case a robot. Factors addressed in this volume's chapters on cognition and perception in HRI (Gillan et al.; Pazuchanics et al., respectively)—including support for operators' situational awareness, task switching, and information integration—can influence effort expectancy. Performance expectancy refers to the extent to which military personnel believe that using a given robot will help them attain gains in their jobs. Social influence involves whether personnel believe that others who are important to them (e.g., superior, teammates) feel they should use an available robotic assistant. Finally, facilitating conditions entail soldiers' beliefs that an appropriate organizational and technical infrastructure is in place to support their use of the robot. Several moderators are said to influence the extent to which these four factors lead to acceptance of a given technology; these include gender, age, experience, and the degree to which using the technology is in fact voluntary (Venkatesh et al., 2003).

At present, much more work is needed to inform interventions designed to encourage technology acceptance and the appropriate utilization of robots. Research in the domain of technology acceptance has begun to show how perceptions of effort expectancy, performance expectancy, social influence, and facilitating conditions develop and unfold. Scientists and practitioners should consider each of these four factors if robots are to be properly integrated into military operations. Such research should consider HRI from multiple perspectives. In other words, it is important to remember that issues related to user acceptance are not limited to interactions between a robot and its operators. At times, remotely operated robots will need to interface with people in "front" of the robot as well. In the context of a search and rescue mission, for example, a given robot may not only interact with its operators; it may also deal with survivors or other people at a disaster site (Murphy, 2004). Thus, the issue of accepting and integrating robots into the battlefield needs to be considered from multiple vantage points.

Acceptance also needs to be considered from multiple levels of analysis. Whereas much research and theory in the domain of HRI has focused on user acceptance at the individual level, the acceptance of robots in military operations is often a team-level phenomenon. It is not uncommon for the types of robots described in this book to operate in a team environment consisting of a variety of human and technological entities. For example, an important use of mobile robots for urban search and rescue occurred during the World Trade Center disaster, where the team consisted of a mix of traditional rescue workers and robots. After-action analyses of this event revealed group dynamics which were at times suboptimal (Murphy, 2004). Since then, the Center for Robot-Assisted Search and Rescue (CRASAR) has increased its focus on integrating robots into human teams to avoid resentment and technology acceptance delays among human operators. Based on this analysis of the World Trade Center operation and other robotic-assisted urban search and rescue missions, Murphy (2004) highlighted opportunities for research to contribute to successful HRI systems, including, importantly, encouraging the acceptance and integration of robots within an existing team structure.

Determining how best to assimilate robots into human teams requires an understanding of the processes and challenges that characterize work teams. Teams, by definition, consist of two or more members who interact dynamically, interdependently, and adaptively toward a common and valued goal/objective/mission (Salas et al., 1992). Teams need not be exclusively human. Researchers have effectively argued that emerging automated entities like robots and expert systems can serve as a new kind of team member—albeit technological—in situations where coordination, shared goals, and interdependence among humans and technology exist (Hoeft et al., 2006).

The interactions that occur within human-robot teams may be influenced by a wide variety of affective, cognitive, social, situational, and system design factors (Cuevas et al., 2007). Research and theory from the traditional, human-centered teams' literature provide insights regarding the variables that are important to attend to as robots are integrated into existing social structures. The Team Effectiveness Model (TEM; Tannenbaum et al., 1992) is one well-known framework which describes the factors (e.g., individual team member characteristics, collaborative processes) that influence important team outcomes such as performance quality, quantity, time, errors, and costs.

The *processes* which contribute to team effectiveness are a critical component of TEM. They include coordination, communication, problem solving, and decision making. Coordination involves managing one's own tasks in conjunction with the tasks of teammates; that is, managing interdependencies among activities performed to achieve a goal (Hoeft et al., 2006; Malone & Crowston, 1990). Malone and Crowston (1990) have suggested that "good coordination is nearly invisible" (p. 357) and that we tend to notice coordination most when it is lacking. As described by Bradshaw et al. (2007), effective coordination requires interpredictability, common ground, and directability. Interpredictability entails being able to anticipate what teammates will do. Common ground involves mutual knowledge, assumptions, and beliefs shared among teammates. Shared mental models and shared situational awareness are important in developing common ground. Finally, coordination requires directability, or teammates' responsiveness to each other's influence. The magnitude of the reasoning and representational differences that separate humans from robots can make human-robot team coordination especially challenging (Bradshaw et al., 2007).

Soldiers must know both how to acquire the information they need from teammates and how to communicate information in return. The complexity of human-robot team processes increases to the extent that the input mechanism for conveying information to a robotic teammate diverges from natural human communication processes. As stated by Hoeft et al. (2006), "The formalities structured into communicating with automated team members may impose restrictions that can interfere with the spontaneous, free-flowing nature of human-human communication" (p. 252).

According to TEM, team performance is also a function of the collective problem-solving and decision-making processes in place. Here, it should be noted that the introduction of robots to the team creates the potential for automation bias—an overreliance in which people treat information from the robot as fact without seeking additional information to confirm or refute the validity of the robotic contribution. Automation bias can take the form of omission errors (i.e., the failure to respond to system irregularities when not prompted to do so by the robot) or commission errors (i.e., inappropriately following a robot's recommendation without verifying it or despite contradictions from other sources of information; Mosier et al., 2001).

Of course, automation bias may well depend on trust. Trust dictates the extent to which people rely on teammates, whether they are human or robotic (Cuevas et al., 2007). In the organizational sciences, interpersonal trust is defined as the degree to which a worker is confident in, and willing to act on the basis of the words, actions, and decisions of another (McAllister, 1995). Interpersonal trust derives from perceptions of a teammate's "track record" of reliability, but reliability is not the only factor of importance. In fact, interpersonal trust has both cognitive and affective foundations. Whereas cognitive trust stems from a teammate's competence, responsibility, reliability, and dependability, affective trust has to do with the emotional bonds between individuals (McAllister, 1995). Both are important during collaboration, suggesting the need to deploy military robots that are not only viewed as technically sound but also trustworthy from an affective standpoint.

Admittedly, the Team Effectiveness Model referenced above was developed for exclusively human teams and may therefore fail to capture the entire set of cognitive processes involved in complex teamwork involving robots (Hoeft et al., 2006; Cuevas et al., 2007). In response to this concern, Cuevas et al. (2007) proposed a theoretical framework for augmenting team cognition with automation technology. Team cognition, which emerges from the interplay between each team member's individual cognition and team process behaviors (e.g., coordination, communication), is essential to effective collaboration. Cuevas et al.'s (2007) model emphasizes how information exchange and updating between humans and robots can affect (1) low-level cognitive processes such as working memory; (2) high-level cognitive processes such as sense making; and (3) higher-order, team-level, metacognitive processes such as performance monitoring.

The bottom line is that robots introduce a new and unique wrinkle to the psychology of team performance by requiring interdependence between two distinct cognitive systems—that of the human and the robot (Freedy et al., 2006). Robots pose challenges for the processes traditionally considered important during teamwork (e.g., information exchange, communication, coordination). Moreover, they add new challenges to team collaboration. According to Freedy et al. (2006), "Foremost among these is the ability of the human team to manage, predict, collaborate and develop trust with unmanned systems that may sometimes exhibit fuzzy responses in unstructured and unpredictable environments" (p. 255).

There are many possible ways to address the acceptance challenges discussed above. One solution involves building robots that more closely mirror human processes, allowing soldiers to interact with a robot as a social entity rather than requiring them to learn entirely new cognitive and communication systems as robots are introduced to the team. This raises questions regarding whether humans can and should view robots in social terms.

Viewing Robots in Social Terms

An ongoing discussion pertaining to the acceptance of robots involves the degree to which robots should be designed to exhibit humanlike forms, social behaviors, emotions, and personalities (Dautenhahn, 2002). Should they communicate empathy through responsive faces, engage in sympathy through fellow feeling, and experience intimacy through contact and disclosure (Lee, 2006)? To what extent should these systems be social, sociable, and humanlike (Thrun, 2004)? At one extreme, robots can be very mechanistic, bearing little visual or behavioral resemblance to a human. A robotic arm, for instance, commonly exhibits a machinelike form (Thrun, 2004). It looks and acts nothing like a person. At the other extreme, roboticists are actively working to build androids, synthetic humans, or robots that are virtually indistinguishable from people. The goal, in many cases, is for the robot to pass a modified "Turing test" of sorts—that is, a test designed to examine whether people can differentiate visually (Ishiguro, 2006) and/or socially (Duffy, 2003) between real and synthetic humans. Already, androids have been passed off as human in research exposing people, for short periods of time, to robots exhibiting an anthropomorphic appearance and lifelike movements (Ishiguro, 2006). From a behavioral standpoint, research has begun to show that people can accurately recognize a robot's pre-programmed "personality" based on its verbal and nonverbal behaviors (Lee et al., 2006).

Aside from the question of whether robots *can* be built to demonstrate a human appearance and social interaction, there remain debates about when and whether a humanlike interface is necessary or even desirable. Two central questions emerging from this debate, which pertain to robots in future military operations, include the following. Are military personnel capable of viewing robots in social terms? And, if so, is encouraging them to do so advisable? As suggested below, the answer to the first question is somewhat more straightforward than the answer to the second one.

Will Operators View Robots in Social Terms?

To some degree, whether operators will view robots in social terms has to do with their capacity to attribute human characteristics, such as cognitive or emotional states, to a nonhuman entity, such as a robot, in order to rationalize the entity's behavior—that is, the operator's ability to anthropomorphize (Duffy, 2003). The question of whether people are capable of viewing nonhuman agents as humanlike has been addressed by scholars from many disciplines over the years. Citing Darwin, Freud, Hume, and others, Epley et al. (2007) conclude that anthropomorphism is "surprisingly common," though this tendency[1] does vary across individuals and situations. According to Shneiderman (1989), referring to computers as if they are people is a "primitive urge" that many adults follow without hesitation. Similarly, Nass and Moon (2000) argue that people effortlessly and even "mindlessly" apply social expectations and rules to computers.

This is not to suggest that military personnel will mistake an inanimate desktop computer for a person, or actually believe that a robot possesses a true human personality and identity. Instead, the anthropomorphism of a robot is probably best understood as a metaphor rather than a literal explanation of a system's behavior. That is, operators are prone to metaphorically ascribing humanlike qualities to a system based on their interpretation of its actions (Duffy, 2003). Why? Epley et al.'s (2007) sociality, effectance, and elicited agent knowledge (SEEK) framework suggests that both cognitive and motivational factors work in concert, prompting an operator to attribute social factors to a robot. Whatever the underlying mechanisms, it seems likely that human operators of robots will view robotic devices in social terms in at least certain situations.

Should Operators View Robots in Social Terms?

Given the human capacity to anthropomorphize nonhuman objects, should we encourage military personnel to interact with robots in "interpersonal" ways by designing robotic assistants that possess social and lifelike attributes? The issue of whether robots should exhibit very humanlike characteristics remains a matter of some debate. Arguments in favor of making robots sociable center on the topics of usability, trust, and acceptance. Potentially, applying anthropomorphic paradigms to robotic design could improve effort expectancy (i.e., usability) by exploiting the rules and conventions that are already familiar to people (Thrun, 2004). Some roboticists maintain that workers will collaborate more easily and naturally with robots designed with a lifelike form (Hinds et al., 2004). As a result, anthropomorphizing technological agents could help people effectively learn how to use those agents, reducing the training time needed to get used to a new robot and possibly affecting performance expectancy (Epley et al., 2007; MacDorman, 2006).

Encouraging anthropomorphism through robot design may increase acceptance of robots not only by helping operators understand the automation but also by

---

1   The distinction between anthropomorphic features and anthropomorphic tendencies is an important one. Whereas features pertain to the robot's form and function, tendencies involve how these are perceived. There is no guarantee that the inclusion of anthropomorphic features on a robot will activate anthropomorphic tendencies among operators (Duffy, 2003).

creating bonds that increase a sense of social connection (Epley et al., 2007). Humanlike interfaces may help people relate to the robot, persuading them to share responsibility and rely on robots when necessary (Hinds et al., 2004). Said another way, giving robots social properties may inspire trust, which, as described above, is an important component of teamwork. In support of this position, research has shown that relative to task-oriented computer agents, relational agents equipped with social-emotional, relationship-building skills are trusted, liked, and respected more (Bickmore & Picard, 2005).

Thus, proponents of equipping robots with lifelike attributes maintain that sociable interactions may foster emotional connections to robotic assistants, thereby inspiring trust. Anthropomorphic cues may also allow people to rely on previously learned social rules and assumptions, which are natural to them, to interact with a robot. Rather than making military personnel learn a brand new interface and set of rules for interacting with the robot, familiar and natural (i.e., social) rules of engagement could be exploited. Presumably, this will reduce cognitive load, improve acclimation to new robots, and enhance the quality of interaction with familiar robots. These potential advantages help explain why social robotics has been proposed as an alternative to ubiquitous computing in the quest to make the human-computer interface disappear (Duffy, 2003).

Despite these potential advantages, the humanization of robots could deter acceptance and effective collaboration. Arguments against providing personnel with robots that are sociable and humanlike have to do with the expectations these attributes engender. Veletsianos (2007) reasons that an agent's visual appearance initially drives people's expectations regarding its abilities. Indeed, research by Kiesler and Goetz (2002) has demonstrated that the presence or absence of machinelike interface devices on a robot affects the assumptions users make about the robot's personality. Further, work by Hinds et al. (2004) reveals that people retain more responsibility for work when collaborating with a machinelike rather than a more humanlike robot, especially when the robot is placed in a subordinate role.

Thus, initial evidence suggests that anthropomorphic cues can inspire trust in a robot's competence. But, what will happen when visual and social cues become truly humanlike in appearance? Will trust escalate beyond a level that is warranted? The concern is that humanizing robots will encourage people to reflect capabilities that do not exist into the robot (Dautenhahn, 2002; Duffy, 2003; Thrun, 2004). In essence, when a robotic assistant is designed to mimic something lifelike, people's expectations of the system's performance may become inflated, setting the stage for disappointment and causing the user to feel poorly treated (Shneiderman, 1989). As Duffy (2003) states, "Incorporating life-like attributes evokes expectations about the robot's behavioural and cognitive complexity that may not be maintainable" (p. 183).

Inflated expectations may also lead to safety problems, in addition to disappointment and a lack of user acceptance. The inaccurate perception that robots have a humanlike awareness of the workspace and those within it could cause personnel to inadvertently engage in unsafe behaviors when working with and around robots (Riley, 2006). Making social robots appear too humanlike may also cause people to overlook robotic strengths, for example by fostering erroneous perceptions that the robot is prone to human weaknesses such as selfishness (Duffy, 2003).

The Potential Role of Task Interdependence

Thus, the issue of whether robots in future military operations should look, act, and socialize like people remains a topic of debate. In all likelihood, whether a humanlike design and social capabilities promote effective HRI in future military operations will depend on the degree of task interdependence between personnel and the robots assigned to their units. Some robots of the future will work alongside personnel in a rather independent fashion. Others will be required to cooperate with humans to complete a mission. Thrun's (2004) description of industrial robots and professional service robots illustrates this distinction. Industrial robots (e.g., a robotic arm for industrial welding) tend not to interact directly with people. Meanwhile, professional service robots (e.g., robots designed for military applications such as bomb defusal, search and rescue, and support of special weapons and tactics—SWAT—teams) work more closely with humans. Compared to industrial robots, service robots require richer interfaces (Thrun, 2004), suggesting that the issue of humanizing robots is more relevant for some systems than others.

Coovert (2006) makes finer grained distinctions which further delineate when robot sociability may have the greatest impact on user acceptance. According to Coovert's framework, robots designed to assist with work-related applications exist on a continuum with task-bots on the low end, co-bots (coworker robots) in the middle, and team-bots (teammates) on the high end. Task-bots perform specific tasks by following a series of programmed steps. They require few if any personality and team skills. Meanwhile, "understanding what will make a successful coworker robot is akin to understanding what makes for a successful human coworker" (Coovert, 2006, p. 873). Personality and critical thinking skills may be important. Finally, team-bots require the characteristics of a co-bot plus team skills (e.g., team knowledge, backing up behaviors, leadership, communication, monitoring, and feedback).

In short, whereas anthropomorphic cues and humanlike attributes are probably unnecessary when a robot is intended as a tool rather than a social entity (Duffy, 2003), the demands for social functionality are likely to rise as the interdependence between humans and robots increase.

Identifying an Appropriate Degree of Humanness

Limiting humanlike robots to interdependent tasks does not deal with the problems described above (e.g., inflated expectations), which can ensue if very humanlike robots are deployed in military operations. If one accepts the argument that anthropomorphic features can enhance HRI when military personnel are required to work interdependently with robots, then the concerns regarding humanlike robots need to be addressed. To this end, it should be noted that humanlike features can be viewed as a matter of degree. As such, one fruitful avenue of investigation involves identifying the optimal level and type of anthropomorphic cues with which robots should be equipped.

As suggested above, the quest for many roboticists today is to build a synthetic human which is virtually indistinguishable from a real person (e.g., Ishiguro, 2006). This is based on a commonly held assumption that the most obvious and straightforward way to integrate robots successfully into human systems is to build androids that look and act precisely as people do (Duffy, 2003). Conversely, perhaps a more appropriate goal would be to target an optimal (rather than a maximal) level of anthropomorphism when designing robots for military operations. In other words, one needs to consider the possibility that "more is not always better" when imbuing robots with humanlike qualities. As described below, there two reasons for this assertion. The first rationale has to do with the possibility of adverse reactions when a robot looks almost, but not quite, human. The second rationale involves the management of users' expectations.

Although it has not gone unchallenged, Masahiro Mori's "uncanny valley" paradigm has dominated robotics design for the past 30+ years and continues to do so (Hanson, 2006). This theory suggests that it is possible for anthropomorphism to go too far when designing robots. According to this view, the more lifelike robots become, the more familiar, believable, or acceptable they become, up to a point at which they seem quite close to human without appearing completely real. At this point, acceptability plummets as subtle deviations from human norms prompt an unpleasant or "creepy" feeling of uneasiness caused by the realization that the robot is not real, thereby violating user expectations (e.g., Dautenhahn, 2002; Duffy, 2003; Ishiguro, 2006; Lee, 2006). Due in part to this uncanny valley effect, it has been argued that more mechanistic, abstract, or iconic robotic interfaces may be most effective when designing robots for social interaction (Duffy, 2003).

While the risks associated with social, anthropomorphic robots (e.g., misattributions, inflated expectations, disappointment) have caused some (e.g., Shneiderman, 1989) to argue against this form of design altogether, others advocate a more moderate solution in which robots are equipped with certain anthropomorphic cues without being made to look fully human. A proponent of this approach, Duffy (2003), maintains that the shortcomings of anthropomorphism occur when roboticists indiscriminately apply humanlike qualities to their design, leading to inflated expectations and disappointment when the robot fails to meet these expectations. Dautenhahn (2002) adds that giving robots their own design which employs anthropomorphic features without seeking to imitate humans helps ensure the robot's own behaviors and functionalities, rather than user preconceptions, largely determine people's assumptions about the automation.

Perhaps the goal of HRI in future military operations, then, is not to try to trick operators into believing that a machine has human reasoning capabilities.

Rather, a more appropriate goal may be to build robotic assistants which exhibit a level of anthropomorphism that fosters realistic expectations regarding the robot's capabilities while exploiting people's natural tendencies and eliciting reasonable social expectations. An important criterion is to design for appropriate reliance, balancing users' expectations and a robot's capabilities (Duffy, 2003). To this end, Hinds et al. (2004) suggest that humanoid robotic designs may be more appropriate when wishing to stimulate delegation, but machinelike robotic designs may be preferable when robots are expected to be unreliable and/or when human responsibility should be encouraged.

Contextual Relevance

Once a given level of humanness is decided upon, choices must be made about precisely what the robot should look and act like to maximize integration into the team. While a detailed discussion of interface design is beyond the scope of this chapter, several points should be mentioned. First, the appearance and behavior of robots has not yet received the research attention it deserves. It is a neglected issue which Ishiguro (2006) identifies as a serious problem for developing and evaluating interactive robots. Second, principles from the animation and cartoon industries may provide useful guidelines when designing robots for military use. These industries have developed notable expertise anthropomorphizing characters in a manner which enables viewers to connect and empathize with nonhuman entities in specific and intended ways (Dautenhahn, 2002). Third, the principle of contextual relevance suggests that robots may need to have a military look and feel to be accepted in some environments. Contextual relevance is the degree to which a nonhuman agent's visual characteristics conform to the content area under which the agent functions (Veletsianos, 2007). For example, according to Veletsianos, an "agent that looks like a scientist may be perceived to be more competent in science-related disciplines such as chemical engineering or nuclear physics, than art-related disciplines such as music or graphic design" (p. 374). Contextual relevance may contribute to perceptions of a robot's relevance, seriousness, and authenticity. However, it should be noted that military organizations may be reinforcing stereotypes by capitalizing on them in robotic design and deployment. Thus, although contextual relevance may aid usability, it could also contradict broader societal or organizational goals. For instance, female soldiers could feel alienated if military robots are consistently designed to take a male form.

Conclusion and Guidelines

In conclusion, the domain of HRI offers researchers a new opportunity to study humans' attributions of social factors to technology (Rogers, 2006). How can and should roboticists best exploit elements of anthropomorphism in social robotics (Duffy, 2003)? At present, more work is needed to identify the design parameters needed to facilitate social interaction and create a robot that can function as an effective teammate for military personnel. Such research can clearly benefit from the groundwork laid in other disciplines, including but not limited to sociolinguistics, communications, and social psychology (with a focus on attraction, persuasion, and relationship development as well as the psychological mechanisms underlying anthropomorphic tendencies); industrial-organizational psychology (with a focus on work team processes[2] and issues such as trust); business and information technology management (with a focus on technology acceptance); and comic designs/the animation industry (with a focus on how to create characters that are appropriately believable; Dautenhahn, 2002). Through a multidisciplinary lens, those interested in integrating robots into military operations can take a proactive rather than a reactive approach to understanding acceptance of robots within the existing social structure.

By way of summarizing the major points of this chapter, we present a set of guidelines for the design of robotic interfaces that take into account the social factors discussed previously.

Interface Design

- Soldier-robot interactions on the battlefield increase the importance of human-robot interface design that considers the characteristics of the robot that will affect social interactions.
- Demands for social functionality of the robot interface will increase with increased interdependence among humans and robots.

Team Coordination

- For effective coordination by teams consisting of humans and robots, designers will need to attend to interpredictability, common ground, and directability.
- Interpredictability involves being able to predict what teammates will do.
- Common ground involves shared knowledge, assumptions, and beliefs among team members.
- Directability involves teammates' responsiveness to each others' influences.

---

2   Another important issue pertaining to team dynamics, which is not addressed in this chapter, occurs when civilian and other robot operators are introduced, along with robots, to assist with a temporary assignment or mission (Murphy, 2004). Such robot operators may come from a training background that is quite different from the background shared by other team members. Research is needed to determine how best to orient, socialize, and integrate these human operators into existing, often close-knit, military teams with established norms, traditions, jargon, history, and hierarchies.

Trust

- Interpersonal trust is based on cognitive factors (e.g., reliability of information) and affective factors (e.g., emotional bonds). Thus, humans in human-robot teams will be most likely to trust information from robotic team members when those cognitive and affective factors are present.
- However, one potential danger of human-robot interaction, at least under certain task conditions, may be a tendency for soldiers to treat information from a robot as factual without sufficient confirmation.

Soldier Acceptance of Robot "Teammates"

- One design approach that is likely to increase soldier acceptance of robots is to create robots that "mirror" at least selected human social processes.

Soldiers' Interactions with Robots as Social Entities

- A variety of cognitive and social factors make it likely that soldiers will attribute at least some social characteristics to robots (i.e., anthropomorphize).
- The design of the human-robot interface through the robot characteristics can increase or decrease the anthropomorphizing of a robot teammate by a soldier.
- Use of some anthropomorphic cues may increase the acceptance and ease of learning to interact with robotic teammates.

Matching Robot Social Characteristics to Functional Capabilities

- Robot interfaces that include too many humanlike, social characteristics may lead soldiers to expect the robot to have humanlike cognitive capabilities.
- Unrealistic expectations of robot functional capabilities due to a humanlike interface may lead to disappointment, low soldier acceptance of the robot, decreased trust in the robot, longer processing times in interacting with the robot, and safety problems.
- A machinelike interface may elicit soldiers to take more responsibility in soldier-robot teams, whereas increased humanlike features in the interface may elicit soldiers to share responsibility with or delegate to the robot.

Contextual Relevance

- Contextual relevance—the degree of conformance of the robot's visual characteristics with a military context—may also be a critical factor in soldier acceptance of robotic teammates.

References

Bickmore, T. W. & Picard, R. W. (2005). Establishing and maintaining long-term human-computer relationships. *ACM Transactions on Computer-Human Interaction*, 12(2), 293–327.

Bradshaw, J. M., Feltovich, P. J., Johnson, M., Bunch, L., Breedy, M., Jung, H., Lott, J., & Uszok, A. (2007, November). *Coordination in Human-agent-Robot Teamwork*. Paper presented at the AAAI Fall Symposium Series: The "Intelligence" in Distributed Intelligent Systems. Arlington, VA.

Burke, J., Coovert, M. D., Murphy, R. R., Riley, J., & Rogers, E. (2006). Human-robot factors: Robots in the workplace. *Proceedings of the Human Factors and Ergonomics Society 50th Annual Meeting*, 870–74.

Coovert, M. D. (2006). "Say partner, pass me a screwdriver" Roles for robots in the workplace. *Proceedings of the Human Factors and Ergonomics Society 50th Annual Meeting*. 872–3.

Cuevas, H. M., Fiore, S. M., Caldwell, B. S., & Strater, L. (2007). Augmenting team cognition in human-automation teams performing in complex operational environments. *Aviation, Space, and Environmental Medicine*, 78, B63–70.

Dautenhahn, K. (2002). Design spaces and niche spaces of believable social robots. *Proceedings of the IEEE International Workshop on Robot and Human Interactive Communication,*. 192–7.

Duffy, B. R. (2003). Anthropomorphism and the social robot. *Robotics and Autonomous Systems*, 42, 177–90.

Epley, N., Waytz, A., & Cacioppo, J. T. (2007). One seeing human: A three-factor theory of anthropomorphism. *Psychological Review*, 114, 864–86.

Freedy, A., Freedy, E., DeVisser, J. E., Weltman, G., Kalphat, M., Palmer, D., & Coyeman, N. (2006, August). *A complete Simulation Environment for Measuring and Assessing Human-robot Team Performance*. Paper presented at the Performance Metrics for Intelligent Systems Workshop, National Institute of Standards and Technology, Gaithersburg, MD.

Hanson, D. (2006, July). *Exploring the Aesthetic Range for Humanoid Robots*. Paper presented at the ICCS/CogSci-2006 Long Symposium: Toward Social Mechanisms of Android Science, Vancouver, Canada.

Hinds, P. J., Roberts, T. L., & Jones, H. (2004). Whose job is it anyway? A study of human-robot interaction in a collaborative task. *Human-Computer Interaction*, 19, 151–81.

Hoeft, R. M., Kochan, J. A., & Jentsch, F. (2006). Automated systems in the cockpit: Is the autopilot, "George," a team member? In C. Bowers, E. Salas, & F. Jentsch (eds), *Creating High-tech Teams: Practical Guidance on Work Performance and Technology* (pp. 243–59). Washington, DC: American Psychological Association.

Ishiguro, H. (2006). Android science: Conscious and subconscious recognition. *Connection Science*, 18, 319–32.

Kiesler, S. & Goetz, J. (2002). Mental models of robotic assistants. *Proceedings of the CHI 2002 Conference on Human Factors in Computing Systems*, 576–7.

Lee, B. (2006). Empathy, androids, and "authentic experience." *Connection Science*, 18, 419–28.

Lee, K. M., Peng, W., Jin, S., & Yan, C. (2006). Can robots manifest personality? An empirical test of personality recognition, social responses, and social presence in human-robot interaction. *Journal of Communication*, 56, 754–72.

MacDorman, K. F. (2006). Introduction to the special issue on android science. *Connection Science*, 18, 313–17.

Madhavan, P., Wiegmann, D. A., & Lacson, F. C. (2006). Automation failures on tasks easily performed by operators undermine trust in automated aids. *Human Factors*, 48, 241–56.

Malone, T. W. & Crowston, K. (1990). What is coordination theory and how can it help design cooperative work systems? *Computer Supported Cooperative Work (CSCW) 1990 Proceedings*, 357–70.

McAllister, D. J. (1995). Affect- and cognition-based trust as foundations for interpersonal cooperation in organizations. *Academy of Management Journal*, 38, 24–59.

Mosier, K. L., Skitka, L. J., Dunbar, M., & McDonnell, L. (2001). Aircrews and automation bias: The advantages of teamwork? *The International Journal of Aviation Psychology*, 11, 1–14.

Murphy, R. R. (2004). Human-robot interaction in rescue robots. *IEEE Transactions on Systems, Man, and Cybernetics. Part C: Applications and Reviews*, 34 (2), 138–53.

Nass, C. & Moon, Y. (2000). Machines and mindlessness: Social responses to computers. *Journal of Social Issues*, 56, 81–103.

Riley, J. (2006). How will robots impact the workplace? *Proceedings of the Human Factors and Ergonomics Society 50th Annual Meeting*, 2006, 871–72.

Rogers, E. (2006). Human factors techniques applied to HRI. *Proceedings of the Human Factors and Ergonomics Society 50th Annual Meeting*. 873–4.

Salas, E., Dickinson, T. L., Converse, S. A., & Tannenbaum, S. I. (1992). Toward an understanding of team performance and training. In R. W. Swezey & E. Salas (eds), *Teams: Their training and performance* (pp. 3–29). Norwood, NJ: Ablex.

Shneiderman, B. (1989). A nonanthropomorphic style guide: Overcoming the humpty dumpty syndrome. *The Computing Teacher*, 16(7), 5.

Tannenbaum, S. I., Beard, R. L., & Salas, E. (1992). Team building and its influence on team effectiveness: An examination of conceptual and empirical developments. In K. Kelley (ed.), *Issues, Theory, and Research in Industrial/ Organizational Psychology* (pp. 117–53). Amsterdam: Elsevier Science.

Thrun, S. (2004). Toward a framework for human-robot interaction. *Human-Computer Interaction*, 19, 9–24.

Veletsianos, G. (2007). Cognitive and affective benefits of an animated pedagogical agent: Considering contextual relevance and aesthetics. *Journal of Educational Computing Research*, 36, 373–7.

Venkatesh, V., Morris, M. G., Davis, G. B., & Davis, F. D. (2003). User acceptance of information technology: Toward a unified view. *MIS Quarterly*, 27, 425–78.

Chapter 6

Robots in Space and Time: The Role of Object, Motion and Spatial Perception in the Control and Monitoring of Uninhabited Ground Vehicles

Skye L. Pazuchanics, Roger A. Chadwick, Merrill V. Sapp and
Douglas J. Gillan
New Mexico State University

Uninhabited ground vehicles (UGVs) have the potential to expand human presence and intent across space and time (Riley et al., 2006). Through the use of cameras, microphones, grapplers, and basic vehicle platforms, UGVs can grant human operators the basic abilities to "see," "manipulate," and "move through" off-site environments. Building on these basic abilities, UGVs can be applied to perform any number of complicated tasks remotely in environments ill-suited for humans. Because of this, UGVs have been increasingly incorporated into a variety of dangerous task domains, including: urban search and rescue (USAR), explosive ordnance disposal (EOD), military operations in urban terrain (MOUT), and space exploration.

Although the potential to perform a number of complicated tasks remotely is certainly appealing, success in such operations is largely reliant on a human's ability to *control* a UGV's actions and to *monitor* the stream of data the vehicle provides about its environment throughout the course of its operation. Controlling a UGV often involves directing the vehicle's course across a remotely located space (i.e. *course control activities*) and/or directing how the vehicle manipulates objects using devices such as a grappler (i.e. *manipulation control activities*). Monitoring a UGV involves actively surveying the UGV's environment over the span of its operation, often via a video camera, to determine the presence, motion, and location of objects of interest as well as the UGV itself (i.e. *monitoring activities*). While some systems automate control and monitoring activities, most systems in current use require a "human in the loop" for supervision and guidance—a situation unlikely to change (Fong & Thorpe, 2001).

Despite the importance of human control and monitoring for successful remote operation, these activities remain notoriously difficult across different task domains. Course control tends to be problematic. For example, avoiding obstacles in a simulated USAR scenario was observed as a common difficulty across different teams and systems during the 2002 American Association for Artificial Intelligence Robot Rescue Competition (Yanco et al., 2004). Manipulation control is also difficult. Extensive training and workload are necessary to use grappler arms to move dangerous objects in current EOD and firefighting systems (Lundburg, 2007). Monitoring, too, can be met with considerable difficulty. In USAR scenarios, multiple human operators are often needed to monitor video for objects of interest within the environment and to determine their location (Riley & Endsley, 2004).

To remedy control and monitoring problems such as these, the role that critical perceptual processes play in guiding everyday control and monitoring activities needs to be considered (Tittle et al., 2002; Chen et al., 2006; Chen et al., 2007). When humans control and monitor their own actions across space and time, fundamental perceptual processes including, but certainly not limited to, *object recognition*, *motion awareness*, and the *formation of global spatial comprehension* allow these activities to be completed with some degree of skill. By identifying and addressing the types of "remote perception problems" (Tittle et al., 2002) human operators experience in UGV operation scenarios and the user-interface and environmental contributors to these problems, controlling and monitoring these vehicles can be facilitated. To assist in this endeavor, this chapter describes the roles that object recognition, motion awareness, and global spatial comprehension play in both everyday and UGV control and monitoring activities. It describes the difficulties UGV operators encounter when remotely engaged in each of these perceptual processes and discusses the potential factors that contribute to these difficulties. Finally, the chapter provides guidance for supporting and enhancing remote perception and proposes suggestions for future research.

Recognizing Objects in Space

Humans must recognize objects in order to survive. To do this, a human observer needs to take in different kinds of information from the environment and use that information to both distinguish objects from the space that surrounds them (acquisition) and identify what the objects are (identification). Through this object recognition process, humans are granted the ability to identify assets and threats as they monitor their surrounding space, which in turn enables them to better control their actions. Object recognition is also critical for mission survival when a human controls and monitors a UGV. Monitoring for both the presence and location of objects in the environment will be difficult if one's ability to perceive and identify these objects is compromised. Likewise, effectively controlling a UGV's course of motion through space and its manipulation of objects will be difficult if one cannot recognize obstacles, paths, or objects to be manipulated.

Despite the importance of object recognition in the control and monitoring of a UGV, operators often experience significant difficulty with it. In USAR applications, for example, victim identification is often error prone and time consuming (Yanco et al., 2004; Riley & Endsley, 2004). In MOUT applications, the quality of UGV video has been considered inadequate for allowing an operator to discriminate between friends, foes, and civilians, and to detect obstacles (Lundberg, 2007). Remote object recognition difficulties occur when UGVs and the scenarios in which they are placed fail to provide operators with information necessary to support this perceptual process. A number of different classes of information are relied upon to support object recognition. Investigating the degree to which each of these classes is available in different UGV operation scenarios can help in diagnosing what may be contributing to remote object recognition difficulties, and in providing guidance for ultimately remedying these difficulties.

Object Properties

Object properties are one class of information necessary for object recognition (Kosslyn, 1994). These static visual features include an object's contours, surface texture, and color. Of the object properties, contour information is often regarded as the most critical for object recognition (Marr, 1980; Biederman, 1985). Surface texture and color information are also considered important (Rossion & Pourtois, 2004), albeit to a lesser degree (Biederman, 1985).

Unfortunately, object properties are often compromised in UGV operation scenarios, and this compromise may contribute to object detection difficulties. In some UGV task environments, this information is sparse or simply unavailable. The chaos of a USAR environment, for example, might disrupt many of the object properties that one would usually rely upon to detect and recognize objects (Riley & Endsley, 2004). Scattered debris may both occlude the contour information of objects of interest and present contour information that might result in false recognition of objects of interest. Thick dust in the environments may render any color and surface information in the environments completely unavailable. Environments like these are common across other UGV applications, as well including MOUT and firefighting (Lundberg, 2007), and contribute to the "fog of war" in military operations. In addition to the UGV's environment, characteristics of the video it provides its operator with can also cut out important object properties. For example, a video's different resolution factors often have to be reduced to minimize bandwidth load. These reductions can affect the availability of different object properties, and, consequently, lead to object recognition problems.

Spatial Resolution

Reducing a video's spatial resolution can cut out details about the textures of objects which, when greatly reduced, can contribute to object detection difficulties. In a simulated passive UGV search task, we found that participants correctly assessed whether a target object (pipe bomb) was present in a residential area in 83–90 percent of trials when using moderate (144×96 pixel) and high (720×480 pixel) spatial resolution video. However, performance significantly declined to 65 percent accuracy when using extremely low (48×32 pixel) spatial resolution video.

Additionally, methods for enlarging low spatial resolution video can further alter object properties, which in turn may also hurt object detection. Different resizing algorithms uniquely distort video (McHugh, 2007). Nearest neighbor algorithms, for example, tend to produce images with a significant amount of aliasing, an image artifact that causes smooth edges and surfaces to appear pixilated. This artifact can both break up object contour information and introduce new, false contour information. Bilinear algorithms, on the other hand, produce a significant amount of blurring which can remove both contour and surface texture information. Bicubic algorithms, too, produce blurring. However, they also produce edge haloing, an image artifact that causes dark edges to be surrounded by a much lighter halo. This artifact may help accentuate edges and minimize some of the contour loss due to blurring. We found that the relative effects of these resizing algorithms on target detection in a simulated, passive UGV search task were dependent on characteristics of the places being searched (Pazuchanics, 2008b). Analyses are currently being conducted to determine how environmental characteristics like clutter, lighting, and target similarity can impact the effects of these algorithms.

Chromatic Resolution

Reducing the chromatic resolution of a video, of course, reduces information about the color properties of objects. Subjective measures have shown that color displays are preferred over monochrome in obstacle detection tasks (Miller, 1988), and expert EOD systems operators have voiced that monochrome displays would make their jobs more hazardous because they make detecting small objects in bright or shadowy areas difficult (Drascic, 1991). However, the importance of color in UGV operation scenarios has not been well demonstrated. Some studies report color video as having no significant overall advantage over black and white video in object recognition (e.g. Murphy et al., 1974), which has led some (i.e. Sheridan, 1992) to conclude that its importance for teleoperation may be "overplayed." However, providing color video has been observed by others to entail certain object detection advantages. For example, one study found color to be important for facilitating an operator's ability to recognize objects from a distance (Miller, 1988). Unfortunately, beyond these few findings, surprisingly little is known about the actual impact of reducing color resolution on a UGV operator's ability to recognize objects.

Providing operators with high levels of both spatial and chromatic resolution seems ideal for supporting object recognition because video resolution is a factor in UGV operation scenarios that can be controlled. However, the need to minimize communication bandwidth load often places constraints on the amount of resolution that can be provided. As suggested above, it appears that a certain amount of each resolution factor can be sacrificed before object detection performance is significantly impacted. Furthermore, object recognition theory suggests that spatial resolution and its associated edge and surface information may be more important to maintain than chromatic resolution. However, additional study of the degree to which reducing spatial and chromatic resolution impacts object detection will allow for more appropriate resolution reductions.

Motion

Though object properties are perceptual psychology's most studied class of information for supporting object recognition, motion can also be very useful for identifying and recognizing objects, particularly when object properties are compromised (Kosslyn, 1994). As was the case with object properties, motion information is often compromised in UGV operation scenarios. Like color and spatial resolution, temporal resolution, a video's frame rate, is often sacrificed to conserve bandwidth. And, with this, motion information is also sacrificed.

Reducing temporal resolution, however, does not necessarily reduce one's ability to recognize objects in UGV operation scenarios (Chen & Thropp, 2007). For example, temporal resolution was not found to significantly impact participants' target identification accuracy across 16, 8, 4, and 2 frames per second (FPS) video conditions in either an active or a passive UGV search task (French et al., 2003). However, such does not appear to be the case when object property information is compromised. In a simulated passive UGV search task, we found that when given video with compromised object properties (i.e. low, 48×32 pixel, spatial resolution video that had been resized to 720×480 pixels with a bilinear resizing algorithm) participants' target detection accuracy significantly dropped when temporal resolution was reduced from 30 to 1 FPS. A similar performance drop was not seen across temporal resolution conditions when participants were given high, 720×480 pixel, spatial resolution video (Pazuchanics, 2008a).

Considering this, maintaining motion information in a UGV's video appears to be quite important for supporting object recognition if object properties are compromised. However, if object properties are adequately preserved, this factor may be relatively safe to reduce in situations in which object recognition is the operator's sole required perceptual process. Be that as it may, and as we will discuss in the following sections, temporal resolution is critical for supporting other perceptual processes, so consideration of the importance of these other processes for the operator's task will have to be taken before deciding to make this reduction.

Supporting an operator's natural object recognition processes by maintaining adequate object property and motion information in video is undoubtedly important. If robot operators could perceive objects as well in remote space as they do in their own space, controlling and monitoring UGVs would likely be easier. However, even with supportive video, some operation scenarios, particularly those that take place in chaotic environments, will make object recognition difficult. In these situations, it is important not only to consider how UGV video can better support natural object recognition processes, but also to consider how a UGV may be able to *enhance* these processes. For example, infrared sensors or video-enhancing algorithms may provide information that can improve object detection in certain circumstances (see Yanco et al., 2004; Darken et al., 2001).

Maintaining Awareness of Motion through Space and Time

Perception of motion is also crucial for the functioning and survival of humans. One must be able to perceive the transitional and rotational displacement, velocity, and acceleration of both oneself and others across the spans of space and time. Just as it is necessary for human survival, motion awareness can have a critical impact on controlling and monitoring a UGV through space. Course and manipulation control require a basic sense of the current motion of the UGV and those objects that it may be piloted through or handling. Monitoring tasks, on the other hand, often require an operator to keep track of the history of motion of both the UGV and other objects of interest in order to later direct resources to their locations.

Despite the importance of motion awareness in UGV control and monitoring, this perceptual process is considered to be problematic in many operation scenarios. For example, during the World Trade Center recovery effort, operators could easily lose awareness of both a vehicle's course of motion across space and its attitude (Blackburn et al., 2002). Also, in general manipulation activities, maintaining an awareness of the velocity, position, and orientation of grappling devices is considered difficult for operators (Sheridan, 1992). If an operator cannot reliably perceive a UGV's motion, a number of consequences, including collisions, roll-over accidents (McGovern, 1990), and reduced manipulation capability can result. As with object recognition, humans rely on different kinds of information to perceive motion with some degree of accuracy. Considering how well UGV operation scenarios provide this information can assist in supporting this perceptual process and any control and monitoring activities dependent on it.

Perceptual psychologists, particularly J. J. Gibson (1979), have observed that a good portion of the information humans need to understand current self and object motion comes from the *optic flow*—the continuous visual pattern of relative object motions that expand from one's focal point as one moves through space. This pattern changes with a mover's speed, gaze perspective, and heading. As such, the optic flow is seen as a great source of information from which the heading and speed of one's self, relative to other objects in the environment, can be derived. However, the optic flow provided by many UGV cameras is often considerably different from directly obtained optic flow, and researchers (e.g. Tittle et al., 2002; Darken et al, 2001) have speculated that these differences may contribute to difficulties in perceiving the current motion of one's vehicle.

One notable difference is that the optic flow provided by a UGV's camera often lacks the *continuity* of a directly experienced optic flow (Darken,et al., 2001). This often comes as a result of reducing temporal resolution to free up communications bandwidth. A lack of continuity in the optic flow is thought to lead to misperceptions of a UGV's current motion. Considering research on both teleoperations and augmented reality systems, Chen and Thropp (2007) concluded that, in general, a frame rate at or above 15 FPS needs to be maintained to support perception of a UGV's velocity and heading. However, the results from some studies deviate from this general finding (e.g. van Erp & Padmos, 2003).

Considering that reduced continuity can have an impact on an operator's perception of a UGV's current motion, maintaining a sufficient level of temporal resolution seems important for supporting an operator's ability to control and monitor a vehicle's motion through space. As with the perception of UGV motion, Chen and Thropp (2007) found evidence from a number of UGV operation studies that providing a frame rate of over 15 FPS was necessary for a variety of monitoring and control tasks. This minimum, however, may differ depending on a number of factors within the user's task. See Chen and Thropp (2007) and van Erp and Padmos (2003) for more in-depth discussion of these interactions.

Aside from reducing the continuity of the optic flow, UGV cameras often *capture an optic flow pattern from a height that is considerably different from an operator's current eye-height*. Some researchers (e.g. Tittle et al., 2002) have speculated that this mismatch might make interpreting optic flow patterns difficult—resulting in motion perception difficulties and, ultimately, control problems and even motion sickness. We speculate that humans may be able to adapt more easily to this mismatch in view height if provided with information about the UGV camera's height. Perhaps providing camera height information to the operator via a display aid, a third-person camera perspective (Pazuchanics, 2006), or training may be of assistance. These potential forms of disambiguating information would be worth investigating.

In addition to information from the optic flow, humans can make use of *vestibular* and *somatosensory feedback* to gain a sense of self-motion, particularly rotational motion (Gillingham & Wolfe, 1986; Loomis & Beall, 1998; Schiffman, 2001). Different sensory organs—including the vestibular organs located within the inner ear and mechanoreceptors in the skin—respond to the different body orientations, vibrations, and pressures that occur as a human moves through space. These responses provide the human with feedback about changes in his body's rotation and displacement. This feedback is consciously experienced as different physiological sensations. It is also unconsciously used to stabilize one's retinal image of the environment and keep it gravitationally centered. In UGV operation scenarios, this form of information is almost always absent, and some (e.g. Tittle et al., 2002) have speculated that this might contribute to motion awareness difficulties.

In response to speculations like these, systems have been developed to provide vestibular/somatosensory feedback to operators. For example, controllers and seats that vibrate in correspondence with vehicle velocity and jostling have been developed (e.g. Everett & Nieusma, 1994), as well as more elaborate systems that place the operator in a virtual cab that rotates and vibrates in coordination with the UGV's motion. Studies suggest that these systems may support an operator's awareness of current vehicle motion. Providing vibration has been found to improve the sensation of motion in virtual reality (VR) applications (Riecke et al., 2005). A driving simulator study also suggests that providing vibration may facilitate velocity estimation in course control tasks (McLane & Wierwille, 1975); however, evidence of this is still on shaky ground. Providing combinations of lateral, roll, or yaw information by physically moving operators in coordination with a UGV may also be beneficial. Such has been found to facilitate path following in a driving simulator study (McLane & Wierwille, 1975). Furthermore, providing flying and driving simulator operators with vestibular/somatosensory feedback has been linked to improved navigational performance in general (MacAdam, 2003).

One of the functions of the vestibular system can be artificially provided without having to physically move an operator. A *gravity-referenced* camera system can provide a gravitationally centered image like that the vestibular system helps provide. It does this by rotating the vehicle's camera such that its image is always centered on gravity instead of the vehicle. Studies have demonstrated the utility of these systems. When provided with gravity-referenced views, operators were better able to determine their vehicle's attitude than when provided with vehicle-centered views (Heath-Pastore, 1994). Furthermore, operators provided with gravity-referenced views were better able to pilot a vehicle through chaotic terrain that posed a risk of roll-over accidents than operators provided with vehicle-centered views and separate pitch angle and roll angle indicators (Lewis et al., 2003).

Unfortunately, more elaborate vestibular/somatosensory feedback systems and gravity-centered camera systems have considerable monetary and spatial costs. Thus, it is important to determine when this information may be most critical, and cases in which operators may be able to succeed without it. Considering vestibular/somatosensory feedback systems first, humans have demonstrated their ability to adapt and perceive their current motion quite well in the absence of vestibular/somatosensory information. For example, many can adequately control the motion of vehicles on visual information alone in video games and driving simulators (MacAdam, 2003). However, in situations with little visual information, vestibular/somatosensory information becomes more important. In low-light situations, drivers and pilots are very reliant on this form of information to maintain a sense of orientation and self-motion (Loomis & Beall, 1998). Thus, providing vestibular/somatosensory feedback may only be critically necessary when visual information from a UGV is poor. This potential interaction is worth investigating. Considering gravity-referenced camera systems next, there is some evidence that these systems may be of most use in chaotic environments that afford little information about surface orientation (Lewis et al., 2003). In less chaotic, more familiar environments, they may not be necessary.

Given the likelihood that natural motion perception information will be impoverished to some degree in UGV operation scenarios, one may also consider providing operators with more *artificial sources of current motion information.*

Tools such as speedometers, clocks, odometers, compasses, and attitude indicators may provide more reliable, lower bandwidth current motion information than more naturalistic sources. However, in order for these tools to effectively convey their information, they need to be designed in such a way that their information is readily interpretable. Attitude indicators, for example, can be designed to be more or less useful. While some have found certain attitude indicators to be problematic in UGV operations (e.g. Lewis et al, 2003), aviation psychology can offer some guidance that may be useful for creating more useful indicators. Aviation studies have shown that *frequency separated* attitude displays are more supportive of attitude awareness and reducing accidental control reversals than traditional moving horizon indicators (Roscoe et al., 1980). Also, attitude indicators are recommended to be placed within the span of an operator's peripheral vision to maximize the accessibility of this information (Gillingham & Wolfe, 1986). Investigation and consideration of methods for making these and other artificial sources of motion information more readily interpretable would be valuable.

In addition to being aware of the current motion of the UGV and the objects within its space, operators also need to form an awareness of their past motion. The types of information discussed above can certainly play a role in the formation of this knowledge, as better support of an operator's awareness of current motion will assist his ability to form an awareness of past motion. However, even in the unlikely scenario that an operator has perfect awareness of current motion, forming an awareness of past motion is fairly workload-intensive and difficult. Thus, system tools that can assist an operator in building a history of the robot's path of motion may be of great value. Great care should be taken to make sure that the motion portrayed in these history systems is readily interpretable. While a history system can do a lot of the memory work for an operator, the information it stores will be without value if an operator cannot interpret it. A number of static cues such as arrows or zip lines can be relied upon to provide motion information at a fairly low computer processing and communications bandwidth cost. See Gillan and Sapp (2005) for some basic guidelines in the appropriate application of these cues.

Comprehending Global Space

While the abilities to recognize objects and maintain awareness of motion are critical, an adequate comprehension of *global space* is also necessary for both human and UGV mission survival and productivity. In UGV operation, global space refers to the greater space beyond the immediately perceivable local space surrounding the vehicle. For example, if one were exploring the interior of a building with a UGV, the global space of primary interest would be the space encompassing the rooms and passageways within the building, not just that space viewable from the vehicle's onboard cameras at any given moment. A comprehension of global space is especially important for monitoring a UGV and directing further action in the environment. Considering the building exploration example, this comprehension would allow an operator to keep track of the portion of the building searched and to direct other resources both to encountered objects of interest and to the UGV itself if it should become inoperable.

Despite its necessity, comprehending global space in UGV operation scenarios tends to be incredibly difficult. For example, in studies in which participants explored and later drew a map of a rather barren area with a virtual UGV, we found that participants' awareness of the global space was often severely distorted (Chadwick et al., 2004). Similarly, when mapping out an area in a MOUT training scenario, others (Lundberg, 2007) found that soldiers relying entirely on video captured from a UGV took an average of 96 percent longer and made 44 percent more errors than they did when physically present in the environment. Problems such as these arise when a UGV does not support or enhance the perceptual and cognitive processes an operator relies upon to comprehend global space. By identifying and understanding these processes, we will be better able to identify and understand how to provide UGV operators with a comprehension of global space relevant to their tasks.

UGV operators, and humans in general, comprehend global space through two primary, often interacting, means (Thorndyke & Hayes-Roth, 1982): (1) they can *assemble their experiences of local space*, that immediately perceivable space surrounding one, into a comprehension of global space; (2) they can *rely on some form of an overview image* such as a map or an aerial shot to obtain this comprehension. Each of these means will be described, as well as issues involved in their support in UGV operation scenarios.

The way that humans and other beings assemble a comprehension of global space from local spatial experiences has been a key area of interest in psychology since Tolman's (1948) early work in spatial cognition. A comprehension of global space by these means evolves across a series of awareness stages (Siegel & White, 1975; Thorndyke & Hayes-Roth, 1982). The first level of awareness is a *landmark* comprehension of clusters of objects within a local space. This awareness is then developed into more spatially connected levels of awareness. A *route* level of awareness, a procedural knowledge of actions necessary to travel between landmarks, is developed. Then a more interconnected *survey* level of awareness of the topographical relationships between landmarks is developed.

To evolve a more advanced comprehension of global space, the basic perceptual processes of object detection and motion awareness need to be appropriately supported. Object detection is necessary to form an awareness of landmark clusters. Motion awareness is necessary to form higher, more spatially integrated understanding of these clusters. Thus, the kinds of basic information necessary to support those perceptual processes essential for the level of spatial comprehension required for an operator's task need to be available. For example, in an object localization task that required only a landmark level of global spatial comprehension, we found that maintaining some form of information supportive of object recognition in a UGV's video improved task performance (Pazuchanics,

2008a). When provided video that maintained *either* object property information *or* motion information/optic flow continuity, participants' localization performance was better than when provided with video that maintained neither, but was no different than when provided with video that maintained both. Providing both forms of information may not have had additional benefit because, as discussed above, object recognition tends to rely upon motion information more heavily when visual information is compromised. However, we speculate that if an operator's task demands higher levels of global spatial awareness, maintaining motion information may be of additional benefit regardless of the quality of object property information. This hypothesis would be worth testing.

While supporting an operator's global spatial comprehension through supporting his ability to assemble local spatial experience can certainly be of assistance, comprehending global space by these means is likely to remain difficult. Even in direct viewing situations, in which humans have full access to all of the forms of object and motion information they naturally draw upon, their comprehension of global space is often inaccurate and low level. Because many current and potential UGV application domains (e.g. reconnaissance, search missions) require an accurate comprehension of global space, some form of an overview image, such as a map or an aerial shot, should be provided to augment the operator's comprehension of global space. However, overview images are not an end-all method for providing higher levels of global spatial comprehension as they too are subject to cognitive errors, especially when they must be integrated with other views.

One of the biggest perceptual-cognitive obstacles in using an overview image to form an awareness of global space is *view integration*—particularly coordinating the UGV camera's ground-based local viewpoint with the overview's exocentric, global viewpoint. In order to integrate information from these disparate viewpoints, operators may make identity judgments (see Bedford, 2001), combining radically different pieces of information into a unitary percept of a single object. This is not cognitively trivial. Objects viewed from the air and ground can be perceived as having different shapes, colors, and sizes. A simple cylinder, for example, appears as a circle from the air and, perhaps, as a rectangle from the ground. In tactical situations, every second counts and good decisions for action must be made rapidly. However, studies of participants' integration of ground and aerial views show that object correspondence judgments can take considerable time relative to tactical concerns and are prone to error (Chadwick & Gillan, 2006; Chadwick & Pazuchanics, 2007). In studies which provided participants with pairs of ground and air images, including ground camera position and orientation information (Chadwick & Gillan, 2006), mean judgments of an object's position on a satellite image map took between 6 and 36 seconds, while some individual judgments took nearly 3 minutes. In categorizing the errors made in localizing objects in the referenced study, 20 percent of responses were categorized as rather gross errors.

There is no need to be overly pessimistic about a UGV operator's ability to integrate overview information in a timely and accurate manner. Our recent studies have demonstrated that operators can effectively integrate ground and overview information to their advantage (Chadwick, 2007). Understanding the factors that contribute to accuracy and error in this task will help us to better facilitate awareness. For example, one's awareness of the position and orientation of the ground-based view with respect to the overview can impact the difficulty of making correspondence judgments. One means of providing this awareness is to place an informative icon on the overview that depicts the position and orientation of the UGV and the view of local space captured by its camera. Providing both position *and* orientation information seems necessary, as indicated by our research (Chadwick, 2006). Providing UGV position information alone is often insufficient in resolving correspondence judgment problems. Furthermore, both the viewpoint invariance (Biederman, 1985; Biederman & Gerhardstein, 1993) of objects and their similarity (see Tversky, 1977) to other objects in the global space can impact correspondence judgment difficulty. Considering this, a display aid that increases the viewpoint invariance and uniqueness of objects of interest could be of assistance—perhaps an aid that allows operators to flag objects of interest in one view with a unique, viewpoint-invariant icon that is automatically displayed in both views.

Each of these methods to support view integration has its potential tradeoffs that should be considered. For example, due to limitations in accuracy and reliability with technologies such as global positioning systems (GPS), that might be used to derive UGV position and orientation information (see Chaimowicz et al., 2004), special consideration should be given to provide an operator with an awareness of the reliability of this information. Furthermore, care should be taken to make sure that these support methods do not impose more temporal and cognitive costs on making view integrations. As with any display of information, information to support view integration should be intuitive and easy to access given an operator's attentional demands.

Conclusion

The very same perceptual processes that humans rely upon to control and monitor their own actions through space and time are also often relied upon when humans remotely control and monitor the actions of a UGV. In order for UGVs to more fully reach their potential for expanding human presence and intent across space and time, these processes need to be appropriately supported and, in some cases, enhanced. Otherwise, control and monitoring problems like those described throughout this chapter will arise. By considering the role these processes play in the control and monitoring of UGV action, we were able to identify some of the difficulties operators have with these processes; some factors in UGV operation scenarios that contribute to these difficulties; and the impact these difficulties have on UGV control and monitoring activities. Ultimately, we were able to provide some basic guidance for better supporting these processes and, consequently, one's ability to control and monitor a UGV. This guidance is summarized as a set of basic guidelines in the following "Quick Reference" section. These guidelines should assist in the design of more human-usable remote vehicles.

This chapter provided just a glimpse of the fundamental perceptual processes necessary for the successful UGV operations. While object recognition, motion awareness, and global spatial awareness are critical, a number of other perceptual processes are required for the sufficient control and monitoring of these vehicles. One process of particular importance is *local spatial comprehension*, one's perception of the space immediately surrounding a vehicle. Difficulties stemming from a poor comprehension of the depth and scale of objects around a vehicle abound in both course control (e.g. Yanco et al., 2004) and movement control (e.g. Draper et al., 1991) activities. Unfortunately we were unable to discuss this important process in the chapter due to space constraints.

Also, more research is necessary to provide better guidance for supporting the perceptual processes described in the chapter. First, additional work needs to be done to establish the *generalizeability* of the guidance provided. As mentioned at different points throughout the chapter (e.g. in the discussion of maintaining temporal resolution to maintain optic flow continuity), measures appropriate for supporting a perceptual process in one operation scenario are not necessarily appropriate for another. For those applying any piece of guidance to support a UGV operator's perceptual processes, it is important to consider factors in the operation scenario that might have an impact on the applicability of that guidance. Further research should be conducted to identify variables that impact perceptual demands in different operation scenarios.

Second, additional work needs to be done to establish the *validity* of the guidance provided. In this chapter and other works that have addressed the perceptual issues inherent in a UGV operator's task, some operator problems and pieces of guidance are identified solely on the basis of perceptual theory. While this is certainly a good starting point for gaining a reasonable understanding, work needs to be done to empirically validate these identifications. One reason empirical validation is important is because it is quite possible that some of the important factors involved in remote perception may not be addressed by current perceptual theory. Validation efforts will help to reveal some of these. For those applying guidance derived solely from perceptual theory, reporting successes or inadequacies of this guidance will contribute to this effort.

Finally, additional work needs to be done to establish the *specificity* of the guidelines. While some fairly specific guidance for supporting some perceptual processes exists—e.g. Chen and Thropp's (2007) recommendation for an appropriate temporal resolution for motion perception and control activities— some pieces of guidance are less specific (e.g. recommendations for appropriate levels of spatial and chromatic resolution for object recognition). Research should be conducted to better specify this guidance. Furthermore, considering that specification and generalizeability go hand in hand, efforts to create specifications across a number of operation scenarios and tasks such as that undertaken by Chen and Thropp (2007) would be of great benefit. Taken together, work to address generalizeability, verification, and specificity concerns will allow us to better understand the UGV operators' perceptual needs and their ability to control and monitor a UGV's actions through space and time.

Quick Reference Guide

Supporting and enhancing remote perception in UGV operations is of the utmost importance. In order for UGV operators to be successful in controlling and monitoring their vehicles, an operator's fundamental perceptual processes of *object recognition, motion awareness*, and *global spatial comprehension* must be supported, or, in some cases, enhanced. The lessons learned in this chapter are presented here as a number of guidelines for your quick reference.

Overarching Guideline for Supporting Perceptual Processes in UGV Operations

- Identify, support, and enhance those perceptual processes necessary for a UGV operator's task.

Guidelines for Supporting Object Recognition

- Maintain a useful level of contour/surface texture information in a UGV's video.
- Maintain color information in a UGV's video to support difficult object recognition tasks (e.g. recognizing small objects in bright/dim light and recognizing objects at a distance).
- Maintain motion information in a UGV's video to support object recognition when object property information is compromised.
- Consider providing "extrasensory" object detection capabilities when information supportive of object recognition is compromised.

Guidelines for Supporting Motion Awareness

- Maintain optic flow continuity in a UGV's video (a temporal resolution of 15 FPS is often adequate).
- Consider providing awareness of a UGV camera's height from the ground to make its optic flow more interpretable.
- Consider providing vestibular/somatosensory feedback to facilitate motion awareness if visual motion information is not adequately available.
- Provide gravity-referenced camera systems to support vehicle attitude awareness in chaotic environments.
- Consider appropriately providing artificial sources of motion information (e.g. speedometer, attitude indicator) to assist motion awareness.

- Provide path history recording tools to assist the formation of past motion awareness.
- Make certain that information from path history recording tools is readily interpretable.

Guidelines for Supporting Global Spatial Comprehension

- Support those perceptual processes necessary for the level of global spatial comprehension required for the operator's task.
- Provide information supportive of object recognition to support both the formation of low-level spatial comprehension and the assembly of higher levels of awareness from local spatial experience.
- Provide information supportive of motion perception to support the assembly of higher, more spatially integrated levels of awareness from local spatial experience.
- Provide an area overview to facilitate global spatial comprehension.
- Support one's ability to make correspondence judgments between local views and area overviews.
- Provide position and orientation information for the local view on the overview.
- Consider display aids that increase the viewpoint invariance and uniqueness of objects in the overview.

References

Bedford, F. L. (2001). Towards a general law of numerical/object identity. *Current Psychology of Cognition*, 20, 113–75.

Biederman, I. (1985). Human image understanding: Recent research and a theory. *Computer Vision, Graphics and Image Processing*, 32, 29–73.

Biederman, I. & Gerhardstein, C. (1993). Recognizing depth-rotated objects: Evidence and conditions for three-dimensional viewpoint invariance. *Journal of Experimental Psychology*, 19, 1162–82.

Blackburn, M. R., Everett, H. R., & Laird, R. T. (2002). After action report to the Joint Program Office: Center for the Robotic Assisted Search and Rescue (CRASAR) Related Efforts at the World Trade Center (Technical document 3141). San Diego, CA: Space and Naval Warfare Systems Center (SPAWAR).

Chadwick, R. A. (2006). Operating multiple semi-autonomous robots: Monitoring, responding, detecting. In *Proceedings of the Human Factors and Ergonomics Society 50th Annual Meeting*. Santa Monica, CA: Human Factors and Ergonomics Society, 329–33.

Chadwick, R. A. (2007). Use of aerial views in UGV operations: Costs and benefits. Unpublished manuscript.

Chadwick, R. A. & Gillan, D. J. (2006, November). Strategies for the interpretive integration of ground and air views in UGV operations. Poster session presented at the 25th Army Science Conference, Orlando, FL.

Chadwick, R. A., Gillan, D. J., Simon, D., & Pazuchanics, S. (2004). Cognitive analysis methods for control of multiple robots: Robotics on $5 a day. In *Proceedings of the Human Factors and Ergonomics Society 48th Annual Meeting*. Santa Monica, CA: Human Factors and Ergonomics Society, 688–92.

Chadwick, R. A. & Pazuchanics, S. (2007). Spatial disorientation in remote ground vehicle operations: Target localization errors. In *Proceedings of the Human Factors and Ergonomics Society 51st Annual Meeting*. Santa Monica, CA: Human Factors and Ergonomics Society, 161–5.

Chaimowicz, L., Grocholsky, B., Keller, J. F., Kumar, V., & Taylor, C. J. (2004). Experiments in multi-robot air–ground coordination. In *Proceedings of the 2004 IEEE International Conference on Robotics and Automation*. Hoboken, NJ: Wiley-IEEE Press, 4053–8.

Chen, J. Y. C., Haas, E. C., & Barnes, M. J. (2007). Human performance issues and user interface design for teleoperated robots. *IEEE Transactions on Systems, Man, and Cybernetics. Part C: Applications and Reviews*, 37(6), 1231–45.

Chen, J. Y. C., Haas, E. C., Pillalamarri, K., & Jacobson, C. N. (2006). *Human-robot Interface: Issues in Operator Performance, Interface Design, and Technologies* (ARL-TR-3834). Aberdeen Proving Ground, MD: U.S. Army Research Laboratory Human Research and Engineering Directorate.

Chen, J. Y. C. & Thropp, J. E. (2007). Review of low frame rate effects on human performance. *IEEE Transactions on Systems, Man, and Cybernetics. Part A: Systems and Humans*, 37(6), 1083–76.

Darken, R., Kempster, K., & Peterson, B. (2001). Effects of streaming video quality of service on spatial comprehension in a reconnaissance task. In *Proceedings of the Interservice Industry Training, Simulation, and Education Conference (I/ITEC)*. Arlington, VA: National Training and Simulation Association.

Draper, J. V., Handel, S., Hood, C. C., & Kring, C. T. (1991). Three experiments with stereoscopic television: When it works and why. In *Proceedings of the IEEE International Conference on Systems, Man and Cybernetics*. Hoboken, NJ: Wiley-IEEE Press, 1047–52.

Drascic, D. (1991). An investigation of monoscopic and stereoscopic video for teleoperation. Master's thesis. Toronto, Ontario: University of Toronto Department of Industrial Engineering. Available: http://vered.rose.utoronto.ca/people/david_dir/MASC/MASC.full.html.

Everett, H. R. & Nieusma, J. M. (1994). Feedback system for remotely operated vehicles. Navy Case #73322, U.S. Patent #5,309,140. Washington, DC: U.S. Patent and Trademark Office.

Fong, T. & Thorpe, C. (2001). Vehicle teleoperation interfaces. *Autonomous Robots*, 11, 9–18.

French, J., Ghirardelli, T. G., & Swoboda, J. (2003). The effect of bandwidth on operator control of an unmanned ground vehicle. In *Proceedings of the Interservice Industry Training, Simulation, and Education Conference (I/ITEC)*. Arlington, VA: National Training and Simulation Association.

Gibson, J. J. (1979). *The Ecological Approach to Visual Perception*. Boston, MA: Houghton Mifflin.

Gillan, D. J. & Sapp, M. V. (2005). Static representation of object motion. In *Proceedings of the 49th Annual Meeting of the Human Factors and Ergonomics Society*. Santa Monica, CA: Human Factors and Ergonomics Society, 1588–92.

Gillingham, K. K. & Wolfe, J. W. (1986). *Spatial Orientation in Flight* (Technical Report USAFSAM-TR-85-31). Brooks Air Force Base, TX: USAF School of Aerospace Medicine, Aerospace Medical Division (AFSC).

Heath-Pastore, T. (1994). *Improved Operator Awareness of Teleoperated Land Vehicle Attitude* (Technical Report 1659). San Diego, CA: Space and Naval Warfare Systems Center (SPAWAR).

Kosslyn, S. M. (1994). *Image and Brain: The Resolution of the Imagery Debate*. Cambridge, MA: MIT Press.

Lewis, M., Wang, J., Hughes, S., & Liu, X. (2003). Experiments with attitude: Attitude displays for teleoperation. In *Proceedings of the IEEE International Conference on Systems, Man and Cybernetics*, 2, 1345–9

Loomis, J. M., & Beall, A. C. (1998). Visually controlled locomotion: Its dependence on optic flow, three-dimensional space perception, and cognition. *Ecological Psychology*, 10(3–4), 271–85.

Lundberg, C. (2007). *Assessment and Evaluation of Man-portable Robots for High-risk Professions in Urban Settings*. Stockholm: KTH School of Computer Science and Communication.

MacAdam, C. C. (2003). Understanding and modeling the human driver. *Vehicle System Dynamics*, 40(1–3), 101–34.

Marr, D. (1980). *Vision*. New York: Freeman & Company.

McGovern, D. E. (1990). *Experience and Results in Teleoperation of Land Vehicles* (Sandia National Labs Report: SAND90-0299). Albuquerque, NM: Sandia National Labs.

McHugh, S. (2007). *Digital Image Interpolation*. Retrieved 11 January 2008 from http://www.cambridgeincolour.com/tutorials/image-interpolation.htm.

McLane, R. C. & Wierwille, W. M. (1975). The influence of motion and audio cues on driver performance in an automobile simulator. *Human Factors*, 17(5), 488–501.

Miller, D. P. (1988). Evaluation of vision systems for teleoperated land vehicles. *IEEE Control Systems Magazine*, 8(3), 37–41.

Murphy, R. L. H., Fitzpatrick, T. B., Haynes, H. A., Bird, K. T., & Sheridan, T. B. (1974). Accuracy of dermatological diagnosis by television. *Archives of Dermatology*, 105, 833–5.

Pazuchanics, S. L. (2006). The effects of camera perspective and field of view on performance in teleoperated navigation. In *Proceedings of the Human Factors and Ergonomics Society 50th Annual Meeting*. Santa Monica, CA: Human Factors and Ergonomics Society, 1528–32.

Pazuchanics, S. L. (2008a). Communications bandwidth limitation issues in UGVs: How do spatial and temporal resolution impact object-detection and object-location tasks? Unpublished manuscript.

Pazuchanics, S. L. (2008b). Does resizing matter? The effects of enlarged low spatial resolution video on object-detection tasks. Unpublished manuscript.

Riecke, B. E., Västfjäll, D., Larsson, P., & Schulte-Pelkum, J. (2005). Top-down and multimodal influences on self-motion perception in virtual reality. In *Proceedings of HCI International 2005*. Mahwah, NJ: Lawrence Erlbaum Associates.

Riley, J. M. & Endsley, M. R. (2004). The hunt for situation awareness: Human-robot interaction in search and rescue. In *Proceedings of the Human Factors and Ergonomics Society 48th Annual Meeting*. Santa Monica, CA: Human Factors and Ergonomics Society, 693–7.

Riley, J. M., Murphy, R. R., & Endsley, M. R. (2006). Situation awareness in the control of unmanned ground vehicles. In N. J. Cooke, H. L. Pringle, H. K. Pedersen, & O. Connor (eds), *Advances in Human Performance and Cognitive Engineering Research. Volume 7: Human Factors of Remotely Operated Vehicles* (pp. 359–71). San Diego, CA: Elsevier.

Roscoe, S. N., Johnson, S. L., & Williges, R. C. (1980). Display motion relationships. In S. N. Roscoe (ed.) *Aviation Psychology* (pp. 68–81). Ames: Iowa Sate University Press.

Rossion, B. & Pourtois, G. (2004). Revisiting Snodgrass and Vanderwart's object pictorial set: The role of surface detail in basic-level object recognition. *Perception*, 33, 217–36.

Schiffman, H. R. (2001). *Sensation and Perception: An Integrated Approach* (5th edn). New York: John Wiley & Sons, Inc.

Sheridan, T. B. (1992). *Telerobotics, Automation, and Human Supervisory Control*. Cambridge, MA: MIT Press.

Siegel, A. W. & White, S. H. (1975). The development of spatial representation of large-scale environments. In: H. W. Reese (ed.), *Advances in Child Development And Behavior* (vol. 10, pp. 9–55). New York: Academic Press.

Thorndyke, P. W. & Hayes-Roth, B. (1982). Difference in spatial knowledge acquired from maps and navigation. *Cognitive Psychology*, 14, 560–89.

Tittle, J. S., Roesler, A., & Woods, D. D. (2002). The remote perception problem. In *Proceedings of the Human Factors and Ergonomics Society 46th Annual Meeting*. Santa Monica, CA: Human Factors and Ergonomics Society, 260–64.

Tolman, E. C. (1948). Cognitive maps in rats and men. *Psychological Review*, 55, 189–208.

Tversky, I. (1977). Features of similarity. *Psychological Review*, 84, 327–52.

van Erp, J. B. F. & Padmos, P. (2003). Image parameters for driving with indirect viewing systems. *Ergonomics*, 46, 1471–99.

Yanco, H. A., Drury, J. L., & Scholtz, J. (2004). Beyond usability evaluation: Analysis of human–robot interaction at a major robotics competition. *Human-Computer Interaction*, 19, 117–49.

Chapter 7

Automation Strategies for Facilitating Human Interaction with Military Unmanned Vehicles

Keryl Cosenzo
U.S. Army Research Laboratory

Raja Parasuraman and Ewart de Visser
George Mason University

Introduction: Human-Robot Interaction (HRI) in Military Operations

New technologies are transforming the future battlefield. Force structure, doctrine, and tactics will change as these technologies are introduced. Robotic systems constitute an integral component of the future force. We use the term robot, i.e., unmanned systems (UMS), in a generic sense to describe systems that are unmanned with some degree of autonomy, to include aerial, ground, subterranean, naval surface, and subsurface vehicles. The major rationale for introducing robotic systems is to extend manned capabilities, act as force multipliers, and, most importantly, potentially to save lives (Barnes et al., 2006). Robots have been and continue to be effective in combat tasks that are "dull, dirty, and dangerous" (Carafano & Gudgel, 2007). Remotely controlled robots, both unmanned air vehicles (UAVs) and unmanned ground vehicles (UGVs), play a vital role in current operations; examples include disarming roadside bombs and performing ground based and air based patrol missions.

More autonomous robotic systems with diverse roles, tasks, and operating requirements are being designed to exploit future battle spaces. These advancements will provide a tactical advantage against conventional and unconventional warfare (Carafano & Gudgel, 2007). The prevailing expectation in the robotics community is that autonomy will enable robots (air, ground, or sea) to function with little or no human intervention. The level of autonomy will vary depending on the size and function of the platform. By 2015, the Pentagon's goal is to replace one-third of its vehicles and weapons systems with robots (*The Economist*, 2007).

Robotic operations are continuous and cognitively demanding. For robotic systems the operator will be a planner, a controller, and/or a supervisor (Chen et al., 2007). More specifically, in military operations the soldier who is responsible for the robots will have a multitude of tasks to complete: route planning for the robot; monitoring the robot (current and proposed route and mission plan) during the mission; monitoring and interpreting the sensor information received by the robot; and communicating the sensor information with others. Embedded in these high-level tasks are smaller tasks, which include:

- assuring the platform is within expected or acceptable geographic and mission parameters;
- assuring payload is best used to meet mission requirements;
- selecting best sensor(s) to use for a particular target/mission;
- determining most effective method to provide sensor coverage;
- determining if object is target of interest;
- determining how munitions can best be used to meet mission requirements;
- determining if target should be fired upon and whether call for fire should be issued;
- determining what information needs to be disseminated to others;
- determining how best to communicate this information;
- monitoring status of UMS;
- assuring that all systems are operating as expected;
- monitoring impact of non-optimal conditions on mission and UMS.

The UMS community has purported that with advancements in robotic intelligence the ratio of operators to unmanned systems can exceed 1:1. In order to decrease the manpower requirements from 2:1 to 1:1, with a loftier goal of 1:2 (or greater), automation will be required (Parasuraman et al., 2007; Wickens et al., 2006). As the ratio of operators to UMS decreases the number of tasks the operator will have to complete will increase, and concomitantly his or her workload.

Operations in the complex battlefield (to include robotic operation) require the soldier to perform simultaneous tasks (robotic and non-robotic) with little room for error (Muth et al., 2006). These types of environments impose multiple demands on the soldier's limited cognitive resources. Consequently, the requirement to perform the robotic tasks can interfere with performance on the soldier's other tasks, such as maintaining awareness of the local environment (i.e., security) and communication with other team members. The priority of the robotic tasks versus other tasks will be determined by the current events during the mission and which task(s) are most critical for mission success and soldier safety and security.

The interaction between the operator and the robot is extremely complex and not well understood. An understanding of the cognitive requirements for operator-robot interaction (i.e., HRI) must be established to fully take advantage of robotic technology. We propose that most of the contemplated systems will require either active human control or, at a minimum, supervision with the possibility of intervention. In the most extreme case, soldiers will be required to operate multiple systems while on the move and while under enemy fire. In all cases, the workload will be variable and unpredictable, changing rapidly as a function of the military environment.

The Need for Automation

Unmanned systems will be given to individuals who already have extensive responsibilities. All levels of the command structure, ranging from the infantry soldier to the officer, will use robotic assets such as UGVs and UAVs. Control of these assets, to include mobility and sensor use, will no longer be the responsibility of a few specially trained soldiers, but the responsibility of many. The ability to incorporate robot control with existing tasks will be an issue and the management of workload will problematic. Because of the likely increase in the cognitive workload demands on the soldier, automation will be needed to support soldier performance. Automation can be implemented at several levels and can involve several functions: select data, transform or integrate information, recommend decisions, or control action (Parasuraman et al., 2000).

There are costs and benefits to automation. While automation can support a human operator in a multitask environment, it also changes the nature of the cognitive demands on the operator (Parasuraman & Riley, 1997). Furthermore, while automated systems can enhance system performance, if poorly designed they can also lead to new performance problems. This is primarily due to problems in their use by human operators or to unanticipated interactions with other subsystems. Costs associated with automation that need to be considered include unbalanced mental workload, reduced situation awareness (SA), decision biases, mistrust, over-reliance, and complacency (Billings, 1997; Parasuraman & Riley, 1997; Sarter et al., 1997; Sheridan & Parasuraman, 2006; Wiener, 1988). For example, Wickens et al. (2006) showed that performance on a UAV task increased roughly linearly with the level of automation reliability. However, the automation aid led to poorer performance than manual control when reliability fell below a threshold of about 0.71; lower reliability than this, according to Wickens et al. (2006), represents no more than a "concrete life preserver."

Adaptive Automation

Adaptive automation has been proposed as an approach to automation that may preserve the benefits of automation while minimizing some of its costs (Parasuraman et al., 1992). In adaptive systems the "division of labor" between human and machine agents is not fixed, but dynamic. Adaptive automation uses mitigation criteria that drive an invocation mechanism to maintain an effective mixture of operator engagement and automation for a dynamic multitask environment (Barnes et al., 2006; Parasuraman et al., 2007). The invocation mechanism in an adaptive system is triggered by a measurement process that represents the current state of the operator and/or the task. There are four types of invocation methods:

1. In the critical-events method, automation is invoked only when certain tactical environmental events occur. This method is tied to current tactics and doctrine during mission planning.
2. In the operator performance measurement method operator mental states (e.g., mental workload or, more ambitiously, operator intentions) are inferred from performance or other behavioral measures. Because performance measures can be sparse in many modern semi-automated systems, whereas physiological measures of operator state can be obtained more frequently, operator physiological assessment provides another method for adaptive automation (Byrne & Parasuraman, 1996). In this method behavioral and/ or physiological measures are used as inputs for the adaptive logic.
3. In the human operator modeling method the operator states and performance are modeled theoretically and the adaptive algorithm is driven by the model parameters.
4. Finally, the hybrid method combines one or more of these different invocation techniques (e.g., critical events and operator performance), so that the relative merits of each method can be maximized in order to minimize operator workload and minimize performance.

The adaptive automation concept was proposed over several years ago (Rouse, 1977) but empirical studies are more recent (Inagaki, 2003; Parasuraman, 2000; Scerbo, 2001) and prototype systems have only recently been developed. One example is the U.S. Army's Rotorcraft Pilot's Associate (RPA), which aids helicopter pilots in an adaptive manner depending on mission context, and which has successfully passed in-flight tests (Dornheim, 1999). There has also been a significant amount of empirical work aimed at examining the effects of adaptive automation on human and system performance in different application domains. Automation design effectiveness is evaluated by looking at its effect on human performance, mental workload, and SA (Parasuraman et al., 2000). For example, Hilburn et al. (1997) examined the effects of adaptive automation (i.e., critical event based automation) on the performance of military air traffic controllers. Controllers were provided with a decision aid for determining optimal descent trajectories of aircraft using the Descent Advisor (DA), which was triggered by level of air traffic load. Hilburn et al. found significant benefits for controller workload when the DA was provided adaptively during high traffic loads, compared to when it was static or at low traffic loads. Kaber and Riley (1999) used a secondary-task measurement technique (i.e., operator performance based automation) to trigger adaptive automation in a target acquisition task. The secondary task was used to assess operator workload. When workload reached a preset level it triggered a computer aid for the target task. Kaber and Riley (1999) showed that adaptive computer aiding based on the secondary-task measure enhanced performance on the primary task.

As mentioned previously, the Army RPA system represents one of the few flight-tested examples of adaptive automation (Dornheim, 1999). The widely promoted Augmented Cognition (AugCog) program represents a more recent attempt to implement adaptive automation in real systems (St. John et al., 2004). AugCog specifically follows up on the operator physiological assessment method (Byrne & Parasuraman, 1996; Parasuraman et al., 1992) by evaluating different physiological measures and applying a decision aid or automation appropriate to the task environment and user state (Muth et al., 2006). These approaches to adaptive automation are also consistent with the neuroergonomic approach to human-system performance evaluation (Parasuraman, 2003; Parasuraman & Wilson, 2008). The neuroergonomic approach takes into consideration the neural bases of cognitive functions and human behavior in complex environments and its impact on technology and functioning in the real world (Parasuraman & Hancock, 2004).

The adaptive automation process is more complex than simply unloading (or engaging) the operator of a task, irrespective of the invocation process. To be effective, the invocation process must be sensitive to the operator's combined tasking environment, which depends on interactions among tasks as well as overall workload, stress, and safety considerations (Wickens & Hollands, 2000). Furthermore, adaptive automation can have different effects depending on the functions to which it is applied (e.g., psychomotor, decision making, planning) (Kaber et al., 2006). Another consideration is that in adaptive systems the automation is initiated by the system (without explicit operator input) on the basis of critical mission events, operator performance, or physiological state (Barnes et al., 2006). The adaptive automation concept has been criticized as potentially increasing system unpredictability by being system and not operator driven (Billings & Woods, 1994). Parasuraman and Miller (2006) proposed adaptable automation, an alternative form of adaptive automation, in which the human operator is the supervisor and delegates tasks to the automation (Opperman, 1994). Adaptable automation can be considered a sixth automation invocation mechanism: operator invoked. This differs from a truly adaptive automation system in which the delegation decision is made by automation itself or some expert system. With adaptable automation the human has flexibility over when and what to automate. In the context of military operations, adaptable automation allows the soldier to define conditions for automation decisions during mission planning and execution (Parasuraman & Miller, 2006), whereas in adaptive systems the automation is applied without explicit soldier input.

Adaptable automation has many advantages. The automation can be applied to more coarse grain tasks or more precise tasks. The timing of the automation can also be flexible. The result of a human-delegation type interface is that the appropriate operating points for the automation are selected based on the task demands and the context. Furthermore, the operator has some level of control of the automation, so he or she can declare goals, instructions, and/or level of automation. According to Parasuraman and Miller (2006) there are three parameters for a delegation system: competency of the system, workload of the operator, and unpredictability of the system. Parasuraman and Miller (2006) have implemented and demonstrated the efficacy of adaptable interfaces in several domains, including UAV flight mission planning, game playing, and ground robot simulations. The challenge for an adaptable automation system is that the operator should be able to make decisions regarding the use of automation in a way that prevents such high levels of workload that any potential benefits of delegation are lost (Barnes et al., 2006). While in this chapter we primarily consider how adaptive automation affects system performance, it is important to keep in mind that adaptable automation may provide an alternative approach with its own benefits.

Research reviews (Parasuraman, 2000) have demonstrated that adaptive automation can serve to reduce the problem of unbalanced workload by evading the attendant high peaks and troughs that static automation often induces. However, performance benefits can be eliminated if adaptive automation is implemented in a clumsy manner, supporting the concerns of Billings and Woods (1994) (Parasuraman et al., 1999).

Experiments on Automation Support for Human Interaction with Unmanned Vehicles

The benefits of adaptive automation have now been empirically demonstrated in a number of studies in several different domains of work performance. However, only a small body of work has investigated its benefits (or potential costs) in the context of robotic systems. Furthermore, the specific case of military UMS has been little examined. Finally, very little is known of the particular invocation methods that are best suited to this application case. Accordingly, we developed a program of research to examine the potential benefits of automation, specifically adaptive automation, on operator performance during complex UMS operations. We first developed a capability for evaluating human operator performance in managing UAVs and UGVs, so that subsequent evaluations of the efficacy of adaptive automation could be conducted. This was done in two simulation environments, the Robotic NCO Simulation and the Robotics Simulation Integration Laboratory.

Robotic NCO Simulation Studies

The Robotic NCO simulation was designed to isolate some of the cognitive requirements associated with a single operator controlling robotic assets within a larger military environment (Barnes et al., 2006). The goal was to create a microworld for evaluating UAV and UGV operations, while providing for a high degree of experimental control. The NCO is a notional environment but the tasks were based on current military ideas for robots in the future. For example, the design of the UGV task (described below) was based on field observations of the U.S. Army's Experimental Unmanned Vehicle (XUV) that currently uses an autonomous navigation system in the manner simulated in our studies. The Robotic NCO simulation consisted of four military-relevant tasks.

1. UAV target identification task: A UAV followed a series of pre-planned waypoints during the mission and received electronic hits from potential targets in the area, displayed as small white squares on the imagery. The participant had to locate and identify targets and enemies or unknowns from UAV video imagery.
2. UGV route planning task: A UGV moved through the area following a series of pre-planned waypoints and during the mission the UGV would stop at obstacles various times and request help from the operator. The participant had to determine the appropriate course of action for the UGV, continue on the pre-planned path or reroute the UGV around the obstacle.
3. Communications task with an embedded verbal SA probe task: The participation was presented auditory and visual messages in a separate communications window. Participants had to monitor the messages and acknowledge the message when they heard their own call sign. To assess SA, the messages also requested updates on the UGV and UAV status and the location of particular targets.
4. An ancillary task designed to assess SA using a probe detection method: A change detection task was embedded within a situation map. At unpredictable times during the mission and after the situation map had been populated to a degree, an icon on the situation map (a target previously identified by the participant) changed its location. Participants were instructed to press the space bar when they noticed the change.

It is important to note that participants were told that the UAV, UGV, and verbal SA communications tasks in this simulation were not independent but were coordinated tasks that supported the overall goal—a reconnaissance mission in which participants had to be aware of friendly and enemy unit movements and of the positions of their UAV and UGV assets. The verbal SA queries that were posed over the communications channel provided an evaluation of how well participants followed these instructions. Furthermore, participants were told that they would be asked at the end of the mission to select the best path for a platoon to follow, so that they could not simply ignore UAV targets or enemy units once identified but had to integrate them into their overall map of the battlefield.

A series of three experiments was conducted with the Robotic NCO simulation (Cosenzo et al., 2006; Parasuraman et al., 2009). The goal of the first experiment was to identify tasks in the HRI environment that were challenging to the operator. Cosenzo and colleagues used the Robotic NCO simulation environment to achieve this goal. In the first experiment, we manipulated level of difficulty for the UAV, UGV, and communications task by increasing the number of targets, UGV stops, and communications events. These manipulations created high and low task load conditions which were verified in a pilot study. The design of the experiment was a 2 × 2 × 2 factorial design with factors of UAV task load (high, low), UGV task load (high, low), and communication task load (high, low). The dependent variables were reaction time (RT), SA, and subjective workload. In the first experiment, Cosenzo et al. (2006) found that participants were good at integrating information received from the UAV and UGV during a simulated reconnaissance mission. However, the multitasking requirements of the Robotics NCO simulation decreased performance of the individual tasks that participants had to perform in addition to supervising the UMS. Participants typically took longer to respond to communications when they also had to identify many UAV targets and UGV requests. Additionally, when high-priority but infrequent communications requiring immediate response were presented, participants took longer to respond when they had many UAV targets and UGV requests at the same time. This result appears to indicate that high-priority but infrequently occurring communications pose a particularly high monitoring load on the operator, as suggested by studies of vigilance and monitoring in semi-automated systems (Warm et al., 1996). In general, Cosenzo et al. (2006) showed that performance (e.g., RT) was degraded when task load was high (i.e., many UAV targets to process and many UGV stops to handle).

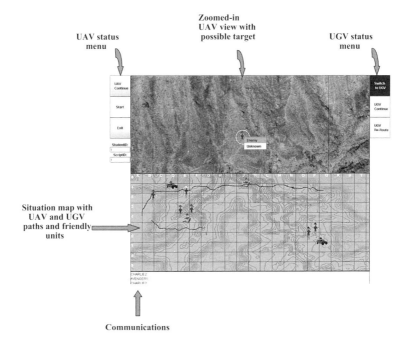

Figure 7.1   Robotic NCO simulation

Having established and evaluated a "baseline" level of UMS and multitask performance in the first experiment (Cosenzo et al., 2006), we examined the potential efficacy of adaptive automation in subsequent experiments (Parasuraman et al, 2009). Based on the results from the first experiment, we implemented an ancillary task, change detection, embedded into the situation map of the Robotic NCO display, as a potential trigger for adaptive automation in Experiment 2 and an actual trigger in Experiment 3 (Parasuraman et al., 2009). The reason for including this task was as follows. People often fail to notice changes in visual displays when they occur at the same time as various forms of visual transients (Durlach, 2004; Simons & Ambinder, 2005; Simons & Rensink, 2005). This "change blindness" phenomenon has been demonstrated in simple laboratory tasks (Simons & Ambinder, 2005) and complex military tasks (Durlach, 2004). In real-life a corollary exists in that in many tactical military operations the operator's situation display may often be updated without warning, so that the operator may miss the change. To capture this phenomenon a change detection task was embedded into the situation map of the Robotic NCO simulation. The change detection implementation in the Robotics NCO simulation was included because reduced SA has been identified as a major contributor to poor performance in robotics missions such as search and rescue (Burke & Murphy, 2004; Murphy, 2004). This experimental manipulation was used to capture transient and dynamic changes in the operators' SA by probing their awareness of the simulated battlefield environment via change detection performance. If automation can improve SA, then change detection performance should be enhanced with appropriately applied automation.

Experiment 2 was conducted to establish the sensitivity of a change detection procedure to transient and non-transient events in a complex multiple UMS mission. At unpredictable times during the simulated mission, and after the situation map had been populated to a degree, an icon on the situation map (a target previously identified by the participant) changed its location. Half of the changes occurred during one of the UGV stops, when the UGV status bar flashed (transient event condition). The other half of the changes occurred while participants were engaged in the UAV task (non-transient event condition). We predicted poorer change detection performance during transient events.

This study also set the levels of low and high task load by manipulating UAV target load which were also used in a follow-up study, Experiment 3. The design of the experiment was a 2 × 2 factorial design with factors of UAV task load (high, low) and communication task load (high, low). We found that change detection accuracy was typically low, ranging from 9.4 percent to 43.8 percent across the various conditions. This result indicated that the *change blindness* effect occurs with more realistic displays relevant to military UMS environments. Despite the low overall level of change detection performance, we found that, as predicted, detection was poorer for changes occurring during transient events (UGV task) than during non-transient events (UAV task). This indicates that the change detection procedure we used was sensitive, and this is encouraging with respect to the potential for using it to assess when adaptive automation might be useful to support the human operator supervising multiple UMS. Also, change detection performance (for non-transient events) was lower when the number of UAV targets was high compared to when it was low. This provides another index of the sensitivity of the change detection task by showing that the increased attentional demand associated with more UAV targets was reflected in poorer awareness of changes in the situation map.

For task loading, the results indicated that a high number of UAV targets and greater uncertainty in the communications task provided a sufficiently challenging level of task difficulty for use in Experiment 3, on adaptive automation. More specifically, high uncertainty on the communications task led to longer RTs for call sign acknowledgment, as predicted. The effects of UAV and communication task difficulty on performance on the UGV route-planning task were also generally as expected, with longer RTs to re-route the UGVs under the more difficult conditions.

In Experiment 3 (Parasuraman et al., 2009), we examined the efficacy of adaptive automation, based on real-time assessment of operator change detection performance, on performance, SA, and workload in supervising multiple UMS under two levels of communications task load (based on the results of Experiment 2). The automation used in the experiment was an Aided Target Recognition System (ATR) that reliably (100 percent) offloaded the responsibility of identifying targets in the UAV imagery from the participants. Parasuraman et al. (2009) implemented three automation invocation methods: no automation, performance-based adaptation, and model-based adaptation. The adaptive automation invocation methods used in this experiment were first developed by Parasuraman et al. (1996). In the performance-based adaptation (i.e., adaptive), operator change detection performance was assessed in real time and used as a basis to invoke automation. The automation was invoked if and only if the performance of an individual operator was below a specified threshold at a particular point in time during the mission. In this experiment the threshold was that change detection performance was below 50 percent; the participant did not detect 50 percent of the icon changes on the SA map. This threshold was based on Experiment 2 results. In the model-based adaptation (i.e., static) the automation was invoked at a particular point in time during the mission, based on the model prediction that operator performance is likely to be poor at that time. The model-based approach assumes that all operators are characterized by the model predictions. In this experiment the predictive model used was that after 50 percent of the changes occurred the automation was invoked; 50 percent of the icon changes on the SA map occurred. Thus, performance-based adaptation is by definition context-sensitive to an extent, whereas model-based automation is not. The design of the experiment was a $2 \times 3 \times 2$ factorial design with factors of communications task (low or high), automation condition (manual, static, and adaptive automation), and block (one and two).

Results showed that, compared to no automation, both the model-based and performance-based automation led to an increase in change detection accuracy and

SA, and a reduction in workload. These benefits of adaptive invocation mechanism for change detection, SA, and workload are summarized in Figure 7.2. Also, in comparison to model-based automation, there was a further increase in change detection accuracy and concomitant reduction in workload with performance-based automation. This last finding is important, because simply demonstrating performance benefits due to automation is insufficient; rather, the specific benefit, if any, of adaptive automation must be shown, over and above that associated with static automation (Barnes et al., 2006; Parasuraman, 1993). Experiment 3 showed that there was a reciprocal relationship between the different operator performance measures in terms of the effects of adaptive automation. Specifically, model-based automation led to an increase in both change detection accuracy and SA and a decrease in workload, with a further increase and decrease in these measures with performance-based automation. However, it should be noted that the additional increase in SA with adaptive automation was not statistically significant. Several studies have documented benefits of adaptive automation for SA (Kaber & Endsley, 2004).

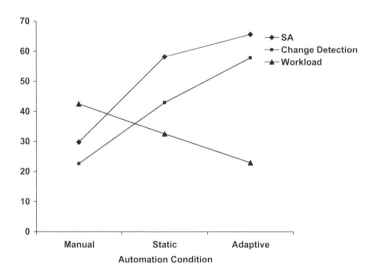

Figure 7.2   Effects of static and adaptive automation on change detection accuracy, situation awareness, and mental workload

Robotics Simulation Integration Laboratory Studies

Another series of experiments was conducted in the Robotics Simulation Integration Laboratory (SIL). The SIL is based on a functional robotic interface used to control an experimental unmanned ground vehicle (XUV). The SIL is a relatively high-fidelity simulation of a planned Army system for mounted Soldiers. The SIL includes the military simulator rSAF, a 3D graphics engine (METAVR), and a Tactical Control User Interface (TCU). The operator can control the vehicles with the TCU using a touch interface.

In the Reconnaissance, Surveillance, and Target Acquisition (RSTA) view screen, snapshots can be taken by pointing and clicking on a target and capturing the picture, while classification was possible through a separate toolbar. The SIL has superior face validity but our ability to manipulate automation and invocation methods was somewhat limited in our initial experiments (but which will improve with new software updates).

The purpose of the first SIL experiment was to investigate the effects of imperfect automation on system performance with multiple UMS under different levels of task load. In SIL Experiment 1 (de Visser & Parasuraman, 2007), we manipulated the reliability of the automation and number of UGVs to monitor or control. The primary task involved monitoring and checking one of the XUVs for target detection accuracy. This XUV performed four RSTA scans during each trial. The ATR capability of the XUV detected targets (soldiers or tanks) during each RSTA scan with one of three levels of detection reliability: 30 percent (low), 70 percent (medium), and 100 percent (high) of targets detected. (The ATR only missed targets when not 100 percent reliable; it never made false alarms.) The participant was instructed to check the ATR system for accuracy and detect any additional targets that were missed by the automation. Furthermore, the participant always had the final say and could either confirm or disconfirm the ATR's decision. Thus, even in the 100 percent reliability condition, while not required to detect targets, participants still needed to check the automation and its decision, and could also acquire additional targets if they wanted to. The second task consisted of classifying all detected vehicles into friendly or hostile units distinguishable by color and shape. Automation assistance was available for the classification task. The third task comprised of monitoring the pre-planned route of several other XUV and UAV assets. Each UMS stopped at each waypoint and waited for the operator's command to continue moving along the pre-planned path. At the waypoints, the operator was also required to note the direction in which each vehicle was facing.

This experiment was a $3 \times 2 \times 2$ factorial design with factors of reliability (low, medium, high), task load (low, high), and block (first, second). The dependent variables were mission completion time, detection performance, classification performance, SA, subjective workload, and subjective trust. Participants received extensive training on the SIL (approximately 60–90 minutes). We have found that this is a minimum level of training that is needed to ensure that participants understand their tasks and show stable levels of performance at the end of training.

We analyzed target detection performance in terms of three components: automation, user, and combined system (user+automation) performance. The mean values of detection performance by the automation at low, medium, and high reliability are shown in Table 7.1. The high automation condition was intended to be "perfect," but in practice the ATR algorithms resulted in performance slightly below 100 percent (97.9 percent) due to the simulation environment. User detection performance in the medium reliability condition increased as task load increased, whereas it decreased for low reliability as task load increased (see Table 7.1). Combined system detection performance was computed by dividing the total number of targets detected by the automation and the user by the total number of targets that could be detected during each trial. For the low level of reliability, combined system detection performance decreased as task load increased. For the other two levels of reliability, there was no effect of task load.

Table 7.2 shows the means and standard errors for target detection collapsed across all conditions for reliability as well as the user and automation detection performance *gains*. These data were obtained by subtracting automation performance and user performance from combined system performance, respectively. That is, "Combined System Gain-User" refers to the gain in system detection performance from including the human user in the system, whereas "Combined System Gain-ATR" refers to the gain from including the automation in the system. As Table 7.2 shows, there was substantial gain to the system of including both the user (23 percent) and the ATR (21 percent).

Table 7.1 **Means and standard errors (SEM) for detection rate (percent) for automation, user, and combined user+automation (system) performance as a function of automation reliability**

Reliability	Task load	ATR detection reliability	User detection performance	Combined system detection performance
High	Low	96.14 (0.87)	-	98.13 (0.65)
	High	99.57 (0.30)	-	99.58 (0.29)
Medium	Low	70.20 (2.21)	72.15 (8.25)	93.40 (1.86)
	High	73.18 (2.29)	77.72 (7.16)	93.75 (1.98)
Low	Low	36.21 (3.81)	75.77 (8.84)	90.14 (2.98)
	High	36.02 (6.53)	66.82 (9.01)	83.65 (4.32)

Table 7.2   Means and standard errors (SEM) of detection rate collapsed across reliability conditions

	ATR Reliability	User Reliability	Combined System Reliability	Combined System Gain-User	Combined System Gain-ATR
Detection rate	67.93 (8.44)	66.33 (7.31)	90.90 (3.14)	22.97 (6.84)	20.94 (6.56)

Our findings showed that system target detection performance was higher than either user or ATR performance alone. In the medium reliability condition, system performance increased about 23 percent in the low task load condition and 21 percent in the high task load condition due to the user's intervention. In the low reliability condition, the user performed most of the target detection task and improved system performance 54 percent in the low task load condition and 48 percent in the high task load condition. In both the medium and low reliability conditions, the user was able to improve on the ATR system, albeit to a somewhat lesser extent in the high task load conditions.

These results are partially consistent with our predictions. Our first hypothesis was supported: there were definite benefits of imperfect automation on performance above a 70 percent level of reliability, consistent with Wickens et al. (2006) (see also Rovira et al., 2007). However, we also found that there were some benefits to imperfect automation with even a reliability as low as 30 percent. This result counters the claim by Wickens et al. (2006) that automation reliability below about 70 percent degrades performance and is worse than no automation at all. This finding may be due to the nature of the task. We instructed participants to detect all targets in the field. Therefore, any targets detected by the ATR would alleviate the workload of the user. In the low reliability condition, the ATR would still find about one target for each RSTA scan. Our results also differ from the previous study by Wickens et al. (2006) in that the ATR did not fully relieve the operator of the task of processing targets, but only assisted him or her in that task. Thus, when automation *extends* human capabilities (Inagaki, 2003) even highly imperfect automation can benefit the user as an "extra pair of eyes." In the studies by Wickens et al. (2006), the automation completely offloaded the task from the operator. Under these conditions, it may indeed be the case that reliability below 70 percent hurts performance because then the human has to expend extra effort to compensate for the poor performance of the automation.

Our second hypothesis was partially supported. In the low reliability condition, system performance was worst in both low and high task load conditions compared to the other reliability conditions, and worse for the high task load condition than for the low task load condition, consistent with Cummings and Guerlain (2007). On the contrary, we found that system performance in the medium condition was about the same in the low and high task load conditions. An explanation for these findings could be that merely having more vehicles to control does not necessarily result in an increase in task load. The secondary task of noting down the direction of multiple vehicles may have failed to create sufficient task load differences in the medium condition, allowing participants to strategically allocate attention to each vehicle. This explanation is consistent with other findings (Levinthal & Wickens, 2006).

Our third hypothesis was partially supported. We did find an interaction between reliability and task load, but not precisely in the direction we predicted. We found that the worst performance occurred in the high task load condition in the low reliability condition. However, performance in the medium condition for both user and system performance remained fairly equal. The reason for the similarity may be due to the possibility of participants with different attention allocation strategies as discussed above.

In the second SIL experiment (de Visser et al., 2008), we manipulated conditions during which the operator performed the task with no automation, static automation, or adaptive automation for the UGV RSTA task. Participants were again supported in the UGV task with the ATR system, thereby offloading them of the responsibility of identifying targets (but not of deciding whether the soldier was actually present). In the static automation condition the ATR automation was always on, and in the adaptive automation condition the ATR automation was on only during high task load phases in the mission. The levels of task load were predetermined based on pilot study results examining the difficulty of the stimuli, which is consistent with the model-based invocation strategy (Parasuraman et al., 1992). In addition to the UGV RSTA task, participants completed three tasks: 1) a UAV target detection task; 2) an ancillary task designed to assess SA using a probe detection method; and 3) a change detection task embedded within the UAV video feed display similar to that used in the Robotic NCO experiments. The experimental design was thus a $3 \times 2 \times 2$ factorial design with factors automation (none, static, adaptive), task load (low, high), and block (first, second).

Sixteen participants were trained for 45 minutes on use of the high-fidelity simulation and completed a total of six missions lasting around 15–20 minutes each. After each mission they filled out an adapted trust questionnaire (Lee & Moray, 1992) and an electronic version of the NASA-TLX workload measure. Analysis of the performance data revealed no significant differences between static and adaptive automation. However, participants reported significantly greater trust in the adaptive automation than in the static automation condition. Participants were also more self-confident in their ability to perform the task using adaptive automation compared to both the no-automation and adaptive automation conditions. Finally, participants reported lower workload when working with adaptive automation compared to static automation. Thus, adaptive automation was associated with equivalent levels of performance as static automation and with lowered workload, greater trust, and greater operator self-confidence. These benefits of adaptive automation complement the performance benefits found previously with the Robotic NCO simulation.

Discussion

The results of five experiments using the Robotic NCO simulation and SIL environments revealed generally consistent evidence for the efficacy of adaptive automation in supporting human operator supervision of multiple UMS. Thus adaptive automation may be a useful mitigation strategy to help offset the potential deleterious effects of high cognitive load on Army robotic operators in a multitasking environment. The adaptive automation was successful not only in supporting the human operator in an appropriate context—when their change detection performance was low, pointing to low perceptual awareness of the evolving mission elements—but also freed up sufficient attentional resources to benefit performance on the sub-tasks.

Future robotics operators are likely to require some automation to complete complex UMS missions. Modeling and simulation studies suggest that the soldier's primary tasks (e.g., radio communications, local security), robotic tasks, and crew safety can be compromised during high workload mission segments (Chen and Joyner, 2006; Mitchell and Henthorn, 2005). As limited capacity individuals, appropriately applied automation can enable operators to perform at greater capacity. The key is to apply automation to the tasks that require assistance or by automating the simple tasks to free up the operators resources to attend to more challenging or cognitively demanding tasks. A second critical component is the temporal nature of the automation; it needs to be applied during the appropriate context or time during the mission. If the mission is at a low operations tempo (i.e., optempo) for a period of time, automation may not be as necessary as when the optempo and the operator's multitasking requirements are high. Some variant of an adaptive system would therefore be particularly well suited to these situations because of the uneven workload and the requirement to maintain SA for the primary tasks as well as the robotic tasks (Parasuraman et al., 2009).

What adaptive automation architectures should be considered? Based on the empirical results discussed previously, we suggest that:

1. Information displays should adapt to the changing military environment. For example, information presentation format (e.g., text vs. graphics) can change depending on whether a soldier is seated in a vehicle or is dismounted and using a tablet controller.
2. Software should be developed that allows the operator to allocate automation under specified conditions *before* the mission (as in the Rotorcraft Pilot's Associate).
3. At least initially, adaptive systems should be evaluated that do *not* take decision authority away from the operator. This can be accomplished in two ways: (a) an advisory asking permission to invoke automation, or (b) an advisory that alerts the operator that the automation will be invoked unless overridden.

4. For safety or crew protection situations, specific tactical or safety responses can be invoked without crew permission.

However, it will be important to evaluate options in a realistic environment before final designs are considered. For example, performance-based adaptive logic has proven to be effective in our laboratory simulations but may be difficult to implement during actual missions because of the temporal sluggishness of performance measures (tens of seconds to many minutes). There is also the danger of adding yet another secondary task to the soldier's list of requirements (although it might be possible to assess performance on "embedded" secondary tasks that are already part of the soldier's task repertoire). Physiological measures could in principle be used to invoke adaptive automation rapidly (Byrne & Parasuraman, 1996; Parasuraman, 2003) because they have a higher bandwidth, but their suitability for rugged field operations has yet to be demonstrated reliably. Also, the type of tasks to automate will depend on engineering considerations as well as performance and operational considerations. Finally, there is the system engineering challenge of integrating different levels of automation into the overall software environment. Fortunately, there are simulation platforms that capture a more realistic level of fidelity for both the crew tasking environment and the software architecture, making the design problem more tractable.

The end goal of our program of research is to design an automation matrix supported by experimental data that lays out the automation–task–mission relationship. The automation matrix prioritizes the tasks that should have some level of automation for UMS operations (Gacy, 2006). More especially it will be a framework for determining what is automated and the impact of such automations on mission performance. A draft three-level framework was developed by Gacy (2006). In the first level of the matrix the soldier tasks are defined. The second level of the matrix lists the potential automation approaches. The third level of the matrix provides weightings for factors such as task importance, task connectedness, and the expected workload. These three levels of the matrix are incorporated into an algorithm which combines the levels and weights into a single number that represents the overall priority for development of that automation. From this prioritized list, various automation strategies (i.e., adaptive or adaptable) can be evaluated. Data is needed to develop the matrix and the weights. We plan to conduct further analytical and empirical studies to investigate the feed the matrix and validate this approach to human-automation interaction design for UMS operations.

This work will continue beyond the current U.S. Army HRI program that it is supporting. As technology changes, the automation and the invocation strategies will have to be modified. As technology becomes more sophisticated, the ability to implement more complex adaptive automation procedures will be realized. We are currently focused on invocation mechanisms that are directly observable and quantifiable. With the advances in neurosensing techniques and empirical research in neuroergonomics, it is foreseeable that future techniques may include nonobservable triggers and non-invasive behavioral probes for adaptive automation human-computer systems.

References

Barnes, M., Parasuraman R., & Cosenzo, K. (2006). Adaptive automation for military robotic systems. In NATO Technical Report, *Uninhabited Military Vehicles: Human Factors Issues in Augmenting the Force* (NATO Publication No. RTOTR-HFM-078, pp. 420–40).

Billings, C.E. (1997). *Aviation Automation: The Search for a Human-centered Approach.* Mahwah, NJ: Lawrence Erlbaum Associates.

Billings, C.E. & Woods, D. (1994). Concerns about adaptive automation in aviation systems: In R. Parasuraman & M. Mouloua (eds), *Automation a Human Performance: Current Research Trends* (pp. 264–9). Hillsdale, NJ: Lawrence Erlbaum Associates.

Burke, J.L. & Murphy, R.R. (2004). Human-robot interaction, situation awareness and search: Two heads are better than one. *Proceedings of the 2004 IEEE International Workshop on Robot and Human Communication*, 307–12. Okayama, Japan.

Byrne, E.A. & Parasuraman, R. (1996). Psychophysiology and adaptive automation. *Biological Psychology*, 42, 249–68.

Carafano, J.J. & Gudgel, A. (2007). The Pentagon's robots: Arming the future. *The Heritage Foundation: Backgrounder*, 2096, 1–6.

Chen, J.Y.C., Haas, E.C., & Barnes, M. (2007). Human performance issues and user interface design for teleoperated robots. *IEEE Transactions on Systems, Man, and Cybernetics. Part C: Applications and Reviews*, 37(6), 1231–45.

Chen, J.Y.C. & Joyner, C.T. (2006). *Concurrent Performance of Gunner's and Robotic Operator's Tasks in a Simulated Mounted Combat System Environment* (ARL Technical Report No. ARL-TR-3815). Aberdeen Proving Ground, MD: U.S. Army Research Laboratory.

Cosenzo, K., Parasuraman, R., Novak, A., & Barnes, M. (2006). *Adaptive Automation for Robotic Military Systems* (ARL Technical Report No. ARL-TR-3808). Aberdeen Proving Ground, MD: U.S. Army Research Laboratory.

Cummings, M.L. & Guerlain, S. (2007). Developing operator capacity estimates for supervisory control of autonomous vehicles. *Human Factors*, 49, 1–15.

de Visser, E., Horvath, D., Parasuraman, R., & Cosenzo, K. (2008). Adaptive automation for human interaction with uninhabited vehicles. *Presented at the Midyear meeting of Division 21 of the American Psychological Association.* Fairfax, VA: George Mason University.

de Visser, E. & Parasuraman, R. (2007). Effects of imperfect automation and task load on human supervision of multiple uninhabited vehicles. *Proceedings of the Annual Meeting of the Human Factors and Ergonomics Society*, CD-ROM, Santa Monica, CA.

Dornheim, M. (1999). Apache tests power of new cockpit tool. *Aviation Week and Space Technology*, 18 (October), 46–9.

Durlach, P. (2004). Change blindness and its implications for complex monitoring and control systems design and operator training. *Human-Computer Interaction*, 19(4), 423–51.

Economist, The. (2007). Robot Wars. *Economist.com, Technology Quarterly*, http://economist.com/printedition/PrinterFriendly.cfm?story_id=9249201.

Gacy, M. (2006). HRI ATO automation matrix: Results and description (Technical Report). Boulder, CO: Micro-Analysis and Design, Inc.

Hilburn, B., Jorna, P.G., Byrne, E.A., & Parasuraman, R. (1997). The effect of adaptive air traffic control (ATC) decision aiding on controller mental workload. In M. Mouloua & J. Koonce (eds), *Human-automation Interaction: Research and Practice* (pp. 84–91). Mahwah, NJ: Lawrence Erlbaum Associates.

Inagaki, T. (2003). Adaptive automation: Sharing and trading of control. In E. Hollnagel (ed.), *Handbook of Cognitive Task Design* (pp. 147–69). Mahwah, NJ: Lawrence Erlbaum Associates.

Kaber, D.B. & Endsley, M. (2004). The effects of level of automation and adaptive automation on human performance, situation awareness and workload in a dynamic control task. *Theoretical Issues in Ergonomics Science*, 5, 113–53.

Kaber, D.B., Perry, C.M., Segall, N., McClernon, C.K., & Prinzell, L.J. (2006). Situation awareness implications of adaptive automation for information processing in an air traffic control-related task. *International Journal of Industrial Ergonomics*, 36, 447–62.

Kaber, D.B. & Riley, J. (1999). Adaptive automation of a dynamic control task based on workload assessment through a secondary monitoring task. In M. Scerbo & M. Mouloua, (eds), *Automation Technology and Human Performance: Current Research and Trends* (pp. 129–33). Mahwah, NJ: Lawrence Erlbaum Associates.

Lee, J.D. & Moray, N. (1992). Trust, control strategies and allocation of function in human–machine systems. *Ergonomics*, 5, 1243–70.

Levinthal, B.R. & Wickens, C.D. (2006). Management of multiple UAVs with imperfect automation. *Proceedings of the Annual Meeting of the Human Factors and Ergonomics Society*, 1941–4. Santa Monica, CA: HFES.

Mahwah, NJ: Lawrence Erlbaum Associates.

Mitchell, D.K. & Henthorn T. (2005). *Soldier Workload Analysis of the Mounted Combat System (MCS) Platoon's Use of Unmanned Assets* (ARL Technical Report No. ARL-TR-3476). Aberdeen Proving Ground, MD: U.S. Army Research Laboratory.

Murphy, R.R. (2004). Human-robot interaction in rescue robots. *IEEE Transactions on Systems, Man, and Cybernetics. Part C: Applications and Reviews*, 32(2), 138–53.

Muth, E.R., Kruse, A.A., Hoover, A., & Schmorrow, D. (2006). Augmented cognition: Aiding the soldier in high and low workload environments through closed-loop human-machine interactions. In T.W. Britt, C.A. Castro, & A.B. Adler (eds), *Military life: The Psychology of Serving in Peace and Combat* (pp. 108–27). Westport, CT: Praeger Security International.

Opperman, R. (1994). *Adaptive User Support*. Hillsdale, NJ: Lawrence Erlbaum Associates.

Parasuraman, R. (1993). Effects of adaptive function allocation on human performance. In D.J. Garland & J.A. Wise (eds), *Human Factors and Advanced Aviation Technologies* (pp. 147–57). Daytona Beach, FL: Embry-Riddle Aeronautical University Press.

Parasuraman, R. (2000). Designing automation for human use: Empirical studies and quantitative models. *Ergonomics*, 43, 931–51.

Parasuraman, R. (2003). Neuroergonomics: Research and practice. *Theoretical Issues in Ergonomics Science*, 4, 5–20.

Parasuraman, R., Bahri, T., Deaton, J.E., Morrison, J.G., & Barnes, M. (1992). Theory and design of adaptive automation in aviation systems (Technical Report No. NAWCADWAR-92033-60). Warminster, PA: Naval Air Warfare Center, Aircraft Division.

Parasuraman, R., Barnes, M., & Cosenzo, K.A. (2007). Adaptive automation for human-robot teaming in future command and control systems. *International Journal of Command and Control*, 1(2), 43–68.

Parasuraman, R., Cosenzo, K.A., & de Visser, E. (2009). Adaptive automation for human supervision of multiple uninhabited vehicles: Effects on change detection, situation awareness, and workload. *Military Psychology*, 21(2), 270–97.

Parasuraman, R. & Hancock, P.A. (2004). Neuroergonomics: Harnessing the power of brain science for HF/E. *Human Factors and Ergonomics Society Bulletin*, 47(12), 1.

Parasuraman, R. & Miller, C. (2006). Delegation interfaces for human supervision of multiple UAVs. In N.J. Cooke, H.L. Pringle, H.L. Pedersen, & O. Connor (eds), *Advances in Human Performance and Cognitive Engineering Research: Human Factors of Remotely Operated Vehicles* (pp. 251–66). Amsterdam: Elsevier.

Parasuraman, R., Mouloua, M., & Hilburn, B. (1999). Adaptive aiding and adaptive task allocation enhance human-machine interaction. In M.W. Scerbo & M. Mouloua (eds), *Automation Technology and Human Performance: Current Research and Trends* (pp. 119–23). Mahwah, NJ: Lawrence Erlbaum Associates.

Parasuraman, R., Mouloua, M., & Molloy, R. (1996). Effects of adaptive task allocation on monitoring of automated systems. *Human Factors*, 38, 665–79.

Parasuraman, R. & Riley, V. (1997). Humans and automation: Use, misuse, disuse, abuse. *Human Factors*, 39, 230–53.

Parasuraman, R., Sheridan, T.B., & Wickens, C.D. (2000). A model of types and levels of human interaction with automation. *IEEE Transactions on Systems, Man, and Cybernetics. Part A: Systems and Humans*, 30, 286–97.

Parasuraman, R. & Wilson, G.F. (2008). Putting the brain to work: Neuroergonomics past, present, and future. *Human Factors*, 50(3), 468–74.

Rouse, W.B. (1977). Human-computer interaction in multi-task situations. *IEEE Transactions on Systems, Man and Cybernetics*, SMC-7, 384–92.

Rovira, E., McGarry, K., & Parasuraman, R. (2007). Effects of imperfect automation on decision making in a simulated command and control task. *Human Factors*, 49, 76–87.

Sarter, N., Woods, D., & Billings, C.E. (1997). Automation surprises. In G. Salvendy (ed.), *Handbook of Human Factors and Ergonomics* (pp. 1926–43). New York: Wiley.

Scerbo, M. (2001). Adaptive automation. In W. Karwowski (ed.), *International Encyclopedia of Human Factors and Ergonomics* (pp. 1077–9). London: Taylor & Francis.

Sheridan, T. & Parasuraman, R. (2006). Human-automation interaction. *Reviews of Human Factors and Ergonomics*, 1, 89–129.

Simons, D.J. & Ambinder, M.S. (2005). Change blindness: Theory and consequences. *Current Directions in Psychological Science*, 14(1), 44–8.

Simons, D.J. & Rensink, R.A. (2005). Change blindness: Past, present, and future. *Trends in Cognitive Sciences*, 9(1), 16–20.

St. John, M., Kobus, D.A., Morrison, J.G., & Schmorrow, D. (2004). Overview of the DARPA Augmented Cognition Technical Integration Experiment. *International Journal of Human-Computer Interaction*, 17(2), 131–49.

Warm, J., Dember, W.N., & Hancock, P. (1996). Vigilance and workload in automated systems. In R. Parasuraman & M. Mouloua (eds), *Automation and Human Performance: Theory and Applications* (pp. 183–200). Mahwah, NJ: Lawrence Erlbaum Associates.

Wickens, C.D., Dixon, S.R. & Ambinder, M.S. (2006). Workload and automation reliability in unmanned air vehicles. In N.J. Cooke, H.L. Pringle, H.L. Pedersen, & O. Connor (eds), *Advanced in Human Performance and Cognitive Engineering Research: Human Factors of Remotely Operated Vehicles* (pp. 209–22). Amsterdam: Elsevier.

Wickens, C.D. & Hollands, J. (2000). *Engineering Psychology and Human Performance*. Upper Saddle River, NJ: Prentice Hall.

Wiener, E.L. (1988). Cockpit automation. In E.L. Wiener & D.C. Nagel (eds), *Human Factors in Aviation* (pp. 433–61). San Diego: Academic Press.

Chapter 8

An Analytical Approach for Predicting Soldier Workload and Performance Using Human Performance Modeling

Diane Kuhl Mitchell and Charneta Samms
U.S. Army Research Laboratory

Over the past several decades the U.S. Army has transitioned from a tactical concept emphasizing heavily armored combat vehicles to a concept of soldiers operating in an information-rich tactical environment. Under this new tactical concept the Army's reliance on armor has been replaced with a reliance on information. Much of this information will be acquired from unmanned systems equipped with multiple sensors. Frequently, soldiers will be expected to process this information concurrent with more routine mission tasks such as moving, shooting, and communicating. Furthermore, the Army has reduced its total force structure, which in turn has minimized crew size. Smaller crews means allocating routine mission tasks across fewer soldiers. Therefore, regardless of the tasks associated with unmanned assets, the soldiers are expected to perform more concurrent tasks than in legacy systems. Adding to these tasks, the tasks associated with unmanned asset use, could potentially overload the soldiers and reduce rather than enhance their performance. To avoid this problem, it is critical to analyze the interaction of soldiers with unmanned assets early in the system design process. The conceptual phase is the optimum time for analysis because changes in tactics, techniques, and procedures (TTPs) and to system design can be made before they become finalized.

The benefits of predicting soldier performance in the concept phase, however, are also the challenges. Analysts must predict how soldiers will use unmanned systems whose interfaces, technologies, and TTPs are preliminary. In some cases, even the platforms the soldiers ride in may be conceptual. To meet these challenges, analysts need an accurate way to represent trade-offs between alternative unmanned system designs, TTPs, and soldier performance. Traditionally, they have compared alternatives by building mock-ups of the conceptual systems or conducting laboratory or field experiments. These traditional approaches, however, were costly and time-consuming and the Army can fund less of these due to economic constraints on the Army's research and development budget. Therefore, Army analysts had to develop fast and cost-efficient alternatives to traditional approaches that are as accurate as, if not better than, traditional approaches for predicting soldier performance. One of these new approaches is human performance modeling prior to experimentation or the model–test–model approach.

Model–Test–Model Approach

With the model–test–model approach, analysts represent soldiers interacting with a conceptual system prior to testing or experimentation using human performance modeling. The human performance models they build identify areas that may contribute to soldier performance problems. During experiments and field testing, these problem areas become key issues for evaluating soldier performance. This process, of model–test–model, results in fewer, more focused studies. Furthermore, soldier performance data can be collected during the studies to validate the models and the data can be input back into the models to be used for additional soldier performance analyses. This approach is efficient in cost and time because it reduces the number of required studies and ensures that the studies conducted focus on relevant soldier performance issues. Barnes et al. (2006) provide several examples of the model–test–model approach to robotic experiments.

Although the model–test–model approach has many benefits, these benefits are contingent upon the human performance models accurately representing how the soldiers will use a conceptual system. For the models to be accurate it is critical that a rigorous analytical process be followed during model development. Following an analytical approach ensures that the predictions from the models accurately identify issues that the experiments then validate as critical soldier performance issues.

Army Research Laboratory (ARL) analysts supporting the ARL program, the robotic collaboration advanced technology objective (RCATO), used the model–test–model approach to guide both laboratory and field tests throughout the entire program. More importantly, however, they used an analytical approach as the foundation for their human performance modeling. Specifically, they used human performance analysis to predict task combinations that are likely to cause high workload for soldiers interacting with unmanned systems. This chapter outlines the steps in the analytical process they used to implement the model–test–model approach. The modeling tool they used in their analytical approach was the human performance modeling tool Improved Performance Research Integration Tool (IMPRINT).

Improved Performance Research Integration Tool (IMPRINT)

IMPRINT is a tool developed by the Army Research Laboratory (ARL) for analysts to use to predict the performance of human operators of a system as they perform a particular set of activities referred to as a mission. In the tool, the analyst represents the mission in the form of a task network. This task network shows the sequence of functions, and tasks within functions, the operators perform as they complete the mission. In addition, environmental factors, such as heat, cold, or vibration that may influence the operators' performance can be included in the model. In addition to these effects, they may become fatigued or experience mental overload. IMPRINT provides the analyst with the capability to predict the effects of these factors on operator performance as well.[1]

IMPRINT Analysis Process

Although IMPRINT provided the ARL analysts supporting the RCATO with the capability for building models and predicting mental overload, they could not begin model building until they had set up their analysis. The first step in this analysis process is to identify soldier performance issues appropriate for both modelers and experimenters to investigate. To ensure that their modeling projects result in recommendations useful to experimenters, ARL analysts use an analytical approach to modeling based on traditional experimental design principles.

Just as an experiment begins with a hypothesis, the analysts must have a clearly defined project objective. This objective is critical in order to accurately decide what key items to incorporate into the human performance models. Therefore, the first step in their analytical process is to format the project goal into a specific question that the models can answer. One of the RCATO questions, for example, was "What are the impacts on soldier workload of a two-soldier function allocation versus a three-soldier function allocation for a future tank system allocated an armed reconnaissance vehicle (ARV)?"

Once the question is clearly defined, the next step in the analytical process is to complete a domain analysis. Just as any good experiment begins with a literature review, the domain analysis for a model building project begins with a review of the current literature on the topic. In addition to the literature review, however, the domain analysis for a modeling project includes interviews with individuals who are using systems similar to the proposed system. This step is critical for obtaining the tasks that will become the task-network models. Basing the task-network models on user input and literature review helps ensure their validity.

For the RCATO projects, the analysts interviewed dozens of combat experienced soldiers. They interviewed soldiers participating in simulations with conceptual equipment as well. Based on these interviews, the analysts developed task analyses to represent the soldiers' mission. These task analyses became their task network models in IMPRINT. They made sure the planned analyses incorporated all potential issues associated with the operation of the proposed system. For example, if the soldiers reported that security was a key factor for crew survivability then the modeler would include local security as a function in the model. In the analysis this function might be allocated to different crewmembers,

---

1   http://www.arl.army.mil/ARL-Directorates/HRED/imb/imprint/Imprint7.htm.

along with unmanned asset operations, and their workload predicted. If workload was high across all allocations, the analysts would recommend an experiment be designed to further investigate this issue.

After the domain analysis is complete, the analysts create an experimental design matrix representing the various conditions to be modeled to answer the analytical question. To set up this design matrix, the modeler, like an experimenter, needs to determine the independent variables that will be manipulated to answer the question. For the evaluation of the crew with an ARV, the independent variable would be the functions the crew is expected to perform and the number of soldiers allocated to each crew design. Dependent measures would be task time, task frequency, and instances of high workload.

By completing the experimental design matrix prior to building any of the models for the RCATO projects, the ARL analysts determined how many models must be built and what factors need to be incorporated into each model. When analysts build models they can model everything associated with a system and operator. However, this is time consuming and some system components may have no impact on the question being evaluated. Therefore, analysts decide what key items to incorporate into their model of the system and the model becomes an abstraction of key system components. For example, military crews initialize and set up equipment prior to combat. Equipment setup probably has little impact on their workload during a combat mission and so would not be incorporated into a model of an armor crew performing a combat mission.

Because each model built for a project represents an abstraction of the operator working with the system, it is important that the modeler have clear guidance on what to put into the model. The design matrices helped the ARL analysts determine what components are critical components of this abstraction for answering the analytical question. It is important that the design matrix determine the key components for the modelers. Without the design matrix, analysts can influence the results by incorporating their own biases into the model. In addition, the combination of a good analytical question, domain analysis, and experimental design matrix helps prevent the "garbage in, garbage out" modeling limitation by providing a strong analytical foundation to the models built rather than the modeler's own personal abstraction of the problem.

With the design matrix as a guideline, the ARL analysts determined the expected output from the models. The output is analogous to the dependent measures or performance outcomes from an experiment. What is output from the model is critical because this is the quantitative data that will be the basis of analytical recommendations intended to change design. The model builder must be certain that the output will answer the project question. Furthermore, the model output should match data that can be collected during an experiment. This ensures that the model–test–model approach can be followed because the model output can be validated by experimental data. Dependent measures collected for the RCATO IMPRINT soldier workload analyses were mental workload predictions, task times, task accuracies, and target detections. Most human performance modeling tools can generate reports of the dependent variables a modeler specifies. The analyst consolidates the data from these reports into a format displaying the overall analytical results. In the analytical process the ARL analysts used for the RCATO projects, they designed an output matrix or data collection sheet prior to building or running the models to ensure that the RCATO models included the correct factors to generate the desired data.

The ARL analysts began building the actual models only after the question, design matrix, and data collection sheet were completed. At this time, they built a model to represent each of the conditions in the design matrix. Creating a separate model for each condition is critical to ensure that the data the model outputs for the different conditions is separated. Analogous to an experiment, each model is a condition and each model run is an experimental trial. The data is analyzed across multiple model runs and collected in the data collection sheet template. Model data formatted in this way are clear, concise, and easy for the project manager to understand. However, the model data will not have impact on system design unless, the model–test–model approach is implemented and the data transitioned into recommendations for experiments. Therefore, for each of the RCATO analyses, the analysts suggested experiments. They recommended an experiment to investigate a gunner acting as robotics controller; an experiment to investigate the effectiveness of an autonomous navigation system (ANS) to reduce soldier workload, and an experiment to investigate the workload of a dismounted squad leader who is controlling multiple unmanned assets. These experiments, and the modeling analysis that initiated them, are discussed in the following three case studies.

RCATO Case Study 1: Gunner-Robotic Operator Analysis and Experiment

The first RCATO project the ARL analysts completed was an IMPRINT analysis of the workload of the crew of a future concept tank platoon leader's vehicle (PLV). In the analysis three crewmembers are controlling an unmanned Armed Reconnaissance Vehicle-Reconnaissance, Surveillance, and Target Acquisition (ARV-RSTA). In addition they are monitoring its sensor feeds, as well as monitoring sensor data from an unmanned aerial vehicle (UAV). The question the analysts were investigating was how to integrate the role of the robotics operator with the roles of the three crewmembers: platoon leader (PL), driver (D), and gunner (G). The analytical objective was to predict their mental workload, identify high workload task combinations, and recommend the function allocation that would result in manageable workload. Table 8.1 displays the design matrix the analysts developed for this project.

Table 8.1   Design matrix for case study 1 IMPRINT analysis

Function Name	Function Allocation 1 Driver-Robotics Operator	Function Allocation 2 Gunner-Robotics Operator
Tactical planning	Platoon leader	Platoon leader
Tactical communications	Platoon leader	Platoon leader
Crew communications	All crewmembers	All crewmembers
Monitor UAV feed	Platoon leader	Platoon leader
Platoon communications	Platoon leader	Platoon leader
Battlefield awareness scan	Platoon leader	Platoon leader
Scan for threats	Gunner	Gunner, platoon leader
Target engage beyond line of sight (BLOS)	Gunner	Gunner
ARV control	Driver	Gunner
ARV payload	Driver, platoon leader	Gunner, platoon leader
ARV security	Driver	Gunner
Vehicle maintenance	Driver	Driver
Driver tank	Driver	Driver

From the design matrix, the analyst built two IMPRINT models of the tank PLV crew, each representing an alternate function allocation. In both IMPRINT models, the tank PLV had a crew of three. One crewmember is the driver (D), one crewmember is the platoon leader (PL), and the third crewmember is the gunner. The two models, however, differed according to who is assigned the additional role of robotics operator. In function allocation 1, the driver is the robotics operator. This driver-robotics controller matched the operational and organizational concept document for this future tank system. In function allocation 2, the gunner is robotics operator. This gunner-robotics controller matched the prime contractor's design concept for the future tank. Table 8.1 displays the functions assigned to each crewmember in each IMPRINT model.

In both IMPRINT models, the platoon peader (PL) performed typical PL functions of tactical planning, battle tracking, and communications, as well as additional functions associated with monitoring information feeds from an UAV and the ARV-RSTA. IMPRINT predicted this combination of functions would result in numerous instances of high workload for the PL. Similarly, in function allocation 1, when the driver acted as robotics operator, IMPRINT predicted s/he would have numerous instances of high workload as s/he attempted to monitor and control the ARV-RSTA as well as drive the tank. In contrast to the PL and driver, the gunner was predicted to have low workload throughout the mission when s/he is performing only the gunner role.

In function allocation 2 when the gunner's role was combined with the robotics role, resolving interventions with the ARV consistently placed the gunner into overload. A study that investigated remote driving and mental workload supports this high workload prediction (Schipani, 2003). This study evaluated operator workload during partially autonomous vehicle operations and found that workload increased significantly during periods when operator intervention was necessary. Because the IMPRINT analysis predicted interventions with the ARV will overload the gunner, s/he will be likely to drop the critical gunning tasks during this time and jeopardize security of the vehicle. Therefore, rather than recommend that ARV-RSTA operation be allocated to the gunner, the analysts recommended an experiment be conducted to investigate whether tank security would be jeopardized by this allocation.

Subsequent to the modeling project, ARL researchers conducted an experiment to investigate the issues associated with the gunner-robotics operator concept and to verify the modeling project's analytical results (Chen & Joyner, 2009). Specifically, the experiment investigated the mental workload and performance of the combined position of robotics operator and gunner. The results of this experiment were consistent with the mental workload and performance predicted by the modeling analysis. In the experiment, the combination of gunning with robotics operations increased the gunner's workload and reduced local security awareness Chen & Joyner, 2009). Thus, ARL recommended that none of the PLV crewmembers should be allocated robotic operations functions. Rather the analysts recommended investigating the possibility of adding a fourth crewmember or allocating operation and monitoring of the ARV-RSTA to the crew of the platoon sergeant's vehicle (PSGTV). To further validate the IMPRINT analytical results and to evaluate the possibility of the platoon sergeant controlling the ARV, the ARL analysts participated in Omni Fusion 06 (OF06), a simulation exercise conducted at the Unit of Action Maneuver Battle Laboratory (UAMBL), Fort Knox, Kentucky.

RCATO Case Study 2: Platoon Leader and Platoon Sergeant Workload Analysis and Experiment

One of the key steps in any modeling project is to validate the model and its results. This step was particularly important for the IMPRINT RCATO analyses because the prime contractor for the future tank vehicle used, and will continue to use, the results of the IMPRINT RCATO analyses in the vehicle design process. To achieve the goal of verifying and validating the models, the ARL analysts collected observational data during Omni Fusion 06 (OF06). Researchers at the UAMBL conducted the OF06 experiment which simulated a future brigade concept team (FBCT) concept of operations in order to evaluate the tactics, techniques, and procedures of the FBCT for the Training and Doctrine Command (TRADOC). The OF06 participants actively engaged in a human-in-the-loop simulation that represented a FBCT in a defense scenario, transitioning to the offense, and an attack to secure a site.

By participating as observers during this experiment the ARL analysts hoped to achieve several goals. Their overall goal was to verify the tasks, task assignments, and task sequences they had incorporated into their future concept tank PLV IMPRINT models by observing the PLV crew performing the modeled tasks, including unmanned asset tasks, during the simulation. The second goal was to validate the impact of high workload on the soldiers' performance predicted by the IMPRINT analyses. In addition they would assess the viability of mitigating the high workload associated with unmanned asset control by allocating the ARV-RSTA to the platoon sergeant's vehicle (PSGTV) by observing the crew of the PSGTV as they interacted with an ARV-RSTA during the simulation.

During the simulation, one of the ARL analysts sat inside the future tank platoon leader's vehicle (PLV) part of the time and the platoon sergeant's vehicle the remainder of the time. Because she was located within the mockups with the crew, the analyst could see and hear all three crewmembers in each vehicle. For her data collection, she had an ARL-designed reference sheet that provided guidelines for the observations to record. Following the reference sheet guidelines, the analyst recorded task start times, task names, interfaces used to perform the task, task end times, and annotated any concurrent tasks the soldiers performed. She recorded all equipment within the vehicle mockups available to the soldiers for performing their tasks. She also annotated any comments the soldier reported about a task while performing it. In addition, she recorded any time a soldier left a vehicle, and the reason for leaving.

As indicators of high workload, the analyst noted when the soldiers repeated a task, dropped a task, or ignored communications. She observed and recorded times of simulation breakdowns, daily start time, lunch breaks, end of test time, network malfunctions, and time and duration of malfunctions. Generally, she annotated any events she observed that might influence the performance of the soldiers participating in the experiment.

A comparison of the observational data from OF06 with the functions in the RCATO future tank PLV model showed that the functions in the models were valid and verified. Table 8.2 shows the complete list of IMPRINT PLV modeled functions. The observed crew performed all of the functions in the ARL IMPRINT models except engagement of a target with the secondary weapon. The crew attempted to perform this function but could not because the simulated secondary weapon did not function. The crew performed a function, Conduct Battle Damage assessment, that was not included in the IMPRINT analyses. ARL analysts will add this function to future IMPRINT tank models so all tank crew functions are included in any future analyses.

Table 8.2   IMPRINT MCS soldier functions with analogous functions observed during OF06

ARL IMPRINT Model PLV Functions	Observed Functions
Tactical planning	Look at common operations picture (COP) and mentally assess situation then give order
Tactical communications	Battalion communications, company communications, unmanned asset controller communications, monitoring, responding
Crew communications	Crew communications within the vehicle
Monitor UAV feed	Monitor UAV and UAV sensor feed
Platoon communications	Within platoon communications, between platoon communications, monitoring, responding
Battlefield awareness scan	Monitor COP, vehicle crew, and platoon vehicles, and company vehicles
Scan for threats	Scanning for threats and lasing
Target engage BLOS	BLOS self, BLOS by others, BLOS of others, BLOS by ARV
Target engage LOS	Shoot line of sight mission
ARV control	ARV control
Engage target secondary weapon	
ARV monitor sensor feed	Monitor ARV and ARV sensor feed
Driver tank	Drive
Monitor driving	Monitor driver and outside field of view (FOV)
	Monitor ammunition status
	Observe round effects via UAV operator's screen

Observations of the driver's performance during OF06 verified the IMPRINT prediction of high workload for the crewmember engaged in the driver function. In addition, the PL experienced high workload when the vehicle was moving, performing local security, engaging targets, and communicating—as predicted by the IMPRINT RCATO analyses. The PL's workload was highest when he was monitoring the UAV sensor and coordinating its output with targeting a threat. This matched the IMPRINT analytical results that had predicted high workload for the PL when monitoring a UAV sensor feed.

In OF06, the PSGT was responsible for moving the ARV-RSTA, monitoring its status, and monitoring information from the ARV sensors. During the experiment he could not perform his role of platoon sergeant effectively while concurrently controlling the ARV-RSTA. He performed either one role or the other. When attempting to perform both roles, he missed communications and targets, which indicated he was experiencing high workload. Because both the PLV crew and the PSGTV crew experienced high workload when simultaneously interacting with unmanned systems and performing routing military tasks such as communications, targeting, and tactical planning, it seems likely that any of the soldiers within the future tank platoon attempting to do these tasks concurrently will be overloaded as well. Because function reallocation could not mitigate the effects of high workload on the soldiers, the analysts focused on how technologies might mitigate the impacts of high workload that are associated with unmanned asset operations. Specifically, they used IMPRINT to analyze the impacts of crew-aided behaviors (CABs) and autonomous navigation systems (ANS) on soldier workload and performance.

Case Study 3: Soldier Workload, Crew Aided Behaviors (CABs), Autonomous Navigation IMPRINT Analysis and Experiments

The CABs and ANS analyzed by the IMPRINT models were developed by engineers working for the U.S. Army Tank-Automotive Research, Development and Engineering Center (TARDEC). To evaluate the effects of these technologies on soldier workload, TARDEC and ARL researchers planned to evaluate them in a series of field experiments (McDowell et al., 2008). Before these experiments were conducted, the technologies were incorporated into IMPRINT models in order to analyze their impacts on soldier workload.

This ARL IMPRINT analysis of the TARDEC technologies had three objectives. The first objective was to use IMPRINT to predict the impact of two specific CABs, mission planning and automated route planning, and ANS on soldier mental workload and performance. The second objective was to compare these IMPRINT predictions to the data from the field experiments. The field experiments planned to have only stationary targets, so the third objective was to extend the experimental data by using IMPRINT to predict the impact of moving targets in combination with the technologies on soldier performance.

The IMPRINT analyst developed a design matrix to represent two conditions for mission planning: manual planning versus planning with an automated planning aid. The IMPRINT analyst built a model to represent each of these conditions. For this analysis, IMPRINT predicted no overload for a soldier planning a mission either manually or with the automated route planning aid. Similarly, the soldiers in the U.S. Army Research, Development, and Engineering Command (RDECOM) field study that evaluated performance with or without an automated route planning aid showed no soldier overload for mission planning.

The ARL analyst set up a design matrix to represent the analysis of ANS and the route planning CAB. This matrix had four conditions:

Condition 1: no CAB, no ANS;
Condition 2: CAB but no ANS;
Condition 3: no CAB ANS;
Condition 4: CAB and ANS.

The analyst built a separate IMPRINT model for each condition in this matrix. For each of the models, the IMPRINT analyst ran a set of runs under the assumption the soldiers would plan a route for a remote follower (RF) vehicle while their own vehicle was moving. She then completed another set of runs under the under the assumption the soldiers would stop their vehicle to plan a route for the RF. In addition, she ran an additional set of runs to simulate moving targets. The moving targets were assumed to be present for 9 seconds only, whereas, the stationary targets in the model were present in the range of 1 to 30 seconds as they would be in the field experiments.

For the analyses of ANS and the CABs, IMPRINT predicted both the CAB and ANS would reduce workload and improve soldier performance. The CAB reduced soldier workload during route planning. When the IMPRINT analysis combined the CAB with ANS, IMPRINT predicted workload to be at its lowest level. Furthermore, IMPRINT predicted both the CAB and ANS would reduce overall mission time and improve target detection performance. Therefore, both these technologies have the potential to increase soldier survivability by reducing workload, reducing mission time, and increasing the number of targets detected.

Despite the reduction in workload due to CAB and ANS, however, the IMPRINT analysis predicted one soldier trying to complete all the mission-related tasks would be overloaded, especially visually and cognitively, throughout the mission. IMPRINT predicted the soldier would be overloaded because in the experiment he is monitoring a remote follower, monitoring communications, monitoring the battlefield situation, and either driving or monitoring ANS. Because of the predicted high workload, the ARL analyst recommends that additional experiments with CABs and ANS consider reallocation of monitoring tasks to a second crewmember as a possible mitigation strategy for overload. Alternative techniques other than visual displays for monitoring battlefield awareness and the status of the remote follower might help mitigate overload as well.

For the second objective of the analysis a comparison of the IMPRINT data with the soldier experimental data, the number of targets the soldiers detected for the non-ANS and ANS conditions of the experiment were consistent with the field test results. It also correctly predicted the reduction in route planning time due to CAB availability. IMPRINT, however, underestimated overall mission time. Therefore, the IMPRINT analyst recommended incorporation of vehicle speed and actual route distance into future IMPRINT analyses to ensure more accurate overall mission time. Table 8.3 displays the comparison of IMPRINT predictions with field test data.

Table 8.3   Experiment data versus IMPRINT model predictions for conditions 1 and 2 with no ANS

	Conditions 1 and 2		Conditions 1 and 2	
	No Autonomous Mobility		No Autonomous Mobility	
	On the Move Planning		Stationary Planning	
	Soldier Data	IMPRINT Model	Soldier Data	IMPRINT Model
Road March Duration (MM:SS.S)				
Minimum:	14:08.43	04:35.00	14:08.43	05:06.90
Median:	16:28.82		16:28.82	
Maximum:	31:38.53	06:51.50	31:38.53	08:02.30
Mean:	18:08.13	05:41.50	18:08.13	06:31.50
Std Dev.:	05:05.88		05:05.88	
Manual Planning Duration (MM:SS.S)				
Minimum:	00:35.01	01:55.00	00:35.01	01:55.00
Median:	01:08.17	01:55.00	01:08.17	01:55.00
Maximum:	03:42.24	01:55.00	03:42.24	01:55.00
Mean:	01:28.00	01:55.00	01:28.00	01:55.00
Std Dev.:	00:54.18		00:54.18	
Planning with CABs Duration (MM:SS.S)				
Minimum:	00:20.53	00:19.92	00:20.53	00:19.92
Median:	00:33.89	00:19.92	00:33.89	00:19.92
Maximum:	02:03.71	00:19.92	02:03.71	00:19.92
Mean:	00:46.46	00:19.92	00:46.46	00:19.92
Std Dev.:	00:32.62		00:32.62	
Percentage of Targets Identified				
Minimum:	18.18%	20.00%	18.18%	30.00%
Median:	66.67%		66.67%	
Maximum:	100.00%	90.00%	100.00%	90%
Mean:	68.42%	60%	68.42%	60%
Std Dev.:	24.97%		24.97%	

A comparison of the experimental workload data with the workload IMPRINT predicted matched across conditions. Both IMPRINT and the experiment showed the highest workload numbers when the soldiers planned a route for the RF in the manual driving condition versus the ANS condition. The IMPRINT analyst could not compare the IMPRINT workload predictions to the soldier's self-report ratings within each mission because the experimenters did not collect workload ratings during the experiment, only at the end. However, the performance decrements due to high workload predicted by IMPRINT were consistent with soldier performance throughout the experiment.

To meet the third objective, the IMPRINT analyst built models representing moving targets and compared this data to the stationary target data from the experiment. The comparison showed no difference in the rate of detection for moving versus stationary targets for any of the conditions. Overall this IMPRINT analysis identified potential high workload task combinations that the TARDEC and ARL researchers could evaluate during the testing. The testing confirmed the risk of high workload when one soldier is performing many monitoring tasks, and confirmed the potential for ANS to enhance performance. Furthermore a comparison of the IMPRINT and experimental data demonstrated that the IMPRINT analyses made valid predictions of workload and its impacts on performance. This analysis and the RCATO human performance analyses discussed so far in this chapter were analyses of soldiers operating with unmanned assets while located with armored vehicles. However, one RCATO analyst used IMPRINT to predict the workload and performance of dismounted soldiers who were interacting with unmanned assets.

Case Study 4: Dismounted Warrior Platoon Analysis and Experiments

The dismounted warrior analysis is unique from case studies 1 through 3 because its focus was on soldiers operating unmanned systems outside of vehicles rather than riding in vehicles. The objective of this case study was to examine mental workload and its impact on the performance of key members of a dismounted warrior platoon (DWP) using unmanned assets to conduct a typical mission.

The DWP is comprised of three rifle squads and one weapons squad. The DWP is part of an infantry company which consists of the headquarters and three rifle squads. Assigned to the company are three assault type armed reconnaissance vehicles (ARV-Assault), three Class II unmanned aerial vehicles (UAVs), six Class I UAVs, and nine small unmanned ground vehicles (SUGVs). Each rifle squad is assigned one of the SUGVs and the remaining unmanned assets can be assigned to groups as needed, depending on the mission they must accomplish.

The model in this analysis was built to depict key members of the DWP using two robotic assets to assist in a clearing mission. Key individuals depicted are: the squad leader (SL), the team leader (TMA), the rifleman (RA1), the platoon leader (PL), and the robotics NCO (RNCO). These soldiers are in the first squad that has been tasked to clear a particular bridge. The company UAV has been assigned to the first squad rather than the other two squads in the platoon because the other two rifle squads and the weapons squad have been sent off to handle other aspects of the mission. They are assumed to be in heavily vegetated areas which are not conducive to UAV use. Therefore the UAV has been slated to assist the first squad with the bridge clearing. Furthermore, the model assumes the UAV has already been launched and sent into loiter mode over the bridge area. Each rifle team from the 1st squad has taken a vantage point near each side of the bridge and is watching the area while a rifleman moves the SUGV in to do recon of the bridge position. The team leader is monitoring the SUGV feed and the RNCO is monitoring the UAV in loiter mode.

Throughout the bridge-clearing scenario the SL, TMA, and RA1 were overloaded as they performed functions associated with the mission. In addition the RNCO and PL were overloaded over a large majority of the scenario, and the PL over a small portion of the scenario. However, the rifleman controlling the SUGV had the highest maximum workload values and instances of overload. The SL had the next highest and the RNCO who was controlling the UAV was next. These results are consistent with an earlier analysis conducted by ARL in which the members of a dismounted warrior squad, the SL, TA, and RA1 were all overloaded with the tasks they were expected to do in a movement to contact scenario (Mitchell et al., 2004).

Since the overload of these platoon members is excessive, it is most important to identify the contributors to the overload in order to develop potential mitigation strategies or aids. Table 8.4 summarizes those tasks which the PL does more than 50 percent of the time in an overloaded condition.

Table 8.4   Platoon leader (PL) tasks conducted in an overload condition

Function Name	Task Name	%
(PL) Monitor platoon	(PL) Adjust platoon position	100%
Platoon Commo	(PL) Communicates	95%
(PL) TacNet	(PL) Send long voice message	88%
(PL) Tactical planning	(PL) Compare courses of action	82%
(PL) Tactical planning	(PL) Analyze courses of action	78%
(PL) Tactical planning	(PL) Issue orders	74%
(PL) Monitor platoon	(PL) Assess platoon position	73%
(PL) Tactical planning	(PL) Estimate situation	70%
(PL) Battlefield awareness	(PL) UAV SA projection	64%
(PL) TacNet	(PL) Receive/comprehend digital message	55%
(PL) Tactical planning	(PL) Change required?	55%
Platoon Commo	(PL) Comprehends	55%
(PL) TacNet	(PL) Formulate long digital response	54%
(PL) TacNet	(PL) Formulate short digital response	54%
(PL) TacNet	(PL) Send short voice message	53%
(PL) Battlefield awareness	(PL) UAV SA comprehension	52%
(PL) TacNet	(PL) Receive/comprehend voice	52%
(PL) Tactical planning	(PL) Decide best course of action	50%

The PL is predicted to experience overload as s/he scans the battlefield to determine if there are any potential threats while assessing the platoon's position and receiving and sending digital and voice messages. These are concurrent activities the PL does continuously throughout a mission which put him into an overloaded state. Since times of overload are indicative of times where soldiers may make mistakes it is important to look at what the PL is doing at these times. For example, when the PL needed to make adjustments to the platoon's position, 100 percent of the time, he was overloaded. When the PL was conducting various tasks of tactical planning, the PL was overloaded over 70 percent of the time. When the PL was sending platoon-level communications, he was overloaded 95 percent of the time. Overload at these times could lead to incorrect or unnecessary adjustment of the platoon's position, flawed tactical planning or miscommunication being sent to the platoon members.

Similar to the workload predictions for the PL, the analysis predicted the RNCO would spend the majority of the mission in mental overload. Table 8.5 displays the tasks that most often lead to overload for the RNCO.

The RNCO spends most of the mission monitoring the flight path of the UAV; scanning and searching a designated sector for important information and potential threats; and sending and receiving communications. When the RNCO is doing these tasks, s/he is in an overload state. For example, when the RNCO is searching and scanning the area for important information and potential threats, 100 percent of the time, he is in mental overload. This is also the case when receiving and sending communications and doing periodic adjustment of the UAV. Overload at these times could cause the RNCO to miss the identification of potential threats, send incorrect communication to the platoon members, or incorrectly adjust the UAV.

Table 8.5   RNCO tasks conducted in an overload condition

Function Name	Task Name	%
(RNCO) UAV control	(RNCO) Alert PL to area of interest	100%
(RNCO) UAV control	(RNCO) Identify potential interest	100%
(RNCO) UAV control	(RNCO) Periodic stick control	100%
(RNCO) UAV control	(RNCO) Periodic throttle to control speed	100%
(RNCO) UAV control	(RNCO) Search/scan	100%
Platoon Commo	(RNCO) Communicates	100%
Platoon Commo	(RNCO) Comprehends	100%
Platoon Commo	(RNCO) Formulates transmission	100%
(RNCO) UAV control	(RNCO) Monitor system status	98%
(RNCO) UAV control	(RNCO) Monitor flight path	95%

Similar to the RNCO, the SL, TMA, and RA1 are predicted to experience overload throughout the bridge-clearing mission. During this mission, the SL focuses his time on scanning the battlefield and determining if there are any potential threats to the squad, and sending and receiving communications. Since the SL is in overload over the whole scenario, it is very likely that he will make mistakes such as missing a potential threat, sending out incorrect communication, or misunderstanding a received message. The TMA is also responsible for scanning the battlefield, monitoring the SUGV feed that comes in, and sending and receiving communication. Due to the constant state of overload, the TMA may make mistakes similar to those of the SL. The RA1, who operates the SUGV in this scenario, also suffers from overload over the whole mission. The RA1 must scan the environment and control the SUGV and monitor its feed. Due to overload, the RA1 may make mistakes such as missing potential threats while scanning the environment, misdirect the SUGV, or miss important notifications from the SUGV.

Overall this analysis shows that key members of the DWP experience significant amounts of mental overload. Most of the overload for each of the operators comes from the need to scan their environment while completing other tasks. Since the DWP is in an exposed environment, the task of scanning the environment is critical to their survivability, and to require them to do other things that distract them from that task could be very dangerous.

The prediction that dismounted soldiers would be overloaded when simultaneously operating SUGVs and performing standard mission-associated tasks is being evaluated during a series of experiments taking place at Fort Dix, New Jersey. These experiments are called the Command, Control, Communication, and Computers (C4) Intelligence, Surveillance and Reconnaissance (ISR)-on-the-Move and took place in 2008 (Bowman and Thomas, 2008).

Summary and Recommendations

As the case studies demonstrate, when a rigorous analytical technique is implemented, human performance modeling can assist experimenters with designing tests that will capture combinations of tasks critical for successful mission performance. The tests, in turn, can be used to evaluate alternative approaches for mitigating the effects of high workload on performance. The test data can then be input back into the models for analysis of factors that could not be incorporated into the tests. This approach is crucial for conducting cost-effective test and evaluation of soldiers interacting with unmanned systems.

Key Points

1. The conceptual phase of the system design process is the optimum time for analysis of the interaction of soldiers with unmanned assets because changes in tactics, techniques, and procedures (TTPs) and to system design can be made before they become finalized.
2. Developing human performance models prior to testing is efficient in cost and time because it reduces the number of required studies and ensures that the studies conducted focus on relevant soldier performance issues.
3. Although the model–test–model approach has many benefits, these benefits are contingent upon the human performance models accurately representing how the soldiers will use a conceptual system.
4. For human performance models to be accurate it is critical that a rigorous analytical process be followed during model development.
5. Following an analytical approach ensures that the predictions from the models accurately identify issues that the experiments then validate as critical soldier performance issues.

References

Barnes, M.J., Hunn, B., & Pomranky, R. (2006). Modeling and operator simulations for early development of Army unmanned vehicles: methods and results. In N.J. Cooke, H.L. Pringle, H.K. Pedersen, & O. Connor (eds), *Human Factors of Remotely Operated Vehicles* (pp. 59–70). New York: Elsevier.

Bowman, E.K. & Thomas, J.A. (2008). C2 of unmanned systems in distributed ISR operations. *Paper presented at the 13th International Command and Control Research and Technology Symposium (ICCRTS)*. Seattle, WA.

Chen, J.Y.C. & Joyner, C.T. (2009). Concurrent performance of a gunner's and robotic operator's tasks in a simulated MCS environment. *Military Psychology*, 21(1), 98–113.

McDowell, K., Nunez, P., & Hutchins, S. (2008). Secure mobility and the autonomous driver. *IEEE Transactions on Robotics*.

Mitchell, D.K., Samms, C.L., Glumm, M.M., Krausman, A.S., Brelsford, M.L., & Garrett, L. (2004). Improved Performance Research Integration Tool (IMPRINT) model analyses in support of the situational understanding as an enabler for unit of action maneuver team soldiers Science and Technology Objective (STO) in support of Future Combat Systems (FCS) (Technical Note ARL-TR-3405). Aberdeen Proving Ground, MD: U.S. Army Research Laboratory,

Schipani, S.P. (2003). An evaluation of operator workload, during partially-autonomous vehicle operations. *Proceedings of the Performance Metrics for Intelligent Systems Conference (PerMIS '03)*, 16–18 September, Gaithersburg, MD. http://www.isd.mel.nist.gov/research_areas/research_engineering/Performance_Metrics/PerMIS_2003/Proceedings/Schipani.pdf.

PART III
UAV Research

Chapter 9

Introducing Cognitive and Co-operative Automation into Uninhabited Aerial Vehicle Guidance Work Systems

Axel Schulte and Claudia Meitinger
Universität der Bundeswehr München

Abstract

Today's highly automated robotic systems are tied into military work processes supporting the human war fighter. This chapter will take uninhabited aerial vehicles (UAVs) as an example. Typically, UAVs are being used to fulfill certain well-defined sub-tasks. The human operator is in the role of the high-end decision component, determining the work process and supervising the automation. With emergent technology highly automated systems can be beneficial on the one hand; but automation may also cause problems on its own. This is the case due to its inherent complexity, which is often not easy to handle by humans. A new way of introducing automation into work systems will be advocated by this contribution, overcoming the classical pitfalls of human-automation interaction and simultaneously taking the benefit as required. This will be achieved by so-called *cognitive automation*, i.e. providing human-like problem-solving, decision-making, and knowledge-processing capabilities to machines in order to obtain goal-directed behavior and effective operator assistance. A key feature of cognitive automation is the ability to create its own comprehensive representation of the current situation and to provide reasonable action. By additionally providing full knowledge of the prime work objectives to the automation it will be enabled to co-operate with the human operator in supervision and decision tasks, thus being intelligent machine assistants for the human operator in the workplace. Such assistant systems understand the work objective and will be aiming to achieve of the overall desired work result. They will understand the situation (e.g. opportunities, conflicts) and actions of team members—whether humans or assistant systems—and will pursue goals for co-operation and co-ordination (e.g. task coverage, avoidance of redundancy, or team member overcharge). Thereby a completely new approach to human-automation interaction will be established, which we call *co-operative control*. Furthermore, cognitive automation can emerge towards being highly automated intelligent agents in charge of certain supportive tasks to be performed in a semi-autonomous mode. These cognitive semi-autonomous systems and the cognitive assistants will be denoted as the two faces of *Dual-Mode Cognitive Automation* (Onken & Schulte, forthcoming).

Introduction

In the field of airborne vehicle guidance, and of course not only there, the general principle of integrating human and automation in the manner of *supervisory control* is very common (Sheridan, 1987). While the automation typically performs fast inner control loops and information management support, the human operator is responsible for mode selection and command value setting as the observable outcome of planning, decision-making, deliberation, and anticipation for the sake of a safe and efficient mission accomplishment. Thereby, a *hierarchically organized work system* is established.

The insufficiency of human mental resources becomes obvious when the supervision of multiple UAVs is required by a single operator. Inevitably, this would lead to erroneous action and performance decline. Nevertheless, such ideas are currently being discussed in the military community under the term *Manned-unmanned Teaming (MUM-T)*. This is an approach to simultaneously control several UAVs and their payload from a manned aircraft or helicopter in order to increase the effectiveness of the manned system in performing its mission. In order to cope with the task of supervising multiple automated processes of vehicle guidance in a co-ordinated manner, while maintaining control over the own vehicle, much more cognition has to be build into the system. How can that be done?

From a purely technology-driven point of view, a superficial answer could probably be to make the UAVs become "autonomous." But, what are the requirements of an appropriate *autonomous system* and what is the difference from the aforementioned automated system? Commonly, an autonomous system would be expected to pursue the work objective of the considered mission, to be reactive to perceived external situational dynamics, to modify the work objective on the basis of the actual situation if necessary, and to generate solutions by means of anticipation, deliberation, and planning and execute them without human intervention; in short to be an *artificial cognitive system* equivalent to the human operator. But of what use might such an autonomous artifact, completely detached from human input, be? On the other hand, is the consideration of human factors in the context of the autonomy debate still adequate? This chapter will show that giving machines full autonomy, in the limited sense of a purely technical treatment, will not be the solution.

In order to embed automation into a highly interactive work environment such as the sketched MUM-T scenario, teamwork is compulsory. The basis of teamwork is the appreciation of the behavioral traits of team members, whether they are humans or machines. Establishing the capability of teaming between humans and automation will be referred to as *co-operative control*, as opposed to the classical paradigm of supervisory control. Interaction will no longer occur on the level of mode selection and command value settings, but through negotiation of requests and commitments on the level of tasks and intents. This will be based upon a common understanding of the current situation by both humans and machines, subject to a common mission objective. Although the final decision authority for high-end decisions will stay with the human operator, there will be established a peer team of humans and *artificial cognitive units* as intelligent systems which assist the human operator in the process of pursuing the mission objective.

The following sections will provide a theoretical approach of, firstly, how to introduce cognitive automation into work; secondly, what is required to make the automation behave in a co-operative manner; and thirdly, how to implement artificial cognitive systems. Finally, an insight into current application-related research will be given.

Introducing Artificial Cognition into the Work System

The Work System as Human Factors Engineering Framework

To figure out a solution to the challenges of future vehicle guidance systems, the first step will be the characterization of conventional automation in the work process, as opposed to the introduction of cognitive automation. Therefore, the consideration of the *work system* as top-level human factors engineering framework will be used. The work system (see Figure 9.1) as a general ergonomics concept, as probably defined for the first time in REFA (1984), has been utilized in a modified definition, and adapted to the application domain of human-machine co-operation in flight guidance by Onken (2002) and to machine-machine co-operation by Meitinger & Schulte (2006).

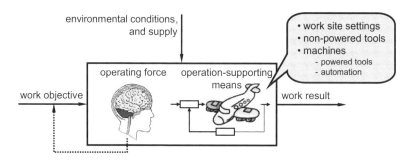

Figure 9.1   Work system

The work system is defined by the work objective, being the main input into the process of work. The work objective mostly comes as an instruction, order, or command from a different supervising agency with its own work processes. Further constraining factors to the work process are environmental conditions and supplies. At its output, the work system provides the work result, including the current state of work and what has been accomplished by the work process as a physical change of the real world (Onken & Schulte, forthcoming).

The work system itself consists of two major elements: the operating force and the operation-supporting means, as characterized in some more detail below.

Operating force   The operating force is the high-end decision component of the work system. It is the only component which pursues the complete work objective. It determines and supervises what will happen in the course of the work process and which operation-supporting means will be deployed at what time. The operating force is the work system component with the highest authority level. One major characteristic of especially a human representing the operating force is the capability of self-defining the work objective himself (see Figure 9.1). Besides operating on the basis of full authority competence this is the decisive criterion for an *autonomous system*.

Operation-supporting means   The concept of the operation-supporting means can be seen as a container for whatever artifacts are available today to make use of in the work process, including basic worksite settings, non-powered tools, and machines. The latter might be a vehicle in the case of a transport work process, but also computerized devices of automation. A robotic system in this sense could be seen as a highly automated machine as part of the operation-supporting means of a work system as well. In the application domain of flight guidance currently used autoflight or autopilot systems including the human-machine control interface can serve as typical examples of automation. Common to the nature of various operation-supporting means is the fact that they facilitate the performance of certain sub-tasks, and only that. By nature, such a sub-task does not form a work system itself, obviously being only a part of another higher-level work task. According to the common ergonomic design philosophy, mostly the operation-supporting means are subjected to the endeavors of optimization in order to achieve overall system requirements and accomplish further improvements.

These elements will be combined with the work system set up in order to achieve a certain work result on the basis of a given work objective. The accomplishment of a flight mission (i.e. the work objective) may give a good idea of what is meant here. In this case, the work system will consist of an air crew being the operating force; and the aircraft, including its automated on-board functions as well as any required infrastructure, represents the operation-supporting means.

The replacement of human work by continuously expanding automated functions, as it could be observed throughout the whole history of industrial mechanization and automation, leads to technical solutions of steadily increased authority. Simultaneously, the human operator is continually being pushed further into the role of supervising more and more machines and, at the same time, more and more complex ones. The process of substituting the responsibilities of the human in a work system might be driven up to a degree where the operator's capabilities would be completely substituted by automated functions. From a purely technology-minded standpoint, the human could theoretically be dropped out of the work system as a consequence. In case this includes the analogue of the human capability to self-assign a work objective, the resulting artifact could be called an *artificial autonomous system*.

In fact, though, such technological artifacts in themselves, not being subjected to any human authority, are of no use, since they are no longer serving the human in accomplishing his work. Therefore, as a consequence such a system, representing a former work system, inevitably will be a part of the operation-supporting means of a higher-level work system whose operating force comprises at least one human operator. According to the principal concept of human autonomy, only this operator is entitled to define the overall work objective. The artificial work system substitute, now being part of the operation-supporting means of that work system, receives its task instructions purely from the operating force of the work system. In this regard we would speak of a *semi-autonomous system*.

Artificial Cognitive Units in the Work System

Traditionally, a human or a human team (see, e.g., Onken, 2002) represents the operating force in the work system. In the conventional sense the human operating force provides the capability of cognition within a work system, whereas the operation-supporting means do not. In order to overcome known shortcomings of such conventional automation (e.g. Bainbridge, 1983; Billing, 1997), often being by far too complex to be handled properly, a configuration of the work system will be suggested whereby so-called *artificial cognitive units* (ACU) are introduced.

Introducing ACUs into the work system, as opposed to the further enhancement of conventionally automated functions or the addition of further humans (first row in Figure 9.2), adds a new level of automation, i.e. the *cognitive level* to the work system (second row in Figure 9.2). The possibility to shape an ACU being either part of the operating force (*operating ACU*) or being part of the operation-supporting means (*supporting ACU*) defines two modes of cognitive automation (see Figure 9.2) (Onken & Schulte, forthcoming). Both modes might be combined together with conventional automation within one work system.

Both modes of automation have in common that they incorporate artificial cognition. Onken (2002, KN5-5) describes the nature of suchlike cognitive automation as follows: "As opposed to conventional automation, cognitive automation works on the basis of comprehensive knowledge about the work process objectives and goals [...], pertinent task options and necessary data describing the current situation in the work process. Therefore, cognitive automation is prime-goal-oriented."

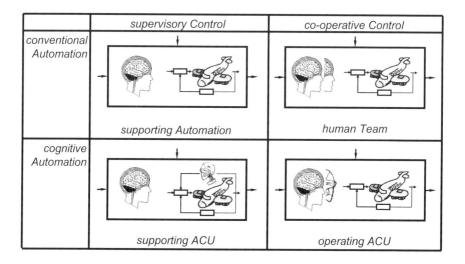

Figure 9.2   Introducing artificial cognitive units in the work system

Particularly due to orientation of its behavior towards explicitly represented goals and its understanding of the current situation, cognitive automation has the potential to avoid problems of conventional approaches, such as complexity and brittleness, which have been thoroughly investigated by Billings (1997) and others. A detailed explanation of this effect will be found in Onken & Schulte (forthcoming).

Concerning the application of cognitive automation as operating ACUs—i.e. on the right-hand side of Figure 9.2, where the human operator and the ACU form the operating force as a team—Onken (2002, KN5-7) comments that in this configuration the ACU has reached "the high-end authority level for decisions in the work system, which was, so far, occupied by the human operator alone."

As a consequence of this consideration both team members—human operator and ACU—have to have the obligation to apply their specific capabilities, which might be overlapping, in order to pursue the overall work objective best. As a consequence, an operating ACU is always characterized by the incorporation of the functionality of what we call an *assistant system*. Such an assistant system can be a *supplement* to the operating force, i.e. to the human operator or the human team. In this case it is mostly of a virtual nature, providing informational assistance. On the other hand it can be embodied in an independent machine (e.g. robot, UAV) *replacing* an otherwise necessary additional human operator. In this case the assistance is also of a physical nature.

Onken (1994) formulates two basic functional requirements for specification purposes of such an assistant system being part of the operating force in the supplementary role. An amended version of these requirements is the following, which will be discussed in Onken & Schulte (forthcoming) in far more detail concerning the different roles taken by assistant systems:

Requirement 1   The assistant system has to assess/interpret the work situation on its own and has to do its best by own initiatives to ensure that the attention of the assisted human operator(s) is placed with priority to the objectively most urgent task or subtask.

Requirement 2 (optional)   If the assistant system can securely identify as part of the situation interpretation according requirement (1) that the assisted human operator(s) is (are) overtaxed, then the assistant system has to do its best by own initiatives to transfer this situation into another one which can be handled normally by the assisted human operator(s).

In these so-called two basic requirements for human-machine interaction the way is paved for automation as part of the operating force of a work system in the sense of cognitively facilitated human-machine co-operation.

Co-operative Automation

In general, working together makes sense, if it is actually possible and useful to work on the given task with several actors, due to one of the following reasons (see Jennings, 1996; Wooldridge, 2002):

- No individual team member has enough competence, resources, or information to solve the entire problem.
- Activities of team members depend on each other with respect to the usage of shared resources or the outcome of previously completed tasks.
- The efficiency of the work process or the utility of the outcome is increased—for example:

  - by avoiding the completion of tasks in unnecessary redundancy; or
  - by informing team members about relevant situational changes;
  - so that task completion will be optimized.

Within the context of the work system both operating ACUs as part of the operating force and supporting ACUs as operation-supporting means need co-operative capabilities for the following reasons:

- Supplementary operating ACUs are supposed to compensate possible human resource limitations for the sake of mission accomplishment. Moreover, substituting operating ACUs are always required to work together with at least one human operator, as it is mandatory that the operating force comprises at least one human.
- The operating force might command a task to several supporting ACUs, which should then accomplish this task together.

Obviously, co-operation becomes a key issue in work systems if there are humans and ACUs involved as several cognitive units. This section will take a look at co-operation and its implications of co-ordination and communication. In this context we will also touch on selected approaches on how ACUs can be enabled to co-operate.

Co-operation

It is considered a fundamental characteristic of co-operation that co-players have a *common objective*. Within a work system there are different levels of co-operation.

On the first level, *human-ACU co-operation* can be present within the operating force, where an ACU representing an assistant system is teaming with the human operator. This level of co-operation allows that the assistant system is co-operating with the human operator as would be the case if there was a human team member. A typical example for a purely human team on this level of co-operation is the pilot team in the two-man cockpit of a commercial transport aircraft. Just like them, both the human operator and the assistant system operate the work process in a co-ordinated way in order to accomplish a common, externally given work objective. This level of co-operation can be called the collaborative level and might go as far as the assistant system taking over full authority in case of incapacity of the human operator. Not all assistant systems on that co-operation level as part of the team constituting the operating force are of that high-end capacity. We can think of others which can be considerably constrained in their authority level and might have no capacity at all to directly operate the work process. Their assistance can be like that of a harbor pilot who advises the captain of a ship, thereby supporting him to bring his vessel safely to the dock, but who would never have his hands on the rudder.

On a second level co-operation exists between supporting ACUs within the operation-supporting means (*ACU–ACU co-operation*), consisting of several cognitive systems working together in order to achieve the sub-tasks assigned to them and monitored by the operating force. This level of co-operation allows for extended capabilities within the operation-supporting means. Although it is in principle desirable to have as much automation as possible within the operating force (Onken & Schulte, forthcoming), there are cases in which this is not possible. An example for such co-operative semi-autonomous systems might be a team of UAVs which is capable of accomplishing a given task, such as the reconnaissance of a certain area, with little or no further intervention by the human operator, including the co-ordination of responsibilities for sub-areas and the appropriate usage of different resources like sensors.

The work system also indicates what we know as *supervisory control*, as opposed to co-operative control, taking place between the operating force and the operation-supporting means. From the engineering point of view, it is important to distinguish carefully between supervision and co-operation. In particular, the requirements and challenges for the realization are considerably different. The main difference between supervision and co-operation lies in the fact that the goals of both the supervisor and the supervised unit are not necessarily the same. It is even not warranted that the goals are compatible under all circumstances. In that sense, the term manned-unmanned teaming (MUM-T) is not necessarily the same as what we have defined here as teaming of co-operating units (Schulte, 2006). There, teaming is used for team structures which can be associated with both supervision and co-operation.

To get an idea of what requirements team members have to fulfill in order to be able to co-operate, Billings's principles of human-centered automation as stated in Billings (1997) can be consulted. They characterize capabilities of both the human and automation which are necessary to enable safe and efficient aircraft guidance or air traffic control. Although these principles refer to a human-machine team they can be generalized for teams consisting of several humans and ACUs working in similar structured task domains, as suggested in Ertl & Schulte (2004). Thus, all team members must:

- be actively involved;
- keep each other adequately informed;
- monitor each other, and therefore should be predictable;
- know the intent(s) of all other team members.

These requirements demand for various capabilities of the ACUs such as comparing expected and observed activities of team members, including the human operator, in order to be able to monitor them in terms of their intents. This requirement has been thoroughly investigated in the context of knowledge-based crew assistant systems by, for example, Wittig & Onken (1993) and Strohal & Onken (1998).

Co-ordination

In order to achieve the desired positive effects of co-operative work, the activities of the participating team members have to be co-ordinated, i.e. interdependencies between these activities have to be managed (Malone & Crowston, 1994). Interdependencies include not only negative ones such as the usage of shared resources and the allocation of tasks to actors, but also positive relationships such as equal activities (von Martial, 1992).

Although there are many approaches to co-ordination (e.g. Wooldridge, 2002), Jennings's (1996) approach will be discussed briefly here, as it is well suited for the application within human-machine teams. He reduces co-ordination to the formula *"Co-ordination = Commitments + Conventions + Social Conventions + Local Reasoning."* Therein, *commitments* are described as "pledges [of agents] to undertake a specified course of actions." Conventions are explained as "describing circumstances under which an agent should reconsider its commitments" and social conventions "specify how to behave with respect to the other community members when their commitments alter." Finally, *local reasoning* stands for the capability to think about own actions and related ones of others.

By choosing conventions and social conventions appropriately, the distribution of commitments in a team can be adapted to the current situation and can consider aspects such as the workload of the human operator, opportunities of team members, or unexpected changes of the environmental conditions.

Communication

The prevailing means for co-ordinating team members is communication, which can be either explicit or implicit. *Explicit communication* in the area of multi-agent research is usually based on the speech act theory founded by Austin (1962) and further developed by Searle (1969). It states that communicative acts can change the state of the environment as much as other actions, i.e. to communicate explicitly means to send messages to team members, usually in order to achieve a certain desired state. *Implicit communication*, in contrast, is based on the observation of team members and an inference of their intentions in order to be able to conclude what content messages could have had, if explicitly sent.

In order to enable machines to communicate with team members explicitly, agent communication languages can be used, one of them being the Foundation for Intelligent Physical Agents Agent Communication Language (FIPA ACL).

Although humans mostly tend to stick to suchlike protocols, they may, in contrast to machines, send redundant messages or use unintended short-cuts. Thus, in cases where machines are supposed to communicate with humans, either the humans have to stick to protocols or—a much more human-friendly alternative—the machines have to be enabled to cope with such imperfect human behavior.

Approaching Artificial Cognition

In the second section the introduction of artificial cognition into work systems was advocated, while the section following gave an overview of the required attributes of an artificial cognitive unit used as a piece of co-operative automation within the work system. This section will provide an engineering approach to artificial cognition.

The concept of automation being a team player in a mixed human-machine team, or having a machine taking over responsibility for work objectives to a large extent, promotes the approach of deriving required machine functions from models of human performance. The following sections will describe the concept of the so-called "cognitive process" and its implementation in a system engineering framework at some more detail.

Cognitive Process

The cognitive process (CP) can be seen as a model of intelligent machine performance which is well suited for the design of ACUs with goal-directed decision-making and problem-solving capabilities based on a symbolic representation of the perceived situation. It aims at the development of technical systems which are capable of exhibiting behavior on all levels of performance, as stated by Rasmussen (1983), i.e. the skill-based, rule-based, and knowledge-based human performance level. Particularly, the possibility to perform on the knowledge-based level makes it possible to develop systems which are very flexible and adaptive to environments, the configuration of which is not exactly known in advance.

From an architectural point of view, the CP follows the approach of knowledge-based systems in computer science, i.e. it separates application-specific knowledge from application-independent processing of this knowledge (inference). Figure 9.3 shows the CP consisting of the *body* (knowledge, inner part) and the *transformers* (inference, outer extremities) (Putzer & Onken, 2003).

The body consists of two kinds of knowledge: the *a priori* knowledge, which is modeled by the developer of the ACU, and the *situational* knowledge, which is created by the CP during runtime by processing information from the environment, already existent situational knowledge, and a priori knowledge.

The above-mentioned transformers are the underlying mechanism for the handling of situational knowledge. They read input data in mainly one area of the situational knowledge, use a priori knowledge to transform or process the knowledge, and write output data ("new" or modified knowledge) to a designated area of the situational knowledge. The following transformation steps work together in order to generate observable CP or ACU behavior, respectively. Although they are described sequentially here, they are performed according to the situation being represented by the situational knowledge as appropriate.

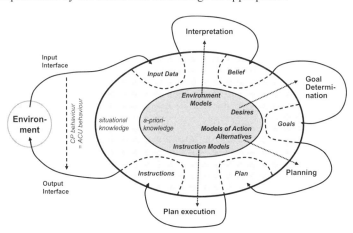

Figure 9.3   The cognitive process

- Information concerning the current state of the environment (*input data*) is acquired via the input interface. These input data include anything the CP is designed to perceive, subject to relevance for further knowledge processing steps. The input data in real-world applications are usually limited by the equipment of the underlying implementation in terms of communication systems and sensors, both concerning the vehicle or robot and the surrounding environment.
- An internal representation of the current situation (*belief*), including the mental model of the external physical world including the comprehension on a higher conceptual level, is obtained by interpreting the input data using environment models. These are concepts of objects, relations, and abstract structures which might be part of the situation.
- Based on the belief, it is determined which desires (potential goals) are to be pursued given the current situation. These abstract desires are instantiated to active *goals* describing the state of the environment, which the CP intends to achieve.
- *Planning* determines the steps, i.e. situation changes, which are necessary to alter the current state of the environment in such a way that the desired state is achieved. For each planning step, models of action alternatives are used.
- Instruction models are then needed to schedule the steps required to execute the plan, resulting in *instructions*.
- Instructions are finally put into effect by appropriate effectors which are part of the environment, resulting in the change of the environment as wanted.

The Framework: From Theory to Implementation

An appropriate architecture for the development of cognitive systems is Cognitive System Architecture (COSA), Putzer & Onken (2003), which offers a framework to implement applications according to the Cognitive Process theory. It supports the developer in two ways: Firstly, COSA provides an implementation of the application-independent inference mechanism so that the development of a cognitive system is reduced to the implementation of interfaces and the acquisition and modeling of a priori knowledge. Secondly, knowledge modeling is supported by the provision of Cognitive Programming Language (CPL), the programming paradigm of which is based on the theoretical approach of the CP. CPL facilitates coding of complex behavior on a high level of abstraction. In this context "programming" means to describe situation-dependent behavior of concepts (knowledge models). These knowledge models are environment models, models of desires, and action alternatives and instruction models as known from the CP.

Thus it differs from conventional programming in the sense of not representing a merely procedural implementation of functions.

The kernel of COSA, which encapsulates the application-independent knowledge processing functionality of the Cognitive Process, is based on Soar (Newell, 1990), the structure of which is similar to that of a production system. Therefore, all knowledge (application-specific knowledge and application-independent knowledge regarding the Cognitive Process being part of the CP library) has to be represented in Soar rules.

Application

In the previous sections the idea of introducing artificial cognition into the work process has been elaborated and discussed to some extent. The discussion resulted in the postulation of co-operative automation as an approach to overcoming human factor-related shortcomings of the current automation paradigm. Some basic considerations were made concerning fundamental features of co-operative automation, and a brief introduction to the concept of artificial cognition as underlying technology has been given. This final section will be dedicated to some application-related issues.

Earlier in this chapter the various forms of co-operation between intelligent agents in the work system have been mentioned—human-ACU co-operation (assistant system in co-operation with human operator) and ACU-ACU co-operation (co-agency of multiple highly automated, i.e. semi-autonomous, entities under human supervision). The following two sub-sections will give some insight into these two aspects. The issue of knowledge-based assistant systems has already been reported on very extensively from our research group in many previous publications. For this reason the elaboration on this will be very brief. The treatment of ACU-ACU co-operation will be elaborated in more detail, though. We expect that there will be massive spin-off for a more human-centered view on how humans and machines might co-operate in the very near future.

Assistant Systems

Researchers at the Universität der Bundeswehr München (Munich University of the German Armed Forces) have been working on co-operative automation technology in the field of aircraft guidance for almost two decades.

Early, but very successful, approaches were on knowledge-based systems assisting airline pilots in instrument flight rules (IFR) flight. The Cockpit Assistant System (CASSY) was successfully flight tested in 1994, being the first prototype of its kind worldwide (Prévôt et al., 1995). The Crew Assistant Military Aircraft (CAMA) followed in the late 1990s, incorporating technology capable of semi-autonomously performing mission tasks (e.g. tactical situation analysis, tactical replanning) on the basis of goal-oriented behaviors while keeping up a situation-adapted dialog with the pilot, in order to balance his workload (Walsdorf et al., 1997; Schulte & Stütz, 1998; Stütz & Schulte, 2000; Frey et al., 2001).

Both systems can be regarded as representing operating ACUs according to the definition given earlier (see Figure 9.2), although not fully following the today's strict design philosophy, which was not available by that time. However, they already bear some crucial capabilities: for example they model expected pilot behavior on the level of control actions (e.g. Ruckdeschel & Onken, 1994; Stütz & Onken, 2001); recognize the pilot's intents and errors through comparison between expected and observed pilot behavior (e.g. Wittig & Onken, 1993); and, finally, manage the human-machine dialog by the use of adaptive speech recognition (e.g. Gerlach & Onken, 1993).

Future work in this field will cover two new application domains: the support of helicopter pilots and assistant systems for the ground-based or even airborne UAV operator. Technology-wise the consideration of user-adaptive automation by use of resources and behavior model based workload estimation (Donath & Schulte, 2006) will be an issue.

Co-operative Supporting ACUs

While work in the field of assistant systems focuses on the introduction of ACUs as artificial team members for the human operator being the operating force in a work system, work in the area of manned-unmanned teaming additionally considers the integration of co-operative semi-autonomous systems into a work system, as discussed earlier in this chapter.

The subject of this investigation is the co-ordination of the UCAVs in a simplified Suppression of Enemy Air Defense (SEAD)/Attack scenario on the basis of dialogs on the knowledge-based level of behavior. The scenario, which is used for the development of the required semi-autonomous and co-operative capabilities, consists of some surface-to-air missile (SAM)-sites and a high-value target in a hostile area which has to be destroyed by the team of UCAVs. Some of the threats are known a priori at the beginning of the mission; others pop up unexpectedly during the course of the mission. One UCAV ("Attack-UCAV") is equipped with a weapon, which can destroy the target. The other UCAVs ("SEAD-UCAVs") have sensors for the detection of pop-up threats and incoming missiles, as well as high-speed anti-radiation missiles (HARMs) for suppression or destruction of SAM-sites on board.

As a first step towards a work system analysis of a manned-unmanned teaming mission the co-agency of the mentioned team of semi-autonomous, co-operating, uninhabited combat aerial vehicles (UCAVs) has been investigated (Meitinger & Schulte, 2006, 2007). The team of UCAVs is supposed to be capable of co-operatively accomplishing a mission of the above-mentioned type as specified by the operating force within a work system setup. Each of the UCAVs is equipped with an onboard supporting ACU responsible for guidance of the vehicle and co-operation with the other ACUs.

In order accomplish the mission, the ACUs have to cover the following capabilities (Platts et al., 2007):

- Use of operation-supporting means, i.e. handling of the conventional automation equipment of the underlying UCAV—namely an autopilot, a flight management system, and a flight planner minimizing threat exposure;
- Safe flight, i.e. ensuring that the UCAV can fly within 3D space without colliding with other UCAVs or terrain;
- Single vehicle mission accomplishment, i.e. the provision of the capability of a single UCAV to actually accomplish tasks such as "suppress SAM-site," to which an individual commitment exists;
- Co-operative mission accomplishment, i.e. the ability to work together with other UCAVs in order to achieve the common objective, namely the mission assigned to the team by the operating force.

All of these aspects had to be considered when implementing a prototype of the ACUs. As the focus of the investigation was co-operative behavior, the first three facets were only covered as much as needed by the co-operative capability. In the following sections, the development of the prototype based on the Cognitive Process and COSA will be described in more detail. First, the most important models of the a priori knowledge relevant to co-operation will be explained. Then, an idea of how different models interact with each other will be given, using communication as an example. Finally, some results will be presented.

A-priori knowledge The behavior of systems which are designed according to the theory of the Cognitive Process is primarily driven by the implemented desires or goals respectively. Therefore, when starting to define the models of the a-priori knowledge of an ACU, the desires have to be modeled at first. In a second step, the action alternatives, which the ACU has in principle at hand in order to achieve goals, must be identified. As it should be possible to execute any action alternative as soon it has been chosen as an element of the plan and its execution is pending, appropriate instruction models have to be modeled next. Subsequently, environment models have to be identified, which are necessary for activation of goals, selection, and instantiation of action alternatives or instruction models. After the declaration of the models and their elaboration with attributes, the behavior of the models, as well as interactions among the models, have to be specified. This approach to cognitive system design is called the "CPL method" (Putzer & Onken, 2003) and was used to identify a-priori knowledge models relevant to co-operation which will be presented in this order in the following.

Figure 9.4 shows the *desires* that underlie co-operative behavior of the ACUs developed. They are arranged in four groups representing the facets of co-operation which have been discussed on pp. 152–153. In order to achieve the common objective, commitments have to be managed and the associated tasks accomplished. Commitment management hereby refers to both accepting and dropping commitments under certain circumstances. Usually, a commitment is accepted by an individual team member or a team if somebody (e.g. operator) requests the accomplishment of a certain task, and the accomplishment of this task is judged as being feasible. A commitment should be dropped if it is achieved, irrelevant, or unachievable (Cohen & Levesque, 1990; Jennings, 1994). As this refers not only to own commitments but also to commitments other team members accepted due to own requests, irrelevant dialogs should not be continued. Moreover, a team or team member should not only consider each commitment apart from the others, but also whether all its commitments are achievable as a whole. Team members as individuals are responsible for the actual accomplishment of tasks associated with commitments; therefore, commitments have to be arranged in a certain order in which they will be considered (so-called "agenda") and actually be complied with at a certain point in time.

Co-operation requires several actors to work together as a team. Following the requirements imposed upon co-operative agents derived from Billings (see p. 153), this includes appropriate information exchange within the team driven by the desires to know the other team members and to keep them informed with respect to, for example, capabilities, resources, and commitments. The distribution of tasks within a team should be organised in a way that all tasks which should

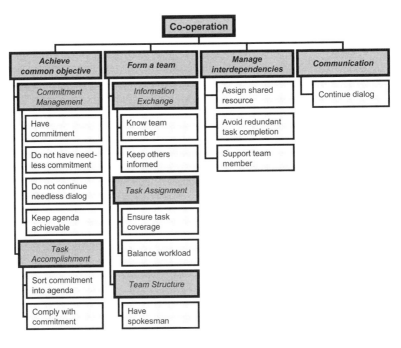

Figure 9.4   Desires for co-operation

be completed are assigned to a team member, and at the same time workload is balanced among the team members. Finally, in the project described here, the team structure is such that all team members are peers, but communication with actors not belonging to the team is performed by a spokesman. Therefore, a desire "Have spokesman" has been formulated.

Within the scenario described above three kinds of interdependencies between team members are relevant, the management of which is triggered by appropriate goals. First of all, shared, non-consumable resources (here a corridor which has to be used by all UCAVs to fly from friendly into hostile area) have to be assigned to individual team members. Secondly, redundant task assignment shall be avoided in order to maximize team effectiveness. Thirdly, team members shall support other team members by completing tasks that facilitate task accomplishment of other team members.

Communication among the ACUs is structured according to the dialog representation described on p. 160. In order to keep dialogs running, the goal "continue dialog" becomes active if the current state of a dialog requires the ACU to send a message next.

In order to be able to achieve these goals, the ACUs may choose among various *action alternatives*, namely to accept or drop commitments, send messages, and initiate dialogs. Within this context, the following different types of dialogs can be commenced, all of them being based on FIPA specifications (p. 154):

- **Request**—The initiator dialog intends that the participant executes the requested action, usually to accept or drop a commitment to perform a task.
- **Propose**—The initiator proposes to the participant to execute a certain action himself, which is usually to accept or drop a commitment to perform a task.
- **Query**—The initiator asks the participant to provide him with some specified information.
- **Subscribe**—The initiator asks the participant to provide him with some specified information and to repeat this information provision every time the value of the specified information changes.
- **Inform**—The initiator provides some information to the participant without the need for a query for that kind of information.
- **Cancel**—The initiator wants the participant to stop the execution of the activities begun due to a prior dialog such as requested task accomplishment or information provision.

As mentioned earlier, appropriate *instruction models* are necessary in order to put selected action alternatives into effect. Unlike "typical" instruction models, most of the instructions within the context of co-operation do not directly influence the environment, but change the situational knowledge first. For instance, the execution of the action alternative "commit" creates a commitment within the belief section, or the instantiation of a dialog actually creates that dialog in the belief. The only instruction being sent to the environment here is sending a message to other ACUs, which in most cases follows a changed situational knowledge.

Within this context, various *environment models* are required, an exhaustive description of which would go beyond the scope of this chapter. However, the following list should give an idea of what has to be modeled. First of all, a representation of actors is needed, which is instantiated for each actor in the setup, here for each ACU guiding an UCAV and for the human operator. Closely related to this model are the ones describing various characteristics of the actors, such as available resources and capabilities. As the actors are organized in a team, a model of what constitutes a team is also necessary. This model, for example, is attributed to the team members who are pointing at the instances of the according actors. With respect to the common objective of a team, commitments have to be represented, which are further described by the task which is related to the commitment and the actor or team which is responsible for the commitment. As commitments refer to tasks, tasks have to be modeled. Furthermore, a representation of dialogs, their states and transitions, is needed, which again are related to actors and/or teams involved in an interaction.

Implementation   After having sketched which knowledge models are necessary in order to implement co-operative behavior on the basis of the Cognitive Process in the previous section, this section describes some of the models actually implemented using CPL and COSA, and how they work together. Here, the request to perform a mission from the human operator to the team of UCAVs is taken as an example. Figure 9.5 shows knowledge models of the a-priori knowledge, as well as instantiations of these classes involved in the representation of such a dialog.

In this example, there are two ACUs (`actor-self`, `actor-4`) involved, which together form a team—i.e. are all members of `team-all`. Moreover, `actor-self` is spokesman of this team, thus an instance of the environment model **spokesman** is present within the situational knowledge of the ACU considered, which is linked to both `team-all` and `actor-self` by appropriate attributes.

Upon receiving a message from the operator, with a request to perform the mission to destroy the target, the representation of an appropriate dialog is created within the situational knowledge. Here, several models are involved, first of all an instantiation of the environment model **dialog-request**, which is inherited from **dialog**, a class encapsulating all knowledge relevant to all types of dialogs. This instantiation is attributed with the protocol and conversation-ID used as well as initiator and participant of the dialog, which are `actor-operator` and `team-all` respectively. It also points to the subject of the request, which is a **mission-order** that requires destroying the target. In addition, several models are needed to represent the states and transitions of the dialog. Directly after the creation of `dialog-request`, the **dialog-state** start is created, being the *current-state* of the dialog at this stage, and pointing to the **dialog-**

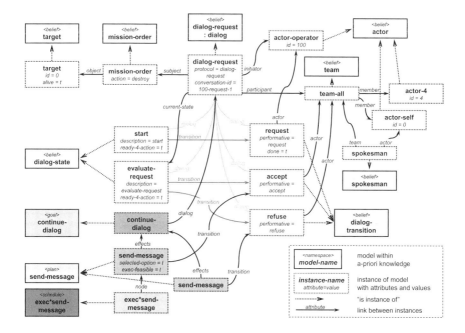

Figure 9.5   Representation of dialog "request"

transition request, which indicates that there is only one possible message, which can be sent when in state start and which points to actor-operator, who is responsible for sending that message. As the operator did already send a message of type ("performative") request, the appropriate transition is attributed with *done=t*, and a new state, evaluate-request, is created and then the current-state of dialog-request. Here, two different messages can be sent, which is indicated by the possible transitions accept and refuse, both of which team-all can initiate.

As the own actor is spokesman of team-all, the desire **continue-dialog** is activated, which can be achieved by sending a message either accepting or refusing the request. In this example, the option to accept the request is selected (instance send-message pointing to transition accept is attributed with *selected-option=t*). Subsequently, this option will be executed by the appropriate instance of an instruction model, which will send a message to the operator.

Results   Figure 9.6 shows, how a sub-problem of the scenario which was introduced at the beginning of this section is solved by a team of two UCAVs based on the knowledge described in the previous sections. Here, one Attack-UCAV (ID 4) and one SEAD-UCAV (ID 0) are commanded to destroy the target, which is being protected by one SAM-site (Figure 9.6, 00:00). Table 9.1 lists some of the dialogs conducted during this simplified mission. At first, the operator (ID 100) requests the destruction of the target from both UCAVs (dialog-request 100-0), which is accepted. The Attack-UCAV commits itself to destroying the target as there is now a team commitment to the destruction of the target and task coverage will be ensured. Moreover, it informs the team about this change in its commitments (dialog-inform 4-15) and asks the SEAD-UCAV to suppress the SAM-site covering the target (dialog-request 4-18), which is accepted. Due to the goal to keep team members informed, the SEAD-UCAV also informs the team about its commitments (dialog-inform 0-30): namely to suppress the SAM-site: to protect the Attack-UCAV from possible pop-up-threats: and finally to return to the homebase.

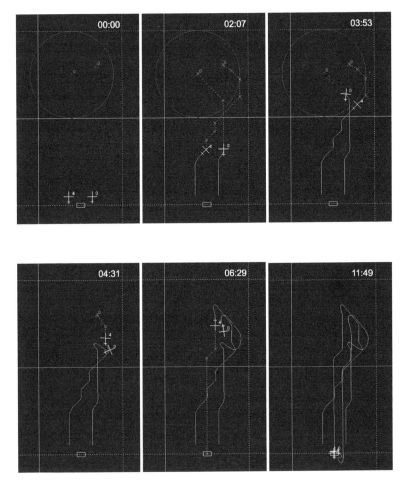

Figure 9.6   Solution of sub-problem by two UCAVs

In the following, both UCAVs comply with their commitments, i.e. the Attack-UCAV flies towards the target, while the SEAD-UCAV starts flying towards the SAM-site (Figure 9.6, 02:07). After the SEAD-UCAV has attacked the SAM-Site (Figure 9.6, 03:53) and this attack is successful, the Attack-UCAV adjusts its flightplan and continues its way to the target, now being escorted by the SEAD-UCAV (Figure 9.6, 04:31).

After having successfully attacked the target, the team continues the request dialog initiated by the operator and informs him about the successful outcome of the activity (dialog-request 0-100, "inform"). Finally, both UCAVs start flying towards the homebase (Figure 9.6, 06:29 and 11:49).

Table 9.1   Excerpt of communication between UCAVs

Protocol			Conversation-ID	
Perf.	**S**	**R**	**Time**	**Content**
dialog-request			*100-0*	
request	100	0, 4	0:42	(*mission-order*(action destroy)(object target))
accept	0, 4	100	0:43	
inform	0, 4	100	5:21	
dialog-inform			*4-15*	
inform	4	0, 4	0:58	(*information*(actor 4)(type commitment)) (*commitment*(responsible(actor 4))(task(name task-destroy)(object target)))
dialog-request			*4-18*	
request	4	0	1:11	(*task*(name task-suppress)(object sam-site)(sam-site 0))
accept	0	4	1:18	
inform	0	4	3:58	
dialog-inform			*0-30*	
inform	0	4, 0	1:24	(*information*(actor 0)(type commitment)) (*commitment*(responsible(actor 0))(task(name task-suppress)(object sam-site)(sam-site 0))) (*commitment*(responsible(actor 0))(task(name task-protect)(ucav 4))) (*commitment*(responsible(actor 0))(task(name task-fly-to)(object homebase)))
dialog-inform			*0-40*	
inform	0	4, 0	4:06	(*information*(actor 0)(type commitment)) (*commitment*(responsible(actor 0))(task(name task-protect)(ucav 4))) (*commitment*(responsible(actor 0))(task(name task-fly-to)(object homebase)))
dialog-inform			*4-27*	
inform	4	0, 4	5:26	(*information*(actor 4)(type commitment)) (*commitment*(responsible(actor 4))(task(name task-fly-to)(object homebase)))

(Perf. = Performative; S = Sender; R = Receiver)

Within the project described here, scenarios consisting of up to ten SAM-sites and five UCAVs as well as a forward line of own troops (FLOT) and a corridor for penetrating into hostile area have successfully been solved in a simulated environment with the implemented prototype, an idea of the development of which has been given in this section.

Summary and Future Prospects

Having in mind our criticism of the often-used marketing term "autonomous system," which implies a purely technology-driven view, this chapter clarifies the articulate necessity of ergonomics research in the field of highly automated systems such as UAVs and other robotic systems. Based upon the human factors engineering framework of the so-called *work system*, clear definitions and distinctions of the terms *autonomous system* and *semi-autonomous system* have been elaborated. According to the given definition a work system consists of an *operating force*, usually a human or human team, and the *operation-supporting means*, being the technology used for work. An autonomous system would emerge if the human function in the work system were completely replaced by a technological artifact, including the ability to self-define its work objective! An artificial system must have to do without this latter capability and, therefore, will be part of the operation-supporting means under supervision of the human operator. In order to counteract obvious problems with human supervisory control of highly automated complex systems, the approach of *co-operative control* has been suggested by this chapter. In this case respective automation, i.e. *co-operative automation* (we call this kind of automation an *assistant system*) will become part of the operating force building a team with the human operator. Co-operative capabilities on the side of the operation-supporting means working under supervision of the operating force have been discussed as well.

The question how to build suchlike co-operative automation has been dwelt on. As a result of our group research the answer is the approach of *cognitive automation*. Cognitive automation is based upon the idea to create artificial knowledge-based behaviour in the computer, being the most flexible aspect of human rationality and cognition and thereby being a guarantee for high integrity. The concept is implementing the cognitive process on the machine side, and here particularly the orientation of behavior towards explicitly represented goals. An approach on how to engineer such an artificial cognitive system has been discussed. In order to enable it to perform as a team member the required capabilities have been specified. These include the capability to perform the relevant mission tasks, supplemented by the capability to co-operate and to co-ordinate on task level.

Currently, two major streams of co-operative automation are under consideration at the Universität der Bundeswehr München: *knowledge-based pilot/operator assistant systems* and *multi-UAV guidance systems*. Whereas the requirements for human-machine co-operation in assistant systems have been studied thoroughly in several recent and current projects (references have been given in the chapter), this contribution has focused more on the systematic engineering of machine-machine co-operation in the domain of uninhabited aerial vehicles.

Although momentarily being focused mainly on the machine aspects this work provides interesting approaches for the modeling of human-machine co-operation by consequently providing models of goals for co-operation, of co-ordination techniques, and of dialog management related knowledge. Future work will join the different aspects of human-machine and machine-machine cognitive co-operation in manned-unmanned multi-agent scenarios. In the field of operator assistance the aspect of *adaptive automation* will be considered. Therefore, the modeling of human mental resources and workload will be a prerequisite for the decision of the assistant system in which cases to intervene and how to do so. In the field of supervisory control of UAVs the question of how to reduce the operator-to-vehicle ratio will be predominant. Cognitive and co-operative automation offers an approach to tackle many forthcoming questions in this field.

References

Austin, J. L. (1962). *How to Do Things with Words*. Oxford: Oxford University Press.

Bainbridge, L. (1983). Ironies of automation. *Automatica*, 19, 6, 775–9.

Billings, C. E. (1997). *Aviation Automation: The Search for a Human-Centered Approach*. Mahwah, NJ: Lawrence Erlbaum Associates.

Cohen, P. R. & Levesque, H. J. (1990). Intention is choice with commitment. *Artificial Intelligence*, 42, 213–61.

Donath, D. & Schulte, A. (2006). Weiterentwicklungsmöglichkeiten kognitiver und kooperativer Operateurassistenz: Ein Konzeptansatz für modellbasierte adaptive Automation. *48. Fachausschusssitzung Anthropotechnik T5.4 Cognitive Systems Engineering in der Fahrzeug- und Prozessführung*. Fraunhofer-IITB, Karlsruhe, 24–25 October.

Ertl, C. & Schulte, A. (2004). System design concept for co-operative and autonomous mission accomplishment of UAVs. In: *Deutscher Luft- und Raumfahrtkongress 2004*. Dresden, 20–23 September.

FIPA (Foundation for Intelligent Physical Agents). http://www.fipa.org.

Frey, A., Lenz, A., Putzer, H., Walsdorf, A., & Onken, R. (2001). In-flight evaluation of CAMA—the Crew Assistant Military Aircraft. In: *Deutscher Luft- und Raumfahrtkongress*. Hamburg, Germany, 17–20 September.

Gerlach, M. & Onken, R. (1993). A dialogue manager as interface between pilots and a pilot assistant system. *Proceedings of the 5th International Conference on Human Computer Interaction* (HCII'93). Orlando, Florida.

Jennings, N. R. (1994). *Cooperation in Industrial Multi-Agent Systems*. Singapore: World Scientific.

Jennings, N. R. (1996). Coordination techniques for distributed artificial intelligence. In: G. M. P. O'Hare & N. R. Jennings (eds). *Foundations of Distributed Artificial Intelligence*. New York: Wiley. pp. 187–210.

Malone, T. W. & Crowston, K. (1994). The interdisciplinary study of coordination. *ACM Computing Surveys*, 26(1), 87–119.

Meitinger, C. & Schulte, A. (2006). Cognitive machine co-operation as basis for guidance of multiple UAVs. NATO RTO HFM Symposium on *Human Factors of Uninhabited Military Vehicles as Force Multipliers*. Biarritz, France, 9–11 October.

Meitinger, C. & Schulte, A. (2007). Onboard artificial cognition as basis for autonomous UAV co-operation. In: Platts, J. (ed.). *Autonomy in UAVs*. Final Report of the GARTEUR Flight Mechanics Action Group FM-AG-14.

Newell, A. (1990). *Unified Theories of Cognition*. Cambridge, MA: Harvard University Press.

Onken, R. (1994). Basic requirements concerning man-machine interactions in combat aircraft. In: *Workshop on Human Factors/Future Combat Aircraft*. Ottobrunn, Germany, 4–6 October.

Onken, R. (2002). Cognitive cooperation for the sake of the human-machine team effectiveness. In: *RTO-Meeting Procedures MP-088, HFM-084: The Role of Humans in Intelligent and Automated Systems*. Warsaw, Poland. 7–9 October.

Onken, R. & Schulte, A. (forthcoming). *Dual-Mode Cognitive Design in Vehicle Guidance and Control Human-Machine Systems*. Heidelberg: Springer.

Platts, J., Ögren, P., Fabiani, P., di Vito, V., & Schmidt, R. (2007). *Final Report of GARTEUR FM AG14*. GARTEUR/TP-157.

Prévôt, T., Gerlach, M., Ruckdeschel, W., Wittig, T., & Onken, R. (1995). Evaluation of intelligent on-board pilot assistance in in-flight field trials. In: *6th IFAC/IFIP/IFORS/IEA Symposium on Analysis, Design, and Evaluation of Man-Machine Systems*. Cambridge, MA, June.

Putzer, H. & Onken, R. (2003). COSA: A generic cognitive system architecture based on a cognitive model of human behaviour. *International Journal on Cognition, Technology, and Work*, 5, 140–51.

Rasmussen, J. (1983). Skills, rules, and knowledge; signals, signs, and symbols, and other distinctions in human performance models. *IEEE Transactions on Systems, Man, and Cybernetics*. Vol. SMC-13, no. 3, May/June.

REFA (1984). *Methodenlehre des Arbeitsstudiums*, vols 1–3. 7. Auflage. Verband für Arbeitsstudien und Betriebsorganisation. Munich: Carl Hanser Verlag.

Ruckdeschel, W. & Onken, R. (1994). Modelling of pilot behaviour using petri nets. In: R. Valette (ed.). *Application and Theory of Petri Nets*. LNCS vol. 815. Berlin: Springer.

Schulte, A. (2006). Manned-unmanned missions: chance or challenge? *The Journal of the Joint Air Power Competence Centre (JAPCC)*, edition 3, 34–7.

Schulte, A. & Stütz, P. (1998). Evaluation of the Crew Assistant Military Aircraft (CAMA) in simulator trials. In: NATO Research and Technology Agency, System Concepts and Integration Panel. *Symposium on Sensor Data Fusion and Integration of Human Element*. Ottawa, Canada, 14–17 September.

Searle, J. R. (1969). *Speech Acts: An Essay in the Philosophy of Language*. Cambridge: Cambridge University Press.

Sheridan, T. B. (1987). Supervisory control. In: G. Salvendy (ed.). *Handbook of Human Factors*. New York: Wiley.

Strohal, M. & Onken, R. (1998). Intent and error recognition as part of the knowledge-based cockpit assistant. In: *Proceedings of the SPIE '98 Conference*—The International Society for Optical Engineering. Orlando, FL.

Stütz, P. & Onken, R. (2001). Adaptive pilot modelling for cockpit crew assistance: Concept, realisation and results. In: *The Cognitive Work Process: Automation and Interaction*. 8th European Conference on Cognitive Science Applications in Process Control, CSAPC'01. Munich, 24–26 September.

Stütz, P. & Schulte, A. (2000). Evaluation of the Cockpit Assistant Military Aircraft *CAMA* in flight trials. In: *Third International Conference on Engineering Psychology and Cognitive Ergonomics*. Edinburgh, 25–27 October.

von Martial, F. (1992). Coordinating plans of autonomous agents. Lecture Notes in Artificial Intelligence 610. Berlin: Springer.

Walsdorf, A., Onken, R., Eibl, H., Helmke, H., Suikat, R., & Schulte, A. (1997). The Crew Assistant Military Aircraft (CAMA). In: *The Human-Electronic Crew: The Right Stuff?* 4th Joint GAF/RAF/USAF Workshop on Human-Computer Teamwork. Kreuth, Germany, September.

Wittig, T. & Onken, R. (1993). Inferring pilot intent and error as a basis for electronic crew assistance. In: *Fifth International Conference on Human-Computer Interaction (HCI International 93)*. Orlando, FL.

Wooldridge, M. (2002). *An Introduction to Multi Agent Systems*. Chichester: Wiley.

Chapter 10

Situation Awareness in Human-Robot Interaction: Challenges and User Interface Requirements

Jennifer M. Riley, Laura D. Strater, Sheryl L. Chappell, Erik S. Connors
and Mica R. Endsley[1]
Systems Applications and Technologies, Inc.

In many of today's complex military and civilian domains, remotely controlled ground, aerial, sea surface, and underwater vehicles are being utilized as tools to extend the sensory and psychomotor capabilities of the human to remote environments. These unmanned robotic vehicles facilitate the projection of the operators' presence and intent upon the distant space. However, the degree to which the use of unmanned robotic systems leads to improved task performance is dependent upon not only how well the operator can task, control, and manage the robot(s), but also upon how well the operator is able to develop and maintain an accurate understanding of the unfolding situation while interacting with the robotic system. Specifically, operators need to maintain a high level of *situation awareness* (SA) with regard to both changes in the environment and the current and future actions of the robots under their control. Indeed, studies investigating issues surrounding robot control have pointed out that most of the problems encountered during robotics navigation tasks, or other more complex tasks such as search and rescue, resulted from operators' lack of awareness during robot control (Burke et al., 2004; Drury et al., 2003). Difficulties such as these limit the utility of robots for enhancing task performance in remote environments.

Given the ever-increasing prevalence of unmanned robotic systems, it is important for us, as researchers and interface designers, to understand the factors

1   We wish to thank two anonymous reviewers for their comments which helped to greatly improve the chapter. We also wish to express our appreciation to Dr Haydee Cuevas for her valuable editorial guidance in finalizing this chapter. This research was supported through participation in the Advanced Decision Architectures Collaborative Technology Alliance sponsored by the U.S. Army Research Laboratory (ARL) under Cooperative Agreement DAAD19-01-2-0009. The views and conclusions contained herein, however, are those of the authors and should not be interpreted as representing the official policies, either expressed or implied, of the ARL or the U.S. Government.

that impact the SA of operators performing tasks that rely on effective human-robot interaction (HRI). Explaining the nature of this impact can facilitate development of user interface designs for supporting SA and, ultimately, lead to improved human-robot task performance. Toward this end, in this chapter we present a theoretical model of SA in HRI with the goal of achieving a better understanding of what it means to have SA in robotics tasks across various platforms and system types, as well as to identify the challenges and kinds of errors that occur in HRI. We begin with a brief overview of the SA construct and the research efforts that laid the foundation for our model's development. We then examine each component of our model in greater detail, citing relevant findings from our empirical research. We conclude with implications for SA-oriented design of HRI interfaces.

Model of Situation Awareness in Human-Robot Interaction

Situation awareness has been formally defined as the "perception of elements in the environment within a volume of time and space, the comprehension of their meaning and the projection of their status in the near future" (Endsley, 1995b, p. 36). This definition highlights three levels of SA, namely perception (Level 1 SA), comprehension (Level 2 SA), and projection (Level 3 SA). Awareness at each level of SA is required for effective use of robotic devices. Operators using unmanned systems need to be aware of the robot's current location and status (Level 1 SA), how elements in the environment (humans, other robots, obstacles, etc.) impact the system (Level 2 SA), and how the robot's current and future behavior impacts the team's goals or supports meeting mission objectives (Level 2 and 3 SA).

A unique aspect of unmanned system control is that the operator's sensory connection is oftentimes completely mediated by an interface (e.g., display components or interaction/control devices). The importance here is that in most contexts, including military operations, human operator control and functioning will continue for some time to be essential in task completion, particularly in very complex domains, instances of robot system failures, or automation indecision. This requirement supports the philosophy of not just extending the technological capabilities of robot mechanics and physics, but also working to improve the quality and sophistication of robotics interface design, specifically with regard to presenting the right information to the operator in the right (highly usable) format.

Accordingly, our theoretical model of SA in HRI identifies several factors that can impact SA in robot tasking and control at different stages of information processing, and thus, serves as a useful introduction to the potential issues in SA related to unmanned system operation. As shown in Figure 10.1, many different types of factors can influence the quality of SA achieved by robot operators. Although these factors are also relevant to other complex task domains, specific aspects associated with certain factors of the model are unique to HRI. In addition, whereas some factors impact SA overall, other elements within a factor category affect different levels of SA at various stages of information processing.

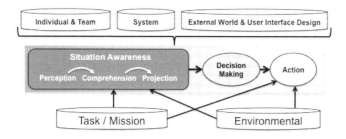

Figure 10.1 Theoretical model of situation awareness in human-robot interaction

The foundation for our model's development is based on our extensive research efforts related to human interactions with unmanned vehicles. These investigations have revealed the importance of SA in human use of robots, the types of SA-related errors observed in robot control, and the potential interface design requirements for ameliorating the performance decrements that result from poor SA in HRI. Next, we briefly discuss the central facets of our research approach, which included conducting cognitive task analyses, field observations, and empirical investigations.

The cognitive task analyses were targeted at understanding SA requirements in robot control across numerous platforms, including unmanned ground vehicles (UGVs), unmanned aerial vehicles (UAVs), and unmanned underwater (UUVs) and sea surface vehicles (USVs). Utilizing a unique form of cognitive task analysis called goal-directed task analysis (GDTA) (Endsley et al., 2003), we interviewed subject matter experts in various task domains, including robot controllers, operators, and mission information stakeholders with experience of using or developing various platform types (Chappell & Craddock, 2006; Riley & Endsley, 2004; Strater et al., 2008). The GDTA provides a hierarchical representation of the major goals and subgoals, related decisions, and key mission-relevant information requirements at all three levels of SA. We used the results of our GDTAs to isolate how various aspects of HRI may negatively influence operators' decision processes and goal attainment, leading to degraded SA. For example, our GDTA for UUV control identified several critical SA information requirements needed by the operator to support the goal of monitoring critical UUV systems status. These included understanding how factors such as changed mission requirements, enemy capabilities, mission demands, system degradation, terrain, and weather will impact projected system capabilities and system requirements (Strater et al., 2008).

In addition to GDTAs, we also participated in training exercises and field observations to conduct contextual inquiries in the operation of ground vehicles for urban search and rescue (SAR) operations and aerial vehicles, specifically Class I and Class II military vehicles (Chappell & Craddock, 2006; Riley et al., 2006). This work involved direct observation of operators (both single user and team control) performing representative domain tasks in realistic missions, followed by after-action review sessions. Our approach involved: (1) documenting interface design and control issues; (2) monitoring team communications to assess the influence of information exchanges on SA and team coordination; and (3) identifying human factors and SA problems during training/field performance. This research allowed us to identify the HRI factors that challenge operators' ability to acquire and maintain SA, which, in turn, supported the development of our model.

Lastly, we conducted empirical studies to assess the effects of specific HRI factors on SA and other dependent measures (e.g., performance and workload) (Riley & Strater, 2006a; 2006b; Riley et al., 2008). These studies involved experimentation using miniature ground robots and scaled-down environments. The ground robots were equipped with devices representative of the current tools on unmanned vehicles, i.e., cameras and (simulated) sensors. In our experiments, operators were provided with computer interfaces that presented, for example, data on task instruction and status, displays for monitoring system status, camera video, mechanisms for collaborative control, and maps. Not all features were present on the interfaces in all studies, but this list represents some of the interface components of interest across studies. In all our studies reported in this chapter, SA was measured using SAGAT queries (for a detailed description of this methodology, see Endsley, 1995a), which assessed participants' SA at all three levels with regard to the task, environment, and robot. These experiments facilitated collection and analysis of empirical data that provided insights into the nature of the impact of HRI factors on SA, further highlighting SA challenges that could be addressed by innovations in interface design. We next describe the central components of our model, citing in greater detail the relevant findings from our research efforts investigating SA in HRI.

Task/Mission Factors

Task or mission related factors impacting SA are associated with the parameters under which operators must complete their tasks (see Figure 10.2). Many of these factors affect SA at the perceptual level (Level 1 SA). For example, in robot tasks remote control is typically non-line of sight, with limited to no opportunity to observe the environment directly; the impact the robot has on the environment and vice versa; or to observe the robot while it is completing various control actions. (There are some exceptions, for example, in SAR and for SUGVs when ground or aerial vehicles are periodically in the visual field.) For the most part, operators must rely on the system interface for information on the robot and distant space. Consequently, the quality of SA achieved is dependent upon how well the system technology serves to virtually immerse the operator's senses in the remote environment through sensory feedback (visual, auditory, and/or tactile). On the one hand, operator SA is hindered when sensory tools designed to mimic direct observation are limited, providing degraded feedback. On the other hand, operator SA is enhanced when technology extends what might normally be perceived (e.g., night vision devices or augmented reality).

Mission parameters (multiple vehicle control, multiple platform control, etc.) can often overload the perceptual and cognitive resources of operators. In particular, multiple vehicle control is an important consideration for current and future robotics operations. A goal for many military unmanned vehicle programs is to increase the operator span of control in robotics (i.e., increase the number of vehicles a single operator can manage). For example, the U.S. Army is projected to transition from the current two operators per UAV to a single operator for multiple UAVs. However, this shift toward multiple vehicle control will inherently result in a corresponding increase in operator mental workload, imposed by the cognitive requirements of concurrently managing more than one system. The amount of information that an operator must process in robotics tasks can be overwhelming. Robotics operators must (1) maintain awareness of data on system location, status, and activities; (2) understand the environment, where the robot has or has not been, dangers in the space, and spatial relationships; and (3) track task progress against time constraints and objectives and relate task states to efforts performed by other people. The amount of data to manage and information to process increases as the number of robots increases. When multiple goals across multiple systems must be met, conflicting system or goal priorities lead to difficult decisions. Requirements for collaborative control, where teams of operators must coordinate to complete a robot-assisted mission, further compound these issues. Not surprisingly, the cognitive resources of the operator can be greatly taxed, and if the operator cannot cope with these demands, SA will be degraded, negatively impacting task performance.

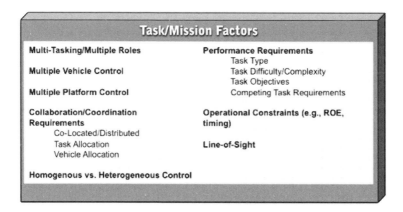

Figure 10.2   Task/mission factors influencing SA in HRI

Indeed, we have observed such degradations in SA during our investigations of multiple robot control. For example, in one study (Riley & Strater, 2006a), single operators were tasked with controlling two ground robots of the same type either serially (manually one after the other) or simultaneously to perform a maze exploration task. Three levels of simultaneous control were manipulated: (1) manual control of both robots at the same time; (2) manual control of one robot and supervisory control of the other robot under automation that mandated operator intervention at decision points; and (3) manual control of one robot and supervisory control of the other robot under full automation. In general, results showed that operators controlling robots simultaneously, whether manually or in mixed manual-automation mode, exhibited poorer SA than operators allowed to control the robots serially. Specifically, results revealed a significant difference with respect to SA projections on operator decisions and actions based upon the overall status of the robots. Participants controlling one robot at a time exhibited the highest SA scores (about 80 percent correct responses to SAGAT queries), while participants controlling one robot under full automation exhibited the lowest SA scores (about 49 percent correct responses to SAGAT queries). These findings suggest that operators allowed to concentrate efforts on one robot at a time had more cognitive processing capacity to track robot state and assess the impact of system status compared to simultaneous dual robotic control, even with automation support.

Such potential degradation in SA is an even greater concern during control of multiple systems of different types and capabilities (e.g., UAVs with UGVs or with USVs) spanning a wide area of operations. The U.S. Navy Littoral Combat Ship (LCS) concept of operations is one notable example. The LCS is one of a family of future Navy surface ships that will play a key role in littoral sea control (e.g., anti-submarine warfare, mine warfare, surface operations, and intelligence–surveillance–reconnaissance). The area to which LCS will be deployed covers a wide region, extending from the open ocean to the shore, and to inland areas that can be attacked, supported, and defended directly from the sea. In the LCS concept, multiple types of unmanned vehicle systems will be employed, including USVs, UUVs, and UAVs. The spatial and temporal distribution of such a situation can strain the knowledge and information exchange that underlie effective collaboration and the cognitive processes that drive good SA. The key to successful human-robot task performance in this domain will be providing the myriad, dynamically changing, team-relevant operational data for effective coordination with these different robots. Achieving this objective will require theoretically grounded, human-centric technologies to support collaborative task and control situations.

Performance requirements—task type (navigation, target detection, object manipulation), task difficulty, and task objectives—also impact operator SA. For tasks that require, for example, complex manipulations of robot parts (e.g., robotic arms and end effectors and other tactical devices), SA can be challenged by attentional tunneling. When focusing closely on the control of the system (e.g., making it move a certain way, maintaining its balance on an incline, or positioning joints for grasping an object), the operator is likely to fixate on a subset of SA requirements to the exclusion of others. Robot operators experiencing attentional tunneling may either intentionally or inadvertently discontinue monitoring behaviors and assessments, resulting in reduced overall SA due to the complexity of the requirements on a portion of the overarching task. This presents a key SA challenge, especially for military robotics, as maintaining awareness of the environment can impact troop safety and, further, robot task control may not be the only tactical responsibility of the operator.

As another example of task requirements, one of the biggest challenges of SA in robot control is detecting an object or a condition that is undefined. In our interviews with military operators (Chappell & Craddock, 2006), they often described seeing something that "didn't look right," which turned out to be an improvised explosive device (IED) or other type of danger. Operators detected these hidden hazards by noticing subtle cues such as soil texture that was inconsistent, a vehicle that was in a different location, or even something that was missing. The ability to detect these important cues stems from operators' experience with the area and terrain, as well as from the shared experiences of others. Technological advances in image analysis, which could be employed on a robot, are of little use for this type of detection. Such detection requires the visual and perceptual skills and experience of the operator. However, we suggest facilitating rebuilding SA by supporting robot operator cue detection through interface design, for example, providing easy access to images of the same area taken at an earlier time, allowing operators to quickly confirm that something is different and alert others of the potential danger. This opportunity for comparison with past states is important when elapsed time necessitates rebuilding or refining one's SA. Additional issues related to interface design will be discussed later in this chapter. This next section addresses system factors such as the use of automation.

System Factors

The way that the system is designed to work—its functional capabilities and limitations, its onboard systems, and mechanical make-up—significantly influences the operator's SA (see Figure 10.3), especially Level 2 (comprehension) and Level 3 (projection) SA. In particular, with the shift toward multiple vehicle control, appropriate system design and judicious use of automation will be required to make the operator's increased span of control possible. When automation is implemented to take over portions of tasking and control requirements, SA can be affected by the level of automation employed and its reliability, as well as the type and logic of the automation. The utility of automation really depends on the environment and the task context. In urban SAR, for example, the environment lends no affordances for the benefit of automated navigation, particularly with the size of the robots being used. Spaces are categorized as messy, cluttered, and jumbled with debris, likely making it hard for automated navigation strategies and algorithms to be much help in finding paths. Nevertheless, because of the perceptual strain associated with visual search tasks, SAR operations might benefit greatly from improved sensors that support automated object detection or from aids that facilitate judging the size and distance of objects which provide clues for finding victims.

In our own research, we have observed specific effects of automation on SA in control of simulated ground vehicles. In one study (Riley and Strater, 2006b), we compared the effect of manual versus automated robot control on the SA of participants performing a maze exploration task that required navigation, detection, and identification of landmarks. Results revealed that participants who controlled the robots under automated navigation exhibited significantly poorer Level 2 (comprehension) and Level 3 (projection) SA (ranging from 43 percent to 57 percent correct responses to SAGAT queries) than participants who utilized manual control (ranging from 62 percent to 75 percent correct responses to SAGAT queries). Given such findings, the decision of when and to what degree to implement the use of automation must consider its effects on the related constructs of performance, workload, and SA.

To guide such automation usage decisions, our user-centered design approach emphasizes the goals of the human operator, not the technology to be built, and, therefore, allows us to address the desired end state of the control tasks. For example, UAVs in theater now have the capability to hover and easily change direction and altitude. During our interviews with military operators regarding UAV control for reconnaissance missions (Chappell and Dunlap, 2006), they repeatedly remarked that they want to "fly the camera," that is, operators do not want to have to devote attention to controlling the vehicle and its systems, but rather they want to position the camera where it needs to be to support the battle objectives. To address this operator goal, we are developing user interfaces for

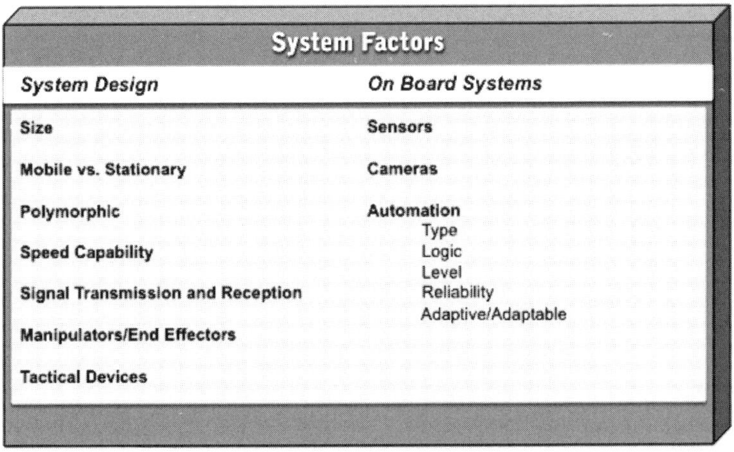

System Factors	
System Design	**On Board Systems**
Size	Sensors
Mobile vs. Stationary	Cameras
Polymorphic	Automation
	Type
Speed Capability	Logic
	Level
Signal Transmission and Reception	Reliability
	Adaptive/Adaptable
Manipulators/End Effectors	
Tactical Devices	

Figure 10.3   System factors influencing SA in HRI

UAVs that eliminate direct control of roll, pitch, and yaw. Instead, operators move the vehicle with a game controller by pointing the joystick in the direction they want to look, and the system automatically follows through. Reducing the SA demands associated with controlling the vehicle's movement allows UAV operators to concentrate on building SA for the primary task of reconnaissance. We believe a great deal of progress can be made in this area as we automate vehicle systems and advance in display technology.

Beyond automation, the actual physical design and features of the system also affect SA requirements. Mobile robotics requires SA on different elements (e.g., mobility and terrain effects) than stationary industrial robots. Embedded devices (e.g., manipulator arms and sensors) and the reliability of said tools can influence what information the operator needs to develop SA and the degree to which he or she can acquire knowledge in the remote environment. Poor signal transmissions can mean lapses in availability of SA-relevant data.

Systems that can change in size or shape require a deeper understanding of how the robot mechanics and configuration can impact, for example, robot mobility on various types of terrain or the current robot viewpoint or perspective provided with onboard cameras and other sensors. For small robots, the point of view is unnatural (e.g., very low to the ground). Shape-shifting robots can have multiple viewpoints that may vary in terms of how optimal they are for detection, identification, fitting through tight spaces, driving over tricky terrain, etc. Drury et al. (2006) explored the potential benefit of providing visual representations of robot shape through an interface. They found that fewer incidences of the robot getting "caught" or stuck occurred when participants could see a representation of the configuration. Also, fewer attempts to change the configuration were made when the representation was available. Drury and colleagues did not assess SA directly, but speculated that the visual representation may have improved SA on that particular element, thus helping to decrease mobility errors.

Along the same lines, our own observations (Chappell & Craddock, 2006; Riley et al., 2006) suggest that military ground and aerial systems might benefit from similar graphical representations of camera direction and orientation, relative to the direction and orientation of the unmanned system. Unmanned vehicles can travel in a direction that is different from the direction of the camera. This ability is important when conducting reconnaissance while en route. The problem lies with collisions that can occur with loss of SA and poor localization (often a function of the operator misunderstanding the deviation between the camera direction and orientation and that of the vehicle). UAVs operating in an urban environment at a low altitude are especially vulnerable when the operator has lost awareness of proximate obstacles. To facilitate SA in these situations, we developed an interface design that displayed a 360-degree scan indicating the direction of travel and the field of view of the camera focus (Chappell, 2007). This scan is updated using a primary camera at intervals specified by the operator based on the level of hazard. As a bonus, the scan has also been useful for detecting unexpected objects.

Environmental Factors

Unmanned systems are controlled from a remote and potentially very distant space, creating two distinct environments that impact operator SA: the *local* environment in which the operator is stationed and the *remote* environment where the robot is located (see Figure 10.4). Operators could conceivably develop SA on one environment to the exclusion of the other. Or, they could develop some level of SA on both environments simultaneously. The use of multiple robots often involves multiple remote locations, with multiple systems sent to different places for completing coordinated, synchronized missions. Developing SA on two (or more) environments simultaneously will be critically affected by the operator's ability to divide attention across multiple locations (Draper et al., 1998; Endsley and Jones, 1997). Consequently, the operator's SA will be constrained by limits in attentional resources as well as a substantial number of factors that diminish attentional capacity. (This issue will be further discussed under "Individual and Team Factors").

Both the local and remote environments present challenges to the operator. For military operations, risks associated with either environment can impact SA. The threat may be in the form of unstable conditions or other actors that pose risks to the human operator or to the robotics system. Fortunately, threats to humans in the remote environment, for the most part, are minimized by the use of robots; indeed, one of the primary reasons robots are deployed is to serve in the place of humans in dangerous task environments that are ill-suited to human insertion. However, threats in the local environment are still a very real and important concern for troops.

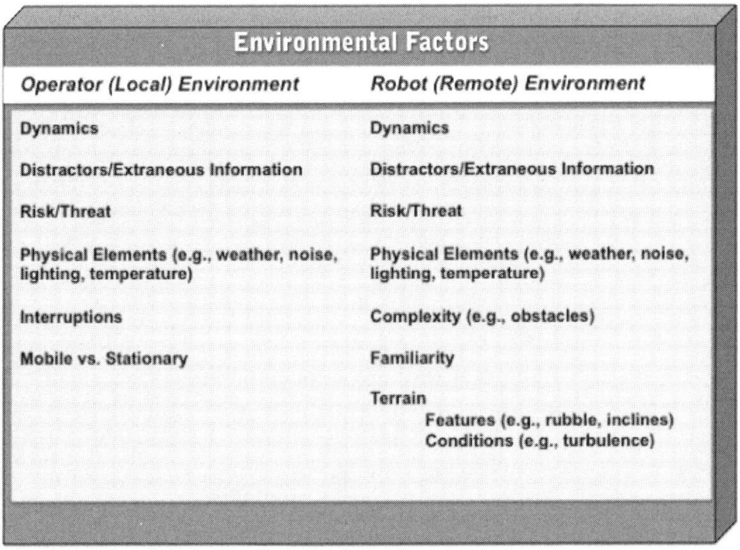

Environmental Factors	
Operator (Local) Environment	*Robot (Remote) Environment*
Dynamics	Dynamics
Distractors/Extraneous Information	Distractors/Extraneous Information
Risk/Threat	Risk/Threat
Physical Elements (e.g., weather, noise, lighting, temperature)	Physical Elements (e.g., weather, noise, lighting, temperature)
Interruptions	Complexity (e.g., obstacles)
Mobile vs. Stationary	Familiarity

	Terrain
	    Features (e.g., rubble, inclines)
	    Conditions (e.g., turbulence)

Figure 10.4   Individual and team factors influencing SA in HRI

The effects of the remote environment on SA can also sometimes be attributed to features of the terrain. For ground, aerial, and sea vehicles, the terrain will vary considerably, but general aspects (e.g., dynamics, complexity, conditions) of the remote environment influence SA across vehicle types. For example, USV and UUV operators both have to contend with meteorological and oceanographic conditions. The dynamics of the sea can greatly impact their understanding of, and confidence in, current target location and/or direction of movement, vehicle ability to traverse to the next objective on time, and sensor coverage. Projections on future locations for linking with other systems, when the system(s) will be unavailable for communications relays, when routes need to be altered, etc. are also hindered. With aerial vehicles, urban skyscrapers and weather conditions (e.g., precipitation and wind speed and direction) can impact visibility and referencing for localization.

In military operations, UAV operators also have to know where they are relative to the ground forces they are supporting and the combat objectives. This can be challenging in both natural and urban environments. To enhance SA, a hybrid of current imagery, previously collected imagery, and a synthetic environment can provide a view of the world with phase lines, road signs, and building numbers (supporting terrain awareness). An interface has been developed for the U.S. Army's Future Combat System that provides a virtual environment framework for the real-time video from the vehicle; the video is georectified on the virtual three-dimensional display. This methodology has been shown to improve search and rescue (Drury et al., 2006).

Individual and Team Factors

As shown in Figure 10.5, human-robot task performance can also be affected by numerous operator characteristics, at both the individual and team level. At the individual level, each operator has personal experiences, training, background, and knowledge bases to access on the job. Certain skills and abilities also come into play to impact SA and robot task performance. Operator skills include both the ability to control robotic systems and specific skills directly related to the task domain (e.g., IED detection). A high level of proficiency in both types of skills is necessary for developing SA on the task and environment and achieving successful performance.

Attentional resources and the ability to effectively divide attention will significantly affect operator SA. Given the potential for attentional tunneling, maintaining global SA on both the operator (local) and robot (remote) environments is absolutely critical. Effective dividing of attention will promote survivability. Experimental results from two of our studies have revealed the effect of divided attention on SA. In the first study (Riley et al., 2004), participants utilized a single robot to conduct a (simulated) minefield remediation task. Participants were required to divide their attention between monitoring one system embedded in the remote robotic task interface and monitoring a separate system associated with the local environment (the second system was on a different monitor and not included as an overlay or component of the remote robotic task display). Attention, measured in terms of signal detection theory, was assessed by calculating the hit-to-signal ratio in a secondary task that required monitoring the local environment system display for changes in status. Results of correlation analysis revealed that effective attention allocation between the remote robotic task interface and local environment system display (as indicated by higher hit-to-signal ratios on system monitoring) was associated with better SA for the robotics task (as indicated by higher percent correct responses to SAGAT queries).

In the second study (Riley & Strater, 2006a), we assessed SA and attention allocation across two robotics environments (two robots in two different locations, both with a secondary monitoring task). Attention and SA were assessed in the same manner as described for the first study. Correlation analysis showed that higher performance on the robotic task was associated with better SA for that robot. This finding is consistent with the hypotheses and statements in other research which maintain that SA is mediated by attention allocation to tasks (Endsley & Jones, 1997). Drawing from these investigations, we designed user interfaces that promote SA of the robot's remote environment and preserve SA of the operator's local environment by continuously providing a graphical representation of the

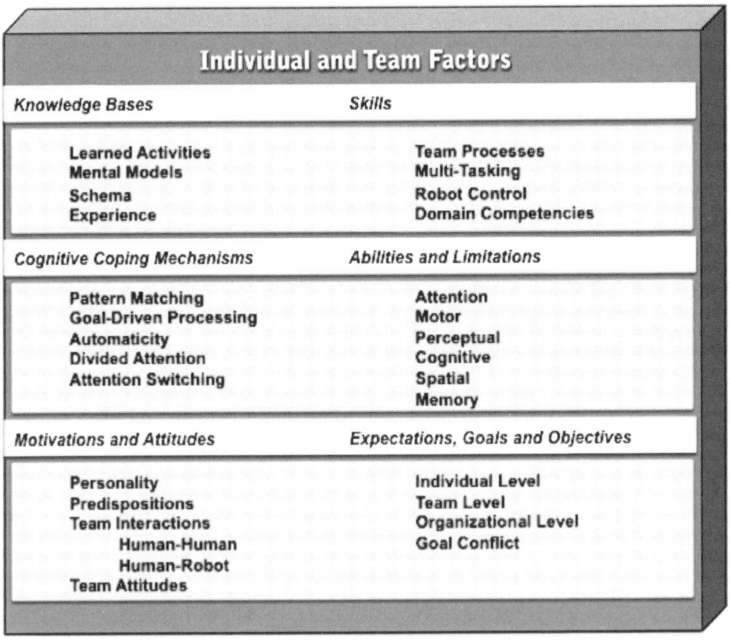

Individual and Team Factors

Knowledge Bases	Skills
Learned Activities **Mental Models** **Schema** **Experience**	**Team Processes** **Multi-Tasking** **Robot Control** **Domain Competencies**

Cognitive Coping Mechanisms	Abilities and Limitations
Pattern Matching **Goal-Driven Processing** **Automaticity** **Divided Attention** **Attention Switching**	**Attention** **Motor** **Perceptual** **Cognitive** **Spatial** **Memory**

Motivations and Attitudes	Expectations, Goals and Objectives

Personality
Predispositions
Team Interactions
    Human-Human
    Human-Robot
Team Attitudes
	Individual Level
	Team Level
	Organizational Level
	Goal Conflict

Figure 10.5   Environmental factors influencing SA in HRI

relative position of the operator with respect to the vehicle (Chappell, 2007). In addition, significant events or changes in the operator's environment are presented as alerts, drawing his/her attention to any immediate personal threat.

Individuals also vary in terms of their perceptual skills, memory capacity, and cognitive abilities, which may differentially influence the ease with which they develop SA (e.g., individuals with poorer perceptual skills may struggle with developing Level 1 SA). Spatial abilities, in particular, have been shown to be essential to robot operations (e.g., Chen & Barnes, 2008), as these involve the ability to: recognize objects when viewed from different perspectives; perceive and comprehend spatial relationships; mentally imagine configurations and space layouts; and perform mental rotations of objects. In our own research, we have demonstrated the importance of spatial abilities to SA and task performance. For example, we evaluated the relation between spatial ability and SA within the context of collaborative robotic control in a study involving 12 teams performing a navigation and search task (Riley et al., 2008). Spatial ability was measured with the Manikin Test (Carter & Woldstad, 1985), which involves determining the directional location of a colored target that matches the color of a cue. Results revealed that greater accuracy on the spatial abilities test was significantly and positively correlated to better SA (percent correct responses to SAGAT queries). Results also showed how participants' ability to spatially orient in their environment affected task performance. Specifically, 70 percent of the 65 total performance errors recorded involved poor robot localization (e.g., attempting to complete tasks in the wrong place, orienting the robot in the wrong direction for task completion).

At the team level, SA is influenced by a variety of team processes and interactions as well as individual and team expectations, motivations, and attitudes. Robot teams may consist of one or more operators assigned as the vehicle operator or driver of the system, while one person or more is assigned as the sensor (or payload) specialist, interpreting the information received from onboard robot devices. Team members sometimes may be required to multitask, taking on both of these roles (or other responsibilities). In any case, the team is likely required to provide synthesized information to multiple stakeholders, who will use this data to support their own SA and decision making. Effective human-robot task performance, therefore, is dependent upon not only each team member's individual SA, but also the team's SA overall as well as their shared SA. *Team SA* represents the degree to which each team member has the SA needed to meet his/her task responsibilities (Endsley, 1995b). Poor SA in one or more team members can negatively impact the performance of the entire team. Similarly, to the extent that the goals of two or more team members overlap, they also need to develop *shared SA*, that is, a shared understanding of the subset of information required to meet their shared goals (Endsley & Jones, 1997).

Effective dissemination of information is essential for supporting the team processes (e.g., collaboration and coordination) that are critical to the development of team and shared SA. In a recent study (Riley et al., 2009), we assessed the communications of teams performing a collaborative robotics task. Twelve teams were required to coordinate to assign tasks to robots based on robot capabilities, collaborate to complete joint tasks involving sensor readings and environment habitability assessments, and, in some teams, coordinate to share control of a third semi-automated robot. In our analysis of the robot teams' communications, we observed that teams varied greatly in the number of verbal statements made during task completion. Across two 35-minute trials, teams uttered between a low of 207 and a high of 1,492 statements. More importantly, teams differed in what they communicated in those statements. We categorized and tallied verbal statements based on content (i.e., what they talked about), type of statement made (e.g., a question or instructional statement), and function of the statement (e.g., to request information, report data, or clarify a previous statement). These categories were based on Burke's (2003) Robot Assisted Search and Rescue Communications Coding Scheme, though we did tailor some categories and groups to fit our military-based scenario.

Our communications analysis revealed that teams differed significantly with regard to the content of their discussions (e.g., some teams talked more about robot functioning than others) and the function of their verbalizations (e.g., some talked more for updating or others talked more for clarifying previous statements). Notably, our analysis also found a number of significant correlations between the number of statements made in a particular category and SA and performance scores. Specifically, more team communications about strategy (content—on task plans, procedures, or decisions) was significantly and positively correlated with increased performance on the robotics task. Similarly, providing more situation reports (function—to share observations about the environment or the tasks/ targets) was significantly and positively correlated with higher SA scores (percent correct on SAGAT queries). Conversely, a significant negative correlation was found between SA scores and the number of uncertainty statements communicated (function—to express doubt, disorientation, or loss of confidence in a state of observation); that is, teams that verbalized more uncertainty statements had lower SA scores. Off-task communications (statement type—unrelated or extraneous subjects) were also significantly and negatively correlated to SA and task performance. These results provide empirical evidence for the significant relation between team communications and SA and human-robot task performance.

External World and User Interface Design Factors

External world and user interface design factors draw attention to interventions that can be implemented to better support human operators as they use robots and contend with the other factors discussed (see Figure 10.6). The location and actions or behaviors of other actors in the external world (e.g., humans, other robots, or targets) can purposely or inadvertently diminish the SA of robotics operators. Conversely, other actors also provide relevant information to support the robot team's SA (e.g., intelligence reports from higher headquarters).

With regard to user interface design, display characteristics, interaction mechanisms, and control limitations all have a potentially huge impact on robot operator SA. Limitations in interface design and functionality (poor resolution, diminished or narrow field of view, etc.) will limit the degree to which operators' can build SA. Poor user interface designs not only hinder operators' ability to develop Level 1 SA (perception of elements), but, more significantly, make it increasingly difficult for operators to transition to higher levels of SA (comprehension and projection) due to the lack of integrated, situation-relevant information provided through the robot interface. Low-level information only serves to slow operators down as they must work to cognitively assimilate data items in order to comprehend what is going on (Riley et al., 2006). In our SAR field observations (Riley & Endsley, 2004), we noted that operators used interfaces that presented many low-level data items, mostly involving system diagnostics on mechanical components of the system (e.g., flipper angles, flipper rotations, multiple motor functioning, and system temperatures). Operators in this domain are primarily focused on the demanding visual search task, which precludes devoting their attention toward integrating all this Level 1 SA system data into a global understanding of the overall situation (Level 2 and 3 SA).

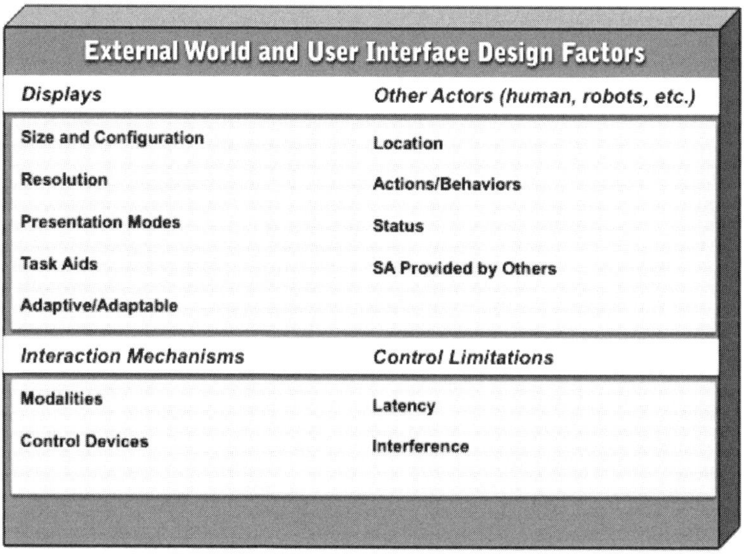

External World and User Interface Design Factors	
Displays	**Other Actors (human, robots, etc.)**
Size and Configuration	Location
Resolution	Actions/Behaviors
Presentation Modes	Status
Task Aids	SA Provided by Others
Adaptive/Adaptable	
Interaction Mechanisms	**Control Limitations**
Modalities	Latency
Control Devices	Interference

Figure 10.6   External world and user interface design factors influencing SA in HRI

In our research, we have found evidence for this difficulty in transitioning to higher levels of SA. In one study (Riley & Strater, 2006a), participants performing a robot-assisted search task provided 70–80 percent correct responses to Level 1 SA SAGAT queries on basic perceptual items such as component status (current robot battery level, communication link status, etc.), but were only about 30 percent correct in answering Level 2 and 3 SA SAGAT queries assessing their situation comprehension and projection. Results showed that participants were unable to integrate status data acquired through monitoring of individual systems into an understanding of the overall system state; comprehend the impact of status on robot functioning; or make decisions on the basis of this understanding. Further, participants were unable to integrate multiple series of control actions completed into an understanding of the environment layout (often failing in attempts to sketch the environment) or knowledge of where and how they were positioned in the space. To address these issues, robot interface designers must first understand what operators in various contexts need to know and then develop presentation methods that display this information in an integrated format to support their SA and decision making.

Control latency (i.e., a lag between control actions at the interface and robot responses in the remote environment) has also been shown to degrade overall task performance, often resulting in a suboptimal "move-and-wait" control strategy. Unmanned vehicles of all types can suffer from control latency, and even small time lags associated with data transmissions to and from the vehicle can hinder SA, particularly exacerbating Level 3 SA (projection) problems (Endsley & Riley, 2004). In UAV operations, for example, communicating critical SA on position, the commanded position, and future locations is important to task performance, but greatly hindered by control lag. With expectations of as much as a 2-second lag in control and a high degree of variability, interface concepts for ameliorating SA and performance degradations are clearly warranted. We have developed an interface design to expressly address this latency issue (Chappell, 2007). The operator specifies a point to where the vehicle is to move using a point-and-click interface with a three-dimensional map. The vehicle location is updated as the transmission latency permits, augmented by projections based on vehicle and environmental dynamics. This interface methodology reduces the impact of the move-and-wait strategy, while mitigating the chances of pilot-induced oscillations that result from latency.

Beyond difficulties associated with developing SA in general, user interface design must also consider how to support individual differences in operator experience and skill level. Newly assigned robot operators may have limited experience in controlling unmanned systems, particularly within a combat environment. The challenge for designers, therefore, is to create a control system and interface for operators with minimal training who are charged with controlling the unmanned vehicle while performing other combat tasks on the move. The user interface must also support more experienced operators performing more complex tasks. To address this challenge, we have developed a user-centered methodology

(Chappell & Dunlap 2006) that is being applied to the development of both UAV and UGV operator interfaces for the Army's Future Combat System. Our design offers different levels of interaction based on the operator's experience, allowing more experienced operators to drill down and specify vehicle and mission parameters in greater detail. These concepts can be extended to support novice operator control for other platforms.

Summary

Robots are designed and deployed to aid humans in task performance by extending their sensory and psychomotor capabilities in remote environments. Good interface designs optimize the potential benefit of robots on the job, while poor designs, unfortunately, constrain the utility of this technology. Given the demonstrated relation between SA and performance, and other important elements of information processing, researchers and interface designers need to understand the effects of technological, task, human, and environmental factors on SA, and explore how to better facilitate SA through interface design. In our research on SA, we have investigated how all data to be considered in dynamic and potentially complex military (and civilian) robot operations need to be integrated and comprehended to improve human-robot interactions and task performance. The model of SA in HRI described in this chapter provides a theoretically grounded, empirically based framework to guide robot interface designers in considering how various factors interact to impact SA and performance. SA-oriented design complements other user-centered design practices by going beyond robot technology and a system-oriented approach to include comparable care for and improvements to interfaces that help operators easily absorb, assimilate, and track relevant information over time.

We have presented our lessons learned here, along with some of our ideas and approaches, because of the observed impact of SA on robot performance and in an effort to stimulate thought on how to ameliorate these challenges. Below we briefly summarize several implications for HRI interface design based on the research reported in this chapter. As we continue to explore SA in HRI, our goal is to extend our empirical research efforts to more fully specify and validate the components of our model and to better understand how augmenting operator SA can lead to improved human-robot task performance.

SA-Oriented Design Guidelines for Enhancing Human-Robot Interaction

Guideline 1: Task/Mission Factors

Our research has shown that task/mission factors impact SA directly (through the ability to perceive objects and events) as well as indirectly (through increased mental workload, attentional narrowing, or extreme attention distribution) (Riley & Endsley, 2004; Riley et al., 2004; Riley & Strater, 2006a). Interface design must, therefore, support the perceptual (e.g., cue detection), cognitive (e.g., attention), and physical (e.g., motor control) demands imposed on the operator by performance requirements (e.g., task type, task difficulty, and task objectives) as well as different mission parameters (e.g., collaboration/coordination requirements, multiple vehicle control, and operational constraints). For example, in situations where ongoing and detailed awareness of the environment, task, or robot status are critical to mission performance, requirements for divided attention should be minimized and the vehicle-to-operator ratio should be reduced to facilitate higher levels of SA.

Guideline 2: System Factors

Automation usage decisions must be guided by a user-centered design approach that emphasizes the goals of the human operator, rather than the technology to be built. Our research has demonstrated why the use of automation must be tempered by consideration for the effects of different automation characteristics (e.g., level, reliability, type, and logic) on operator workload, SA, and task performance (Riley & Strater, 2006a; 2006b). In addition, varying configurations of system design characteristics (mobility, size, shape-shifting, signal transmission speed, manipulators, etc.) may inherently increase or decrease the number of elements required by operators to develop and maintain SA. Therefore, system factors need to be considered in relation to task factors and mission objectives and goals. For example, interfaces should be designed to reduce the SA demands associated with controlling the vehicle's movement to allow operators to concentrate on maintaining SA for their primary task (e.g., reconnaissance, IED detection) (Chappell & Dunlap, 2006). Similarly, to enhance SA as well as improve robot control and navigation performance, interfaces must support operators' localization and orientation in their environment, providing for direct representation of vehicle position, orientation, and direction relative to other mission-critical areas, environmental features, other actors, or systems (Riley & Endsley, 2004; Riley et al., 2006, 2008; Riley & Strater, 2006a). This information presentation should also include timely and accurate awareness of camera (and other onboard systems) positioning, relative to the direction and orientation of the unmanned system under operator control.

Guideline 3: Environmental Factors

Well-designed user interfaces will enable operators to maintain SA for both their own *local* environment and the *remote* environment where the robot is located. When multiple robots (and multiple remote locations) are involved, interfaces will also need to support operators' prioritization of environments or tasks, such as through the proper use of alerts and goal-based alarms. Additional design considerations include determining how to help operators maintain or rebuild SA when distracted by other tasks, interruptions, extraneous data, alarms or alerts, threats, etc. Since operators will need to maintain awareness of the impact of certain environmental factors (risks/threats, physical elements, terrain, etc.) over time, operators will require interfaces that promote the development of higher levels of SA (comprehension and projection) (Endsley et al., 2003).

Guideline 4: Individual and Team Factors

Effective human-robot task performance is dependent upon each operator's individual SA, the team's SA overall, and the team's shared SA. Our research has shown how individual-level factors such as attention allocation (Riley et al., 2004; Riley & Strater, 2006b) and spatial skills (Riley et al., 2008) interact with task/mission and system factors to significantly impact SA and robot-assisted task performance. Thus, interfaces should be designed to flexibly adapt to individual differences in operators' innate abilities, skills, and experience levels. In addition, consideration for task/mission factors calls for reducing operators' performance requirements in the local environment in order to maximize their ability to control the robot in the remote environment. This allows operators to effectively manage the cognitive demands imposed by having to divide their attention across multiple environments.

At the team level, our analysis of robot-team communications highlighted how the nature of the information exchanged among team members is directly related to SA and task performance (Riley et al., 2009). Accordingly, interfaces should be designed to present the appropriate level of task-relevant information to foster the development of team and shared SA. Specifically, interfaces need to promote comprehension and communication of key shared SA information requirements that relate to improved human-robot task performance (Riley et al., 2009; Strater et al., 2008). Robot teams also need context-specific collaborative tools to help them meet their overlapping goals and make joint decisions.

Guideline 5: External World and User Interface Design Factors

Other actors (humans, robots, targets, etc.), display characteristics, interaction mechanisms, and control limitations interact with other HRI factors (task/mission, system, etc.) to influence operator SA. For example, in our research, we have found that control limitations due to delays in data transmissions to and from the vehicle can hinder SA (Endsley & Riley, 2004). Fortunately, we have demonstrated that these negative effects can be mitigated by designing interfaces that expressly address problems associated with control latency (Chappell, 2007). Delays in the control loop can be mitigated with predictive displays that support operator projections on future vehicle locations and/or task activities or progress. SA-oriented design concepts for display characteristics such as presentation modes and task aids will be required to promote higher levels of SA. Specifically, beyond simply presenting basic perceptual data to support Level 1 SA, interfaces need to facilitate operators' ability to integrate this information to achieve a global understanding of the overall situation (Level 2 and 3 SA) (Riley et al., 2006; Riley & Strater, 2006a). Interfaces should directly present integrated data on, for example, system status and projected impact of status on robot functioning or task objectives. Finally, interfaces that offer different levels of interaction based on the operator's experience will increase the utility of unmanned systems for both novices and experts (Chappell & Dunlap 2006).

References

Burke, J. (2003). *Moonlight in Miami: A Field Study of Human-robot Interaction in the Context of an Urban Search and Rescue Disaster Response Training Exercise*. Unpublished master's thesis. University of South Florida, Tampa.

Burke, J., Murphy, R.R., Coover, M., & Riddle, D. (2004). Moonlight in Miami: An ethnographic study of human-robot interaction in USAR. *Human-Computer Interaction*, 19(1–2), 85–116 (Special Issue: Human-Robot Interaction).

Carter, R. & Woldstad, J. (1985). Repeated measurements of spatial ability with the Manikin Test. *Human Factors*, 27(2), 209–19.

Chappell, S.L. (2007). UAV control with pictures and pointing. *Proceedings of the American Institute of Aeronautics and Astronautics Conference and Exhibit*. Reston, VA.

Chappell, S.L. & Craddock, C. (2006). *Robotech Goal-directed Task Analysis*. Internal Report for U.S. Army Future Combat Systems Program. Report No. WMIS-GDTA-0035 (limited distribution).

Chappell, S.L. & Dunlap, K.L. (2006). Incorporating operator situation awareness into the design process: A key ingredient for future combat unmanned rotorcraft. *Proceedings of the American Helicopter Society 62nd Annual Forum*. Alexandria, VA.

Chen, J. & Barnes, M. (2008). Robotics operator performance in a military multi-tasking environment. *Proceedings of the 3rd ACM/IEEE International Conference on Human Robot Interaction*. The Netherlands.

Draper, J.V., Kaber, D.B., & Usher, J.M. (1998). Telepresence. *Human Factors*, 4(3), 354–75.

Drury, J.L., Richer, J., Rackliffe, N., & Goodrich, M.A. (2006). Comparing situation awareness for two unmanned aerial vehicle human interface approaches. *Proceedings of the SSRR 2006 Conference*. Gaithersburg, MD: National Institute of Science and Technology (NIST).

Drury, J.L., Scholtz, J., & Yanco, H.A. (2003). Awareness in human robot interaction. *Proceedings of the IEEE Conference on Systems, Man, and Cybernetics*. Washington, DC, October.

Drury, J.L. Yanco, H.A., Howell W., & Minten, B. (2006). Changing shape: Improving situation awareness for a polymorphic robot. *Proceedings of the 2006 ACM Conference on Human-Robot Interaction*, Salt Lake City, UT.

Endsley, M.R. (1995a). Measurement of situation awareness in dynamic systems. *Human Factors*, 37(1), 65–84.

Endsley, M.R. (1995b). Toward a theory of situation awareness in dynamic systems. *Human Factors*, 37(1), 32–64.

Endsley, M.R., Bolte, B. & Jones, D.G. (2003). *Designing for Situation Awareness: An Approach to User-centered Design*. New York: Taylor & Francis.

Endsley, M.R. & Jones, W.M. (1997). Situation awareness information warfare and information dominance (No. AL/CF-TR-1997–0156). Wright-Patterson AFB, OH: United States Air Force Armstrong Laboratory.

Endsley, M.R. & Riley, J.M. (2004). Supporting situation awareness in human operator–UAV/UGV collaboration. Presentation at the Army AMRDEC Science and Technology Workshop, Moffet Field, AFB, CA.

Riley, J.M. & Endsley, M.R. (2004). The hunt for situation awareness: Human-robot interaction in search and rescue. *Proceedings of the Human Factors and Ergonomics Society 48th Annual Meeting*. Santa Monica, CA.

Riley, J.M., Kaber, D.B., & Draper, J.V. (2004). Situation awareness and attention allocation measures for quantifying telepresence experiences in teleoperation. *Human Factors and Ergonomics in Manufacturing*, 14(1), 51–67.

Riley, J.M., Murphy, R.R., & Endsley, M.R. (2006). Situation awareness in control of unmanned ground vehicles. In N.J. Cooke, H. Pringle, H. Pedersen, & O. Connor (eds), *Advances in Human Performance and Cognitive Engineering Research, Vol. 7: Human Factors of Remotely Operated Vehicles* (pp. 359–71). San Diego, CA: Elsevier.

Riley, J.M. & Strater, L.D. (2006a). Effects of robot control mode on situation awareness and performance in a navigation task. *Proceedings of the Human Factors and Ergonomics Society 50th Annual Meeting*. Santa Monica, CA.

Riley, J.M. & Strater, L.D. (2006b). Assessing effects of robot control mode on performance and situation awareness in a maze navigation task. Lecture presentation at the 3rd Annual Workshop on Human Factors of Unmanned Aerial Vehicles. Mesa, AZ, 24–25 May.

Riley, J.M., Strater, L.D., Davis, F., Strater, S., and Faulkner, L. (2009). Situation awareness and team communication in robot control. *Proceedings of the Human Factors and Ergonomics Society 53rd Annual Meeting*. San Antonio, TX.

Riley, J.M., Strater, L.D., Sethumadhavan, A., Davis, F., Tharanathan, A., & Kokini, C. (2008). Performance and situation awareness effects in collaborative robot control with automation. *Proceedings of the Human Factors and Ergonomics Society 52nd Annual Meeting*. Santa Monica, CA.

Strater, L.D., Connors, E.S., & Davis, F.C. (2008). Collaborative control of heterogeneous unmanned vehicles. Lecture presentation at the Texas Regional Human Factors and Ergonomics Society Conference. Austin, TX, 18 April.

Chapter 11

Imperfect Reliability in Unmanned Air Vehicle Supervision and Control

Christopher D. Wickens, Brian Levinthal and Stephen Rice[1]
Alion Science and Technology

The goal of this chapter is to present and synthesize the results of two experiments that have examined issues of unreliable automation in the control/supervision of unmanned air vehicles (UAVs). These are results that can be seen to generalize to a broader class of mobile robots. The focus on automation is a logical outgrowth from the continuing pressures to reduce pilot workload (via automation) in order to increase the ratio of UAVs to pilots. This ratio currently stands at 1:2 for the U.S. Army's Hunter/Shadow class. To the extent that a single pilot could control/ supervise a single UAV (1:1), or even extend to a 2:1 ratio, greater surveillance coverage could be obtained without increasing personnel requirements. Because automation can replace many aspects of routine piloting, sufficient automation should enable this gain to take place. Indeed, Dixon et al. (2005) have shown how, in some circumstances, automated UAV systems can enable a 2:1 ratio to be as effective as a 1:1 ratio in the absence of automation.

Unfortunately, there are constraints in substituting automation for human functioning, to the extent that the automation is *imperfectly reliable* and hence in need of human supervision to catch the instances when some aspect fails. To the extent that the human must monitor automation for these failure events, many of the benefits of offloading human cognition and control are offset (Parasuraman, 1987; Wickens & Dixon, 2007). Asserting that such imperfection exists is quite plausible on two grounds. First, interviews by Parasuraman of Hunter/Shadow pilots revealed the extent to which they complained that "things go wrong," and, in fact, intuitively estimated that this failure rate was "around 10 percent." Second, there are certain types of automation, which will be the focus of the current chapter, in which such a failure rate is quite plausible. In particular, these

---

1   This research was sponsored by subcontract #ARMY MAD 6021.000-01 from Microanalysis and Design as part of the Army Human Engineering Laboratory Robotics CTA, contracted to General Dynamics. David Dahn was the scientific/technical monitor. Any opinions, findings, conclusions, or recommendations expressed in this chapter are those of the authors and do not necessarily reflect the views of the U.S. Army. The authors also wish to acknowledge the support of Ron Carbonari and Jonathan Sivier (in developing the UAV simulation).

are what Parasuraman et al. (2000) label as "stage 2 automation," or automation information integration and inference. This is where automation is tasked to diagnose or predict a particular state of the world, which may be based upon fuzzy data. Such instances might involve diagnosing a target as friend or foe, diagnosing the cause of a particular engine failure, or making a prediction of enemy activity. In all such cases, particularly those involving prediction, the data upon which the diagnosis is based are fuzzy and uncertain. Hence a substantial error rate in diagnostic accuracy is anticipated.

One critical feature of any such imperfect diagnostic automation system, whether in UAVs or any other system, is the *threshold setting*, or decision criterion, for the diagnostic judgment. A low setting is one that allows few automation misses, but many false alarms (FAs)—when the alert sounds, or a target is identified but there is no dangerous event, or a target is not present, respectively. In contrast, a high threshold setting will reduce the FA rate, but increases the miss rate or, in the case of predictive alarms, will increase the frequency of late alarms (alarms offered too late for users to take an appropriate response in light of the forecast event).

A key design question addresses the relative costs of high versus low threshold settings. Were only a single system responsible for the diagnostic judgment, then the optimal setting of the threshold could be determined by establishing the costs and benefits of the two types of automation failures (misses and false alarms), along with the frequency of the forecast events as dictated by signal detection theory (Green & Swets, 1966; Wickens & Hollands, 2000). However, when such diagnostic automation is coupled, with the monitoring performance of the human operator, monitoring the "raw data" in parallel with the output of automation (see Figure 11.1), and when the human also depends upon the automation to do a substantial amount of this monitoring, because his/her attention must also be allocated to concurrent tasks, the issue of threshold effects is more complex., as we explain below.

Performance of both agents can be represented by the signal detection theory outcome matrix, as shown at the bottom of Figure 11.1. The human may be more or less dependent on automation in making a total system diagnosis. This dependence is governed in part by the reliability of automation (R), shown at the bottom. Finally, human dependence is influenced by, and influences performance on, concurrent tasks to which the human must allocate some attention.

We consider four ways in which human-system performance in the parallel monitoring environment depicted in Figure 11.1 can be influenced, and how these can be associated with two important categories of human dependence on diagnostic automation proposed by Meyer (2001, 2004; see also Chen & Barnes, 2008, Dixon & Wickens, 2006; Maltz & Shinar, 2003). *Reliance* is the state of human cognition when the automation is silent. It is characterized by trusting the automation to alert when there is a signal or dangerous event; hence when there is *no* signal, it is characterized by reallocating resources from monitoring the raw data, to concurrent tasks. *Compliance* is the state when automation sounds the alarm, and leads to a rapid switch of attention from those concurrent tasks to the

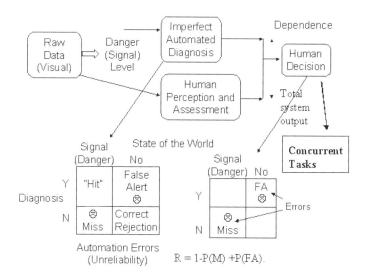

Figure 11.1   Parallel monitoring of human and automation

automated domain. (e.g., ascertaining the source of alarm, and/or preparing actions to take as a consequence). As we expand this general description of cognitive responses to specific behavioral measures, we identify the following four classes of behavior or hypothesized "threshold effects".

1. *Compliance effects—cry wolf.* Increasing false alarm rate as the threshold is lowered will reduce compliance, as manifest in the so-called "cry wolf" effect (Breznitz, 1983). Events will be missed, or responded to late.
2. *Reliance effects—automation misses.* Increasing automation miss rate as the threshold is raised will decrease reliance, causing the operator to spend more time monitoring the raw data in the monitoring domain. As a result, on those occasions when the automation *does* miss a target, the operator will actually be more likely to detect it. Conversely, with the lower threshold in an FA-prone system, the automation miss rate will be lower and reliance will be greater. During those (now) very rare occasions when the automation does miss a signal, such a miss will be less likely to be detected by the human. Hence there will be a greater likelihood of a total human-automation system miss. We can refer to this phenomenon as low threshold-induced complacency.
3. *Reliance effects—dual task resource allocation.* With a miss-prone threshold, the reduced reliance, forcing the operator to monitor the raw data more, will force a reallocation of attention away from concurrent tasks. Hence performance will be degraded on these tasks during the typically long intervals of time when the alarm is silent.

4. *False alarm disruption.* This fourth effect of alerting system reliability falls in neither the reliance nor compliance category, but characterizes the manner in which false alarms tend to disrupt concurrent task performance in a qualitatively different way than does miss-prone automation (as the latter is reflected in a reliance decline described above). Here, false alarms appear to have two effects (Dixon & Wickens, 2006; Dixon et al., 2007). First, their repeated occurrence often requires cross-checking the raw data to establish that they *are* indeed false (so long as they are not ignored altogether as mentioned in (1). Second, because they are very salient sorts of automation errors, they tend to lead to a general distrust of the alerting system (Madhavan et al., 2006), hence leading to the loss of reliance (and its negative effect on concurrent tasks) as well.

We integrated the results of 13 studies carried out at the University of Illinois in which the threshold of imperfect diagnostic automation was varied in a multitask setting (reported, collectively, in Wickens & Colcombe, 2007; Dixon & Wickens, 2006; Dixon et al., 2007; Wickens et al., 2005; Levinthal & Wickens, 2005; Wickens et al., 2005). This integration revealed the general robustness of the compliance cry wolf effect (10 out of 13 studies show an effect as the threshold is lowered, and none contradict this), and the low threshold complacency effect (6 out of 7, with no contradiction). Both of these influences penalize event detection performance when the threshold is low. Examination of dual task effects, however, reveals an ambiguous pattern of results, in terms of which threshold setting (low or high) is more detrimental. Either a miss-prone or FA-prone threshold degrades concurrent tasks compared to more reliable automation, but the nature of the two dual-task effects, (3) and (4) described previously, appear to somewhat offset each other. It is important in this regard to note that one other such study, performed in a UAV context (Chen & Terrence, 2008), also noted an ambivalent pattern of dual task interference within the experiment, with FA-prone automation disrupting concurrent (UAV) performance more for subjects with low attention control, and less for subjects with high attention control. This dimension of individual differences also modulated the extent of compliance with FA-prone automation (low compliance for high-attention control subjects).

The goal of the current chapter is to describe two of the Illinois studies in detail, integrating their collective message on alerting thresholds. Both have the use of a UAV simulation in common, and both have the threshold varied within a dual task setting. In Experiment 1, the simulation is of modest fidelity, and we use recorded eye movement to infer the allocation of attention between the alert domain and other concurrent tasks when controlling a single UAV (1:1). In Experiment 2, a high-fidelity UAV simulation is employed, and we explicitly examine the workload issue by comparing performance of a single operator with two and four UAVs.

Experiment 1: Method

Forty undergraduate and graduate students at the University of Illinois were placed in groups of eight pilots, and received $8 per hour, plus bonuses of $20, $10, and $5 for 1st, 2nd, and 3rd place finishes, respectively. Figure 11.2 presents a sample display for a UAV simulation, with verbal explanations for each display window and task.

As seen in Figure 11.2, the experimental environment was subdivided into four separate windows. The top left window contained a 3D egocentric image view of the terrain directly below the UAV. During regular tracking periods, the operator could only view straight down to the ground at a 20-degree angle. During a loiter pattern, in which the simulated UAV would automatically fly a racetrack pattern around a suspected or located target, the operator was able to zoom in and pan for a closer inspection of the ground. During this inspection, the operator was tasked with detecting camouflaged targets, as discussed below.

The bottom left window contained a 2D top-down map of the 20×20-mile simulation world to support navigation to command targets (CTs). The bottom center window contained a Message Box, with "fly to" coordinates and mission-related command target questions. These flight instructions were present for 15 seconds, and could be refreshed for another 15 seconds by pressing the Repeat key.

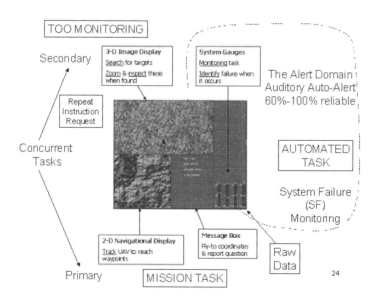

Figure 11.2   The pilot's UAV interface in Experiment 1

The bottom right window contained four system failure (SF) gauges acting as the automation-support alert domain. Each gauge represented a different onboard system. White bars oscillated up and down continuously, each driven by sine waves ranging in bandwidth from 0.01 Hz to 0.025 Hz. A system failure occurred when one of the white bars moved into a red zone.

Eye-tracking data were collected from each pilot during the mission, to examine scanning between these four areas of interest on the display and therefore infer how attentional strategies supported reliance and compliance as the threshold of the automated task was varied.

Participants used a Logitech Digital 3D joystick to manipulate the aircraft/ camera and an X-Key 20-button keypad with which to indicate responses. The joystick had controls for turning the UAV, manipulating the camera on the x and y axes, zooming, detecting targets, loitering around targets (to the left or right), and detecting system failures. The keypad was used for indicating which system failure occurred, the ownership coordinates for that system failure, and for typing in mission coordinates during the automation condition.

Each pilot flew one UAV through ten different mission legs, while completing three goal-oriented tasks commonly associated with UAV flight control: mission completion, ground target search, and systems monitoring. At the beginning of each mission leg, pilots obtained their flight instructions for that leg via the Message Box. Once pilots arrived at the CT location, they loitered around the target, manipulated a camera for closer target inspection, and reported back relevant information to mission command (e.g., *What weapons are located on the south side of the building*?). Around each CT were one to three tanks and/ or helicopters, located within 10–30 feet of the building. These weapons were always located on the north, south, east, or west sides. Location was to be specified in cardinal directions, thereby forcing a relatively high level of spatial-cognitive activity (e.g., Gugerty & Brooks, 2001).

Along each mission leg, pilots were also responsible for detecting and reporting randomly located low-salience targets of opportunity (TOOs), a task similar to the CT report, except that the TOOs were much smaller (1–2 degrees of visual angle) and were camouflaged. TOOs could occur during simple tracking (low workload) or during a pilot response to a system failure (high workload). These two types of TOOs occurred, respectively, with a ratio of roughly 4:1.

If the participant detected a CT or TOO, he or she was required to indicate detection by pulling the joystick trigger. The duration of time between when the target entered the 3D display and when the pilot pressed the detection button was recorded as TOO target detection time. The participant then pressed the loiter button (loiter would be selected either left or right) on the joystick. This put the UAV into an automated oval pattern around the target. After making the report, the participant could then depress the loiter button again, which would unloiter the UAV and unzoom the camera, returning the egocentric view to 6,000 feet altitude.

Concurrently, pilots were also required to monitor the system gauges for possible system failures (SFs). This was the automated task. When a gauge went "out of bounds" (i.e., the needle moved from the green zone to the red zone), they had to press a button to detect the SF, indicate which SF gauge had failed, and then report the current location of the UAV during the SF. SFs were designed to fail either during simple tracking (i.e., easy concurrent task) or during TOO and CT inspection (i.e., difficult concurrent task). The SFs lasted only 30 seconds, after which the screen flashed bright red and a harsh auditory alarm announced that the pilot had failed to detect the SF (the UAV was considered to have "crashed" if pilots did not detect the failure quickly enough). There were a total of ten SFs, with never more than two SFs occurring during any mission leg.

Automation aids, in the form of auditory auto-alerts during SFs, were provided for three out of the four conditions in this between-subjects design. The A100 condition (A = automation; 100 percent reliable) never failed. The A60f condition (f = false alarm; 60 percent reliable) was created by imposing three automation false alarms and one automation miss (four automation failures) out of the ten system failures that actually occurred; hence the reliability is considered to be 0.60. The A60m condition (m = miss; 60 percent reliable) resulted in more misses than false alarms (3:1 ratio). The final condition was a baseline condition (Man), with no automation aid to assist pilot performance.

Pilots were not aware of the precise level of reliability provided by each automation aid; however, depending on the experimental condition, they were told that the automation was either "fairly reliable" or "not very reliable," as well as, in the latter case, the bias setting (i.e., more false alarms or more misses).

Results

We describe the results of experiment 1 in terms of the four classes of threshold effects, as outlined in the introduction to the chapter:

1. *Compliance: system failure detection.* Analysis of the system failure detection data (the automated task) revealed no effects of automation reliability at lower workload, but did show response times (RT) that were faster when any type of automation was present, $F(3, 26) = 5.40$, $p < .01$. Importantly, highly significant effects of reliability emerged at high TOO surveillance workload, as revealed by the significant load X condition interaction in both accuracy, $F(3, 26) = 7.91$, $p < .01$, and RT, $F(3, 26) = 9.65$, $p < .01$. Our particular interest was in the increased cost of FA-prone relative to miss-prone automation where, in the high workload condition, both accuracy, $t = 1.96$, $p < .04$ (a drop from 58 percent to 42 percent detection rate) and RT, $t = 3.53$, $p < .01$ (an increase from 15 to 23 seconds detection time) demonstrated a greater cost in the FA-prone condition. Thus the compliance effects of the cry wolf phenomenon were robust.

2. *Reliance: rare automation misses and low threshold complacency.* In those trials on which the automation missed a failure, and the alert was therefore silent, RT was long in general (around 23 seconds), but was significantly longer in the FA-prone condition, in which this event was quite rare (28 seconds) compared to the miss-prone condition (18 seconds), hence demonstrating the complacency effect. The manifestation of these effects in visual attention will be discussed below.
3. *Reliance: concurrent task performance—TOO monitoring.* We focused our analysis here only on TOO events that occurred during the long periods of low workload in which a system failure had not occurred (i.e. during the period of reliance), and observed the trend of both accuracy and speed to be degraded with less reliable automation, particularly in the miss-prone condition (although this trend was not significant for RT, a 2.5-second, 33 percent time increase, and only marginally so for detection rate, $F(3, 26) = 2.31$, $p = .10$, a decrease in accuracy from 75 percent to 61 percent).
4. *FA disruptions of concurrent tasks.* We were unable to directly measure FA-disruption except to note that the effects in the predicted reliance direction of greater interference with miss-prone automation (effect 3), were muted (see previous paragraph), as if these were offset by the general distrust and disruption of the FA-prone system.

Visual scanning analysis: compliance   We inferred that compliance would be reflected in the speed with which visual attention moved to the SF gauges, from wherever it was located at the time that the automation alert sounded. These measurements were computed by hand from a time-file of scanning across the four display windows depicted in Figure 11.2. The data for these "scan RTs" are shown in Table 11.1 when the alerts occurred during the high workload period while the pilot was engaged in image scanning of the 3D ground display.

A one-way analysis of variance (ANOVA) on these data revealed a highly significant effect of condition, $F(3,29) = 5.806$, $p < .01$, revealing that looks were as rapid in the miss-prone condition as in the perfect automation condition (pilots perfectly complying with the alerts), but were as slow in the false-alarm condition as they were the baseline (unaided) glance response times. Thus pilots' low compliance when the workload is high reflecting cry wolf is therefore directly reflected in the slower attention switch.

Table 11.1   Scan RTs in seconds (baseline scans represent the delay between the SF and the first look at the display; all others represent the delay between the auditory alert and the first look)

Baseline	A100	A60F	A60M
19.0	4.50	16.0	4.00
(2.00)	*(1.13)*	*(2.78)*	*(2.48)*

Visual scanning strategies: reliance   We measured the proportion of time that the eye was focused on the different areas of interest, a measure of the percentage dwell time, or PDT, data shown in Table 11.2.

Table 11.2   **Percentage dwell time that visual fixation is spent for the four experimental conditions within each area of interest (AOI): 3D image display where the TOOs were located, the 2D navigation display, the system failure (SF) monitoring gauges, and Message Box (MB)**

AOI	Baseline	A100	A60F	A60M
3D (TOO)	50.0 (3.21)	58.7 (2.83)	56.4 (3.59)	45.5 (5.29)
2D	36.7 (2.74)	39.2 (2.45)	32.2 (5.19)	35.1 (4.38)
SF	13.0 (1.35)	5.7 (1.26)	11.3 (1.62)	18.6 (3.18)
MB	4.1 (0.75)	6.6 (1.34)	9.0 (1.64)	11.9 (2.64)

A two-way (AOI X condition) ANOVA carried out on the PDT data revealed a significant effect of areas of interest, $F(3, 78) = 155.75$, $p < .001$. Not surprisingly, the 3D surveillance image window, hosting the most demanding TOO detection task (the concurrent task), commanded the most visual attention. The 2D navigational display, hosting the most important task (command target location information) required around a third of the pilot's attention. The two remaining AOIs demanded the least amount of attention. Importantly, the significant AOI X condition interaction, $F(9,78) = 2.41$, $p = .05$, reflected the attentional characteristics underlying automation reliance. Here we see that visual attention to the TOO window benefited (relative to the baseline value of 50 percent) from having auditory alerts, whether these were fully reliable (59 percent) [100A, $t(14) = 2.05$, $p < .03$], or imperfect, but having few misses and more false alarms [A60F condition, with 56 percent detection, $t(13) = 1.34$, $p = .10$]. However miss-prone automation drew as much, if not more, visual attention away from the 3D TOO image window (45.5 percent) as this window received in the baseline condition (50 percent). While this decrease in looking at the image window relative to the baseline was not significant, the difference between miss-prone and false-alarm prone automation was marginally significant [$t(13) = 1.7$, $p = .06$], reflecting the shift in visual attention away from concurrent tasks. Such a shift would be fostered by a designer's decision to increase the threshold in a way that produces more misses. That is, visual attention to the concurrent task during periods of reliance underlies the effects on TOO detection speed and accuracy, described above. Scanning of the 2D navigational display did not differ significantly between conditions, indicating how pilots treated this display, which hosted primary task information of utmost priority. However, scanning to the SF gauges themselves reflected an expected pattern, opposite to that of the 3D image window. While perfect automation (A100) greatly reduced the visual attention required (5.7 percent) relative to baseline (13 percent) [t(13) = 3.97, p < .01], the miss-prone automation condition required far more visual attention to this SF display (18.6 percent). This was expected given that pilots, with their reliance reduced, were paying more attention to the "raw data" of the system gauges compared to the false-alarm prone condition (11.3 percent) [t(13) = 2.05, p = .03], whose attention allocation to the gauges did not differ from baseline. An additional feature is that pilots paid even more attention (18 percent) to the SF display in the miss-prone condition, than in the non-automated baseline (13 percent) [t = 1.71, p < .05], a cost that, as we saw above, bought them nothing in terms of better SF detection performance. There was no difference in scanning to the Message Box across conditions.

One might not have expected the false alarm rate to influence automation reliance, and indeed it did not appear to influence the measures of the residual attention to the 3D image window where the TOOs appeared. However, somewhat surprisingly, the higher FA rate *did* compel more attention to the SF display than the fully reliable automation condition, and induced no less attention there than the baseline condition. Thus no attention was "saved" by FA-prone automation relative to the baseline, in spite of the fact that nearly all failures were alerted. Thus, the general distrust induced by false alarms may have led to pilot suspicion that such a system requires further monitoring, a visual attention effect reflecting influence. This is threshold effect (4) presented previously.

Experiment 1: Discussion

The results of Experiment 1 clearly demonstrated the first three of the effects of unreliable stage 2 automation detailed previously, and indirectly suggested the influence of the fourth. Also these results showed how the performance effects were manifest in visual attention allocation (reliance) and attention switching (compliance) strategies, as inferred from visual scanning measures.

We also note the influence of workload. In particular, compliance effects on system failure detection were most manifest at high workload, reflecting a general conclusion reported elsewhere (Wickens & Dixon, 2007), that both the benefits of highly reliable automation and the costs of lower reliability are amplified when resources are scarce.

In Experiment 2, we examined many of these same issues (workload, automation reliability, and threshold setting) with a high-fidelity UAV simulation; and here we vary workload in a quantifiable fashion, as dictated by the number of assets or UAVs under supervision (see Levinthal & Wickens, 2005 for full details).

Experiment 2: Method

Forty-two students aged 19–29 from the University of Illinois at Urbana-Champaign participated in a 2-hour study, and were compensated approximately $10 per hour for their time.

A Systems Integration Lab (SIL) UAV simulator (from Micro-Analysis and Design, Colorado) was used for this experiment. Figure 11.3 shows the arrangement of two displays for this experiment, which provided raw data and controls relevant to the pilots' two tasks. For the UAV management task (here the "concurrent task"), pilots were presented with an interface designed to replicate those currently in use by the U.S. Army. Pilots interacted with the controls via touch-screen monitor. The displays were arranged such that monitoring each task required an overt shift of attention (via head movement), and critical events such as the appearance of a tank or initiation of "idle mode" were only visible when the corresponding display was viewed directly. The UAV display provided a visual icon of a single UAV (alternative UAVs had to be selected manually), and indicated the aircraft's current progress as well as the subsequent path that pilots had selected.

Pilots navigated each UAV through a series of branching waypoints. UAV flight between waypoints was automated, but pilots were required to perform a decision rule upon reaching each waypoint, in order to select the next correct branch. To choose the correct waypoint, pilots subtracted the last two digits of the x and y coordinates of the UAV, and selected the northernmost waypoint when the difference was greater than 50 (otherwise, they chose the southernmost waypoint). Thus, the demands associated with UAV flight were predominantly cognitive, and not physical. When UAVs reached a waypoint, they circled indefinitely in an "idle" mode until the pilot selected the next waypoint. The accumulation of idle time served as a dependent measure of UAV task performance. In two experimental sessions, pilots were responsible for either two UAVs (low workload) or four UAVs (high workload). UAVs reached a waypoint (i.e. pilots were required to perform the decision rule) once every 15 seconds during low workload sessions,

Figure 11.3   The tank-detection (automated task) display (left) and touch-screen UAV management interface (concurrent task, right)

and every 7.5 seconds during high workload sessions. Thus the UAV task involved keeping track of how close each UAV was to its next waypoint, as well as cognitive computation and choice based upon the location of the waypoint once reached.

While performing the UAV task, pilots monitored the left display in Figure 11.3, depicting raw terrain data, with the goal of finding enemy tanks as quickly and accurately as possible (pilots responded by saying "TANK" and pointing to the target). This surveillance task was the automation-supported task. The terrain scrolled continuously across the screen at a rate of one complete refresh every 30 seconds, thus tanks, when present, were on screen for a full 30 seconds. There were 20 tanks, which appeared pseudo-randomly within a range of 30 seconds and 5 minutes.

In some cases, pilots were provided with an automated target recognition aid to assist in the tank-detection task. In a between-subjects design, the aid functioned at one of four levels: a non-automated baseline (BL), a 90 percent reliable aid that committed equal false alarms and misses (A90), and a 60 percent reliable aid with a 3:1 likelihood of committing either false alarms over misses (false-alarm prone or FAP) or misses over false alarms (miss-prone or MP), as in Experiment 1. Correct alerts sounded no earlier than 3 seconds after the tank appeared, and no later than 7 seconds.

Four dependent variables were recorded to measure compliance and reliance in the context of multiple UAV control. UAV task (concurrent task) performance was measured via accumulated idle time. Compliance was measured by the pilots' response times following an automated alert. Reliance was measured by response times when the automated aid committed a rare miss (low reliance, fast response), and was also measured by the *proportion* of responses made after, rather than prior to, the sounding of an alert. The latter measure provides an indirect measure of reliance as pilots, who, mistrusting the alert system to inform them and thereby frequently viewing the terrain monitor, would be more likely to detect a tank before the alert sounded.

Results

There was no effect of automation condition on the accuracy of tank detection, which was essentially perfect across all conditions. Instead, variability in tank detection performance was demonstrated in response times.

> *(1) Compliance: post-alert response time.* We observed a significant main effect of automation condition on post-alert response time (our primary measure of compliance), $F(2,56) = 27.7$, $p < .001$, indicating that the FAP condition yielded a near doubling of response time (6.0 vs 3.3 seconds) compared to either the 90 percent reliable condition or the miss-prone condition (which did not differ from each other); thus reflecting the cry wolf effect. There was neither an effect of workload, nor a workload × automation level interaction.

(2) Reliance measures: rare automation miss. There was no significant main effect of automation condition on response times when the alert was silent, though there was a non-significant (and anticipated) trend indicating that pilots in the MP condition were faster (8.8 seconds) than those in the FAP condition (11.4 seconds). For the second measure of reliance, the proportion of responses made after the sounding of an alert, there was a significant main effect of workload. As workload increased, this measure of reliance increased as well, from 0.53 to 0.65; $F(1,56) = 14.3$, $p < .001$. There was also a significant main effect of automation condition, $F(1,56) = 8.63$, $p < .002$, indicating that pilots in the FA-prone and 90 percent reliable conditions responded after an alert (i.e. were more reliant) more frequently (0.70) than pilots in the miss-prone condition (0.48). The latter group, with less reliance and hence more raw-data monitoring, made proportionately more responses *prior to* the alert.

(3) and (4) Reliance-based and disruption workload effects on concurrent task. As workload increased from low to high, idle time increased substantially and significantly, $F(1,72) = 248.3$, $p < .001$. Doubling objective task load from two to four vehicles to be supervised resulted in a tripled cost to task efficiency: accumulated idle time for the entire 2-hour simulation increased from 10 to 30 minutes. The differences between automation conditions did not effect UAV performance ($p > .10$), nor did condition interact with workload in a way that might suggest that higher levels of automation could mitigate the effects of higher workload. It is thus apparent in this experiment that pilots truly "protected" the primary UAV task from influences caused by unreliability of the automated task.

Experiment 2: Discussion

As in Experiment 1, the effects of automation reliability threshold setting on both reliance and compliance were again strong and pronounced, in this higher-fidelity simulation. The compliance effects—a doubling of response time, adding a full 3 seconds to this response—are large enough to be of operational significance. The reliance effect was most prominently expressed in the increased proportion of early (anticipatory) detection of tanks before the alert sounded, suggesting that more visual attention was allocated to the raw data when there was miss-prone automation (although such attention was not directly measured here, as it had been in Experiment 1).

However here, in contrast to Experiment 1, these automation reliability effects were manifest at low as well as high workload. Furthermore, also in contrast, automation reliability effects did not influence concurrent task performance; even high reliability (90 percent) automation failed to mitigate the substantial performance decrements associated with the higher workload of doubling the UAVs under supervision (a tripling of non-productive idle time).

Two explanations can be offered to account for these null results. First, we have observed elsewhere that reliability effects are generally more profoundly manifest in the automation task itself (here tank detection) than in concurrent tasks (here UAV supervision; Wickens & Dixon, 2007), although it is surprising here that they were not manifest at all. The event rate or bandwidth of the UAV task was between 10 (low workload) and 20 (high workload) times that of the automated task. Given the strong effect of visual bandwidth on attention allocation (Senders, 1964; Wickens et al, 2008), it is not surprising that subjects would have treated the UAV task as implicitly of much higher priority, despite instructions to give each task equal attention. Second, the idle time period in the current simulation was one in which UAVs were circling the same terrain rather than covering new ground. Hence this was not a period during which tank search (supported by automation) would have yielded many new targets, and so the automation benefits on reducing this idle time would be expected to be muted.

General Discussion and Conclusions

Robotics systems often involve two main automation components. Always they invoke automation of control, whether in travel or manipulation. But they often involve automation of surveillance, and this is often of fuzzy or non-salient data sources which will, therefore, be imperfect. In the two studies reported above, we have examined this second major human-automation interaction component that is well represented, in different forms, within the set of robotic systems that are the focus of this book. The influence on combined human-system performance as the classification threshold setting is varied is both reliable and non-trivial, with effect magnitudes in the order of several seconds, or several percentage points of accuracy. Sometimes (experiment 1), threshold changes had a substantial effect on the attention available for, and resulting performance of, concurrent tasks.

The four classes of diagnostic reliability effects identified above are by now well modeled by the joint attention-cognitive factors of reliance and compliance, along with the additional influence related to the particularly disruptive influence of false alerts (Dixon et al, 2007). As the majority of negative influences of threshold setting appear to be a consequence of FA-prone settings, and as designers generally tend to set the threshold to a low value (producing FA-prone automation), these data raise the possibility that this value may sometimes be too low. Hence it may be better to allow a slightly higher miss rate (Wickens & Colcombe, 2007), even if a truly "miss-prone" system will probably not be optimal. Such threshold adjustment, however, should only take place in situations where the human operator clearly has perceptual access to the same raw data processed by the automation. Furthermore, under these circumstances, even automation reliability as low as 80 percent can generally support an improvement of performance above a non-automated baseline (Wickens & Dixon, 2007).

Finally, we note the influence of workload. Certainly the benefits of high-reliability automation are more manifest as workload increases (Wickens & Dixon, 2007), just as are the influences of threshold setting, as shown by Experiment 1. Yet neither perfect nor imperfect automation can completely offset the costs of high workload (Experiment 2; Dixon et al., 2005); and the research discussed here and elsewhere indicates that automation, while sometimes useful, cannot be seen as the panacea for addressing the workload increases that come, for example, with responsibility for progressively more robots.

Key Points

- In systems in which human and automation monitor raw data in parallel, shifting the automation alerting threshold by which dangerous events are discriminated from safe conditions exerts four qualitatively different effects:

  1. A low threshold, yielding false-alarm prone automation, yields a reduced or delayed compliance to all alerts, both true and false—the so-called "cry wolf" effect.
  2. A higher threshold, designed to reduce false-alert rate by now producing more miss-prone automation, causes the human operator to rely less on the automation and shift attention to more closely monitor the raw data, hence better detecting events when the automation fails to do so.
  3. The higher threshold and shifting attention in 2, above, will also cause a decrease in performance of concurrent tasks, since these will now receive less attention.
  4. The performance cost to concurrent tasks shown in 3, will sometimes be offset in a performance benefit of miss-prone (over FA-prone) automation because it will reduce the disrupting and annoying characteristics of false alarms.

- Effects 3 and 4 often offset each other regarding which threshold setting is worse for concurrent tasks, and indeed concurrent tasks themselves suffer less from the imperfection of automation misses and false alarms than do the automated surveillance tasks to which those automation errors directly apply.
- As long as the overall automation error rate remains below about 20 percent (e.g., 80 percent reliability), such imperfect automation can still provide a benefit to human performance.
- In a mid-fidelity UAV simulation, we demonstrate the above effects of automation threshold setting on both the automated and concurrent task.

- In a high-fidelity UAV simulation, effects 1 and 2 are shown, but concurrent task performance reflects none of the properties of imperfect diagnostic automation because of the high priority that participants assign to the non-automated task.

References

Breznitz, S. (1983). *Cry Wolf: The Psychology of False Alarms*. Hillsdale, NJ: Lawrence Erlbaum Associates.

Chen, J.Y.C. & Barnes, M.J. (2008). Robotics operator performance in a military multi-tasking environment. *Proceedings of the 3rd ACM/IEEE International Conference on Human-Robot Interaction* (pp. 279–86). 12–15 March, Amsterdam, the Netherlands.

Chen, J.Y.C. & Terrence, P.I. (2008). Effects of Imperfect Automation on Concurrent Performance of Gunner's and Robotics Operator's Tasks in a Simulated Mounted Environment. ARL Technical Report. Army Proving Ground, MD: U.S. Army Research Laboratory.

Dixon, S., Wickens, C.D., and McCarley, J.M. (2007). On the independence of reliance and compliance: are false alarms worse than misses? *Human Factors*, 49, 564–72.

Dixon, S.R. & Wickens, C.D. (2006). Automation reliability in unmanned aerial vehicle flight control: a reliance-compliance model of automation dependence in high workload. *Human Factors*, 48, 474–86.

Dixon, S.R., Wickens, C.D., & Chang, D. (2005). Mission control of multiple UAVs: A quantitative workload analysis. *Human Factors*, 47, 479–87.

Green, D. & Swets, J. (1966). *Signal Detection Theory and Psychophysics*. New York: Wiley (reprinted 1988, Los Altos, CA: Peninsula).

Levinthal, B. & Wickens, C.D. (2005). *Supervising Two Versus Four UAVs with Imperfect Automation: A Simulation Experiment* (AHFD-05-24/MAAD-05-7). Savoy, IL: University of Illinois, Aviation Human Factors Division.

Madhavan, P., Wiegmann, D., and Lacson, F. (2006). Automation errors on tasks easily performed by humans undermine trust. *Human Factors*, 48, 241–6.

Maltz, M. & Shinar, D. (2003). New alternative methods of analyzing human behavior in cued target acquisition. *Human Factors*, 45(2), 281–95.

Meyer, J. (2001). Effects of warning validity and proximity on responses to warnings. *Human Factors*, 43, 563–72.

Meyer, J. (2004). Conceptual issues in the study of dynamic hazard warnings. *Human Factors*, 46(2), 196–204.

Parasuraman, R. (1987). Human computer monitoring. *Human Factors*, 29, 695–706.

Parasuraman, R., Sheridan, T.B., & Wickens, C.D. (2000). A model for types and levels of human interaction with automation. *IEEE Transactions on Systems, Man, and Cybernetics. Part A: Systems and Humans*, 30(3), 286–97.

Senders, J. (1964). The human operator as a montior and controller of multidegree of freedom systems. *IEEE Transactions on Human Factors in Electronics, HFE-5*, 2–6.

Wickens, C.D. & Colcombe, A. (2007). Dual-task performance consequences of imperfect alerting automation associated with a cockpit display of traffic information. *Human Factors*, 49(5), 839–50.

Wickens, C.D. & Dixon, S.R. (2007). The benefits of imperfect diagnostic automation: A synthesis of the literature. *Theoretical Issues in Ergonomics Science*, 8(3) 201–12.

Wickens, C.D., Dixon, S., Goh, J., & Hammer, B. (2005). *Pilot Dependence on Imperfect Diagnostic Automation in Simulated UAV flights: An Attentional Visual Scanning Analysis* (AHFD-05-2/MAAD-05-2). Savoy, IL: University of Illinois, Aviation Human Factors Division.

Wickens, C.D., Dixon, S., & Johnson, N.R. (2005). *UAV Automation: Influence of Task Priorities and Automation Imperfection in a Difficult Surveillance Task* (AFHD-05-20/MAAD-05-6). Savoy, IL: University of Illinois, Aviation Human Factors Division.

Wickens, C.D. & Hollands, J. (2000). *Engineering Psychology and Human Performance* (3rd edn). Upper Saddle River, NJ: Prentice Hall.

Wickens, C.D., McCarley, J.S., Alexander, A.L., Thomas, L.C., Ambinder, M., & Zheng, S. (2008). Attention-situation Awareness (A-SA) Model of Pilot Error. In D. Foyle & B.Hooey (eds). *Human Performance Models in Aviation*. Boco Raton, FL: Taylor & Francis.

Chapter 12

Remotely Operated Vehicles (ROVs) from the Bottom-Up Operational Perspective

Tal Oron-Gilad and Yaniv Minkov[1]

Ben-Gurion University, Department of Industrial Engineering and Management

Introduction

The advantages to military commanders offered by using remotely operated vehicles (ROVs) in general, and unmanned aerial vehicles (UAVs)[2] in particular, are numerous. The most noticeable advantages are that UAVs can be used for "the dull, the dirty, and the dangerous" missions (Riebeling, 2004). UAVs are perceived as potent force multipliers, directly releasing manned aircraft and human ground forces for other missions. UAVs also offer increased efficiencies in operation and support costs due to the reduced need to fly actual pilots. The life cycle cost savings from the use of UAVs, in addition to the decreasing risk to human operators/soldiers, promises to be substantive. Clearly all the advantages of ROVs and UAVs will increase and encourage the military to find more and more applications and missions to be performed by ROVs (Wilson, 2002).

The main concern of this chapter is in two critical issues for successful implementation and usage of UAVs: (1) the interaction with the human operator and (2) the integration of UAVs into the various operational missions and military units. The first issue deals with the extent to which the design of the UAV interface is consistent, intuitive, and utilizes multi-modal channels, and the degree to which the mode of operation can be applied consistently across more than one type of UAV platform. The second issue reveals the importance of task definition, delegation of responsibilities within a unit/mission, and the interaction among humans involved in the mission, UAV operators, other soldiers, and the UAVs themselves (as semi-autonomous systems).

The majority of UAV systems today require multiple operators to control a single UAV. The future goal is to reduce this operator to vehicle ratio, with the vision

1   This work was supported in part by the U.S. Army Research Laboratory through the Micro-Analysis and Design CTA Grant DAAD19-01C0065, Michael Barnes, PhD, Technical Monitor. The views expressed in this work are those of the authors and do not necessarily reflect official Army policy.

2   The term unmanned or uninhabited aerial vehicle (UAV) refers specifically to aircraft that are operated without onboard pilots (see also Fahlstrom & Gleason, 1998).

of a single operator controlling multiple semi-autonomous UAVs. Furthermore, in current designs each unmanned system has its own unique workstation and interface, which is not acceptable when operators have to interact with multiple systems. UAVs that are designed to accompany troops in the battlefield are a particular challenge, as are those that are required to have intuitive interfaces. Intuitive interfaces must be easy to learn and to operate while the operator is in the field and performing other tasks. Hence, these UAVs have to serve the dismounted soldier and integrate well with other technologies he or she may use. In this chapter we wish to further examine the utility of UAVs that accompany troops in the battlefield. Our study summarizes interviews conducted with operators of UAVs on their operational experience during the Second Lebanon War in the summer of 2006.

The Top-Down Approach vs. the Bottom-Up Approach

The Top-Down Perspective

Previously, Oron-Gilad et al. (2005) distinguished between the top-down and bottom-up perspectives of human-robot interaction (HRI) design. The top-down perspective requires being very explicit about the goals, intentions, requirements, aspirations, and limitations of the technology at hand. From this perspective, the key to successful implementation of this technology for use in stressful environments is to ensure that the information presented to operators is structured in a manner amenable to direct interpretation, with a minimum of cognitive processing demands (Hancock & Szalma, 2003). If top-down ideas differentiate the feasible from the possible, bottom-up approaches seek to describe the actual from the highly probably. We now turn to the bottom-up perspective.

The Bottom-Up Perspective

The space–time taxonomy, initially introduced by Ellis for network categorization (Ellis et al., 1991) captures the basis of physical and temporal relations between machines and humans. This categorization has already demonstrated the capacity to have an impact in automation implementation (see Yanco & Drury, 2002). The space-time taxonomy divides user-machine interaction into four categories based on whether the humans or machines are using their computing systems at different times, or in different places. Oron-Gilad et al. (2005) broadened this space and time taxonomy by defining space in two dimensions, physical size (nano to macro) and physical proximity, and time as time scale (momentary and extended) and synchronization.

From a data-driven perspective, the expectation is that the "level" at which the individual (inter)actor engages with the system represents the crucial arbiter on performance. Those in greatest proximity to the theater of operation are most liable to respond under conditions of stress and fatigue. They are therefore more likely to adopt a data-driven (bottom-up) approach, as immediate changes in threat or identification of unknown objects may have a direct effect on their safety and the safety of their unit of action. In such circumstances the immediate goal changes from the higher-level tactical operational goals to a more self-centered safety goal. More distal operators are liable to have access to facilities which will mitigate such effects (although in times of great demand such as combat they will be omnipresent in the theater of operation). The goal-driven (top-down) strategy recommendations will depend upon the hierarchy of operational contexts. This taxonomy is given in Figure 12.1.

When we have a model that explains the time, space and perspective aspects of operational remote control, we can then further investigate the linkage between technology limitations and the operational strategies of using these technologies in action.

This study portrays typical operational scenarios of UAVs that were observed in the Second Lebanon War during the summer of 2006. Those scenarios can prototypically be categorized into two types: (a) active control scenarios (as shown in Figure 12.2), i.e., cases in which the UAV control and the use of the information derived from the UAV feed are both done by the same dismounted soldier; and (b) passive control scenarios (as shown in Figure 12.3), i.e. situations where there is a physical separation between the UAV control team and the information utilization team. In such cases the information consumers are considered passive controllers since they cannot directly control the UAV. Instead they have to provide both guidance to and requests for information from those who control the UAV (see also Minkov et al., 2007).

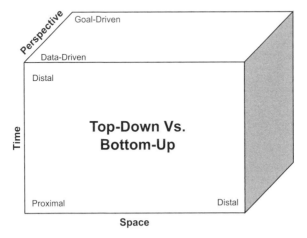

Figure 12.1   A descriptive three-dimensional model which incorporates the operational perspective with the space-time taxonomy (adapted from Oron-Gilad et al., 2005)

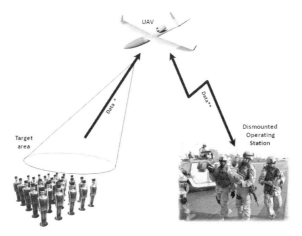

Figure 12.2   Active UAV control. In this scenario the infantry company carries a micro UAV and launches it when achieving the target area. Then, the soldiers who control the UAV receive video images from it and use this information for the operational mission

Note: * Video image uplink from the target area to the UAV and **Guidance controls uplink vs. sensors report and video images.

Interviews and Field-based Findings

Our information is based on interviews that we conducted with Israeli Defense Forces (IDF) soldiers, both users and UAV operators, on their experience with UAVs during the Second Lebanon War (2006). During this war IDF were operating several UAV systems. Our interviews focused on two types of population: active operators of a new tactical, micro-UAV system that was piloted during the war (active users); and passive operators[3] of large, fully operational, remotely operated UAV systems. Both types of operator were present in the interviews and all had operational combat experience. Our goal was to identify the main issues influencing the use of field systems for UAV control and information extraction in both positions. We met with the UAV operators shortly after the operations occurred, which ensured that we collected "fresh/unrefined" descriptions of the operational situations. All of the interviewees were micro-UAV operators or UAV information consumers. This section summarizes impressions and opinions presented to us by both operators and information customers.

---

3   Passive users/information consumers are passive users who get the information provided by the remote vehicle but have no ability to directly control its movements. That information can be video or still images, or textual data processed from these images.

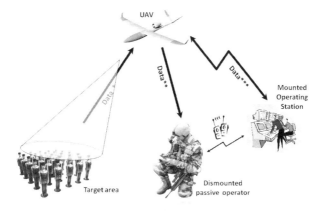

Figure 12.3   Passive UAV control. In this scenario the dismounted soldier carries a data receiver. A ground control station (GCS) controls the UAV according the soldiers reported operational needs. Thus, the dismounted soldier and the GCS team have to communicate during the mission. This scenario is less demanding for the dismounted soldiers since they do not need to attend to the UAV control or to the physical mounting and transport of the UAV. However, the burden of communication with the GCS is larger than when the operators are in proximity

Note: * Video image uplink from the target area to the UAV, ** Sensors report and video images and *** Guidance controls uplink vs. sensors report and video images.

Participants

Active Operators

We interviewed five crew members (UAV operators) who were considered active operators. The five operators varied in their military experience and in their operational background. The common ground of the five was that they all went through a dedicated period of UAV operation training for the system. The UAV that was used was a micro-UAV system manufactured by ELBIT Systems Ltd. An operating team consisted of two to three operators.

Information Consumers/Passive Operators

Four users were interviewed. All four were combat officers. They were familiar with battlefront units and their information needs. Their duty was to connect the infantry battalions to the division headquarter and to process video images that were taken by the UAV and transmit processed information to the battalions. All four interviewees went through professional training. Nevertheless, they all pointed out that actual duty demands in the battlefield were much different from what they had been taught in training.

Experimental Procedure

We met the interviewees personally for 1–2 hours of discussion. The discussion was driven by a questionnaire (given in Table 12.1). Participants were encouraged to suggest and raise new topics and ideas during the interview. The questionnaires included general descriptions of the soldiers' working environment, missions, relationships, and user-interface questions. This was followed by a discussion of theoretical and hypothetical issues such as the use of head-mounted displays (HMDs) for certain missions, and reducing the amount of information displayed to the operator and operating systems (active or passive) during participation in other tasks or missions. The later topics focused on operators' opinions based on their operational conception rather than on actual operational experience. All discussions referred to the specific task (active / passive) of the interviewee.

Table 12.1   UAV Operators and passive users' questionnaire*

Describe the system you used:
  What missions did you operate?
  What information did you need during the mission?
  Did the system provide the needed information? What was missing? What extra information chunks did you get?
  What other missions could you operate with the system?
  Do you have any experience using touch screens? Can it be applicable for such tasks?
	Mission preferences

What (if) other task can one do during operating the system (e.g. using communication devices, walking, fighting, defending his/her standpoint)? Rank the order of the tasks.	**Attention**

What is more important for each of the task types mentioned above?
  Focused image of the target;
  Wide image of the target's environment;
  Area map;
  Textual information;
  Other.
	Priorities

We think that there should be a difference when designing displays for different command levels or different operating environments. If you agree or disagree with this statement, please refer to these differences.

Table 12.1   *Concluded*

What device size will be the most suitable for:
  dismounted rifleman;
  mounted soldier (e.g., tanker);
  mounted operator (operates the UAV from a mounted cabin);
  company coordinator?
	Operational issues

Which dimension is more important: display size or the variety of information displayed for:
  dismounted rifleman;
  mounted soldier (e.g., tanker);
  mounted operator (operates the UAV from a mounted cabin);
  company coordinator?	

What are the main tasks in which the UAV system *can* mostly aid the fighting team? Refer to:
  dismounted rifleman;
  mounted soldier (e.g., tanker);
  mounted operator (operates the UAV from a mounted cabin);
  company coordinator?	

What are the main tasks in which the UAV system *really helped you*? Refer to:
  dismounted rifleman;
  mounted soldier (e.g., tanker);
  mounted operator (operates the UAV from a mounted cabin);
  company coordinator.	

What display size is the most suitable for:
  dismounted rifleman;
  mounted soldier (e.g., tanker);
  mounted operator (operates the UAV from a mounted cabin);
  company coordinator?
	Conformability matters

What is the proper training duration for a UAV operator?
  Should it be the soldier's main training or can it be his secondary task together with "traditional" duty?
  Do you think that using simulation games can help the operator learn the task?
	Training, experience

How do you foresee the ultimate UAV control system? Do you have any experience using HMDs? Can you evaluate the advantages and disadvantages of this use?	**Imagine**

* The same questionnaire was used for both passive and active users but each group referred differently to the items.

Results

The data collected from all nine operators was aggregated into two main topics: technology-related issues and methodological issues. This summary portrays the ideas and suggestions of the interviewee. Topics are presented in no particular order or rank of importance.

Technology Related Issues: Controls and Human–Machine Interaction (HMI)

All active users we interviewed used the same operating system. The interaction devices included a touch screen and a dedicated operating panel in which a tablet PC was docked (see Figure 12.4). The panel included a keyboard, a trackball, and a mouse/joystick. The panel was designed in the Hands on Throttle and Stick (HOTAS)[4] concept of operation, in which operators can keep their hands on the tray and stick for all operating needs without moving their gaze from the screen. Operators suggested adding an alternate interaction device that can be detached from the operating panel (e.g., a portable mouse). Hence, the HOTAS concept is not the most appropriate system for this type of mission, since the operator needs to perform additional tasks that cannot be performed from the dedicated panel. The addition of a portable/detachable control device can add the necessary flexibility to the operation by enabling the operator to take his hands off the panel (for example when using a hard-copy map or a communication device) while maintaining control of the UAV through the portable device. This requirement

Figure 12.4   The interface used for the operation of the UAV system

---

4   Hands on Throttle and Stick (HOTAS) is a human–machine interaction (HMI) style which allows the pilot to access the cockpit functions and fly the aircraft. All switches are placed on the stick and throttle, and hence pilots can keep their hands on throttle and stick without having to take off their eyes off the screen.

deviates from the traditional aircraft HMI design, but is a necessity in an infantry operational environment where the combatant is multitasking.

Displays

Displays related to system state   Some of the UAV's physical status reports, such as vehicle altitude and engine revolutions per minute (RPM), were displayed using alpha-numeric displays. This requires the operator to divert attention to reading the values instead of focusing on the flight mission. Most of the time (as long as nothing exceptional occurs), attending to the system status indicators provides no benefits, but still demands attention from the operator. Operators indicated that they would much rather have the sensor control information presented in a graphic colored-coded display. This would allow them to pay attention to specific system value changes only when they become important, i.e., when they pass a predefined limit and changes in color happen. With the current alpha-numeric black and white displays, the operating method forces the operator to continually monitor the data on the display. This effort increases when technical changes occur frequently (e.g. changes in engine temperature or RPM problems). The operators suggested a conceptual change. They would like to rely more on the automated system and increase their trust in the automation. In order to do so they should get the appropriate warnings at the right time.

Analyzing the level of information needed   There are differences in the amount of information that should be presented to the active operator and the passive operator (who receives the video feed). However, passive tasks still require some technical information about the UAV. One still has to maintain some of the main indicators on the screen, predominately the video itself and the image footprint on a map. What remains to be determined is the minimal amount of information that is crucial for either the active or passive operator.

Innovative technologies   Two technologies concerning graphic dialogue interface were mentioned by the operators as potential technologies that can upgrade the quality of information transferred between the operator and the information consumer:

- For improved collaboration, the use of a dual pointer in which both the operator and the information consumer can mark a specific point on the screen was suggested.
- To decrease team dependence on verbal and communication abilities, it was suggested to add text messages as an additional common communication tool.

Level of Automation (LOA)

Determining the optimal automation level for an unmanned system is a challenge. The interviewees marked it as a key factor for their allocation of attention. For the missions that were performed, the task of operating the UAV was of secondary importance relative to the main tactical/strategic task, and thus the amount of autonomy that the UAV became of substantial importance. Most of the flight control components (e.g. wind sweeping and bumping, altitude, and acceleration or velocity changes) should be controlled automatically by the system, allowing the operator to focus on the tactical aspects of the task. The automation logic rules should know if and when the operator should remain "in the loop," and act accordingly (i.e. management by exception/management by consent). The automation control system should rely on flags, indicators, and operator-adjusted defaults that will produce alerts according to the significance of the situation or problem.

Technical Issues of Displays

In addition, issues related specifically to the technical characteristics of the display and the input devices were specified.

Display size   The 14-inch tablet display (as shown in Figure 12.4) was accepted as a satisfactory size. In "ideal conditions," all interviewees agreed that they would prefer a bigger display (a few of them referred to a 30-inch display they had previous experience with as a preferable size), but they also indicated that "tablet"-size was the optimal size for military environments where weight and portability should be considered. Note that the missions discussed here were performed in an urban environment, which differs from an open rural area. Some examples of these differences include movement distances, accessibility for vehicles, and the ability to move in hidden paths.

Area on the screen   We asked the operators about the distribution of space among the video image, maps, and other controls and sensors. All interviewees thought one should give the video the majority of screen space, since that is in the operator's main interest. Specifically, operators mentioned that they would like to be able to enlarge the video window while reducing the map window.

Screen resolution   Inappropriate screen resolution can reduce the operation quality by preventing retrieval of detailed information (e.g. road names, datum cards) from operators and forcing them to use the hard-copy mapping accessories. Operators did not give any specific quantifiable resolution measures, and we presume that they do not have the ability to estimate this aspect in the same way they can quantify size.

Automatic tracking   Auto-tracking can be applied at two different levels of integration. The first level is related to the hardware, and that must ensure that the communication devices connecting the UAV and the operator control unit (OCU) (e.g. "dish"–direct link antenna) automatically tracks the UAV to keep the communication link alive. For this level of tracking, the current solutions were acceptable. The second level is related to the software and the algorithms that maintain the camera tracking on its target. This is a complicated task even for static targets (and moving vehicles), and becomes even more complicated when moving targets are involved. The UAV system used by the interviewees did not have auto-tracking abilities, and the operators all agreed that this was an ability that should be improved. The lack of automatic tracking ability forces the operator to attend to tracking and depletes the operator's mental effort geared toward the main task.

Working Environment: Physical Environment, Shields, Covers, and Tools

The physical environment surrounding the operator and the tools that the operator has directly affects the ability of the crew to perform its mission. Two environment-related parameters were suggested by the interviewees to have substantial influence on mission accomplishment.

1. *The ability to physically see the details on the system's display.* Thus, variant lightning conditions, such as full daylight, make it difficult to see the display clearly. Furthermore, there are many accessories that can be given to the operator to improve the ability to work under differing daylight conditions (e.g. dark cover).
2. *The level of the display's illumination.* Using the display in combat conditions where the team has to be camouflaged can be problematic. In such cases, the illumination that is caused by the display can expose and thus harm the team. Therefore, we should suggest technologies that enable the operator to use the display without radiating the screen light.

Other elements that affect performance were related to tools and their availability.

Paper-based mapping tools   The current system's configuration forced the operators to use hard-copy maps to complete their mission. Using paper-based maps interferes with the main task. Operators marked the activities related to the hard-copy map as major "time killers."

Digital mapping tools   The operators relied solely on hard-copy mapping accessories. Reliable digital mapping tools could have provided substantial performance enhancement. Good mapping tools can release some of the operator's attention toward the main tasks. Even simple tools can improve performance. Suggestions for additional functionalities/tools are divided into five categories:

1. Basic mapping and interpretation tools:

   – digital zooming (scaling the map and switching between detail levels);
   – map-rolling (roaming abilities);
   – windows synchronization (matching the map and video image in three different levels: scaling, rotating, and positioning);
   – switching map to aerial photo and back;
   – auto-switching (flickering) between the image and the map/aerial photo.

2. Mapping infrastructures for common language between UAV operators and their counterparts (other troops, remote commanders).
3. Fitting the map to the video in all three dimensions: datum point, roll, and scaling.
4. Geographic Information System (GIS) abilities which enable the operators to use merged geographical information with the map/photo infrastructure.
5. Dimming the map over the image.

Examples of commercial applications can be found in several commercial websites such as Tiltan Systems Engineering (2008). These tools, among others, can direct the operator's attention from peripheral tasks to the main tasks, and possibly enhance performance. Some interviewees claimed that for complicated tasks they will always have to use the hard-copy accessories. This assumption needs to be further examined in order to determine whether its origins are based on habit, lack of experience with novel tools, or the information resolution. It is possible that the information given by a hard-copy map cannot be replaced and therefore is a real need. The technology aspects of any unmanned system take an important role supporting the operator's focus on the operational main task. These main tasks are what usually produce the necessary intelligence.

Methodological Issues

Distribution of Tasks among Team Members, Crew and Team Communication

The operators made a distinction between two types of tasks: simple and complex. Simple tasks can be performed by one individual operator. Typically, these include routine area scanning or routine occupancy checkups (the operator is looking at specific targets/areas to see the status of their occupancy by enemy troops). Complex tasks are tasks that are performed under time pressure and/or require non-routine actions and/or coordination with other parties. For simple tasks the team head's (commander) role depends on the accessories and devices installed in the operating station. If the tasks require using hard-copy maps and aerial photos, communication devices, telephones, and/or other peripheral equipment then it demands the team head's presence. Otherwise, if the environment conditions and physical layout of the station enable the operator to access all of the information from his/her workstation, then the commander's involvement in executing the task is not essential. As the tasks become more complex, e.g., weapon guiding or coordinating with other parties, the role of the team head becomes absolutely necessary and cannot be ignored. In these cases, the commander takes part in the task itself by taking responsibility on some of the mission's components, e.g. communicating with other crews or control stations or just monitoring the UAV technical condition (vitality). In addition, the commander monitors the operator's immediate actions in order to have a complete view of the entire mission's progress. Thus, on simple missions, the UAV operator can perform the main tasks by him/her self, and provide the appropriate information to the team head. In fact, in this situation, the team head is acting more like an operation supervisor responsible for setting the operating policy. In the more complicated situations there has to be an additional fourth person on duty, specifically a mission commander whose focus is on strategy and coordination. This is due to fact that the operational mission requires the full attention of both the operator and the team commander. Different working methodologies as specified above lead to different technical needs (e.g. displays) and tools to support each specific methodology.

Training

As in other complex tasks, operator training can be vital. Training can influence the quality of the task's execution, but also improve communication between the crew and the information customers. Two main points were raised in the discussions:

1. Should the operator's main training focus on operating the system or, alternatively, should training focus on acquiring other military skills such as marksmanship or infantry? Most interviewees thought that UAV operation can be the soldier's secondary training, but only on the simple operational tasks. In their collective opinion, it is essential to specifically train for UAV operation.
2. Basic training should be twofold and include training on how to *operate the system* (aviation basics, software, and hardware operating), and training on how to *conduct a mission* (the tactical and strategic elements of the mission, including observation, photo interpretation, and situation analysis).

Conclusions

Five main design guidelines can be drawn from the interviews and discussions that we have conducted. They are all relevant for active and passive operators, with a few limitations described below. Most of the conclusions that are relevant for passive operators are even more crucial for active ones, mostly because soldiers are required to be self-supporting even at large distances from base.

Modularity and Flexibility

The variety of missions, environments, and situations in which UAV systems can be applied is vast and diverse. One of the main requirements for such systems is to be intuitive and to require minimal training. As such, operating uniformity is a major issue. The interface should fit or adapt to different mission properties. This requires technical modularity and flexibility. For example, we would like to see a common interface which can automatically complement and adapt to different display sizes (e.g., 4, 7, 14 inch). The modularity factor mainly affects the active operator, who usually acts in a more complicated environment, under many limitations. Operators should be able to use the most suitable configuration for the specific situations they act in (e.g. observing a terrorist house vs. walking 30 km to reach a bridge).

Automation

Some of the main design efforts should be geared toward helping the operators to focus on the main task, which is usually video interpretation and guiding troops. Automation tools should be developed in order to separate operators from any technical details that could be controlled and supervised by the system itself, leaving the human decision space mainly for operational issues. Although defining the optimal level of automation (what to automate, when, and how) can be a complicated issue, there seems to be no other way but to automate components of the task in order to be able to operate such systems by lightly trained operators. The complexity and amount of information caused by multitasking, which characterizes active operators, makes this principle more crucial for them.

Training

Operating UAV systems is feasible even when based on short training. Short training will enable operators to implement simple tasks but not complex tasks. The question is: To what degree can the interface design bridge this gap? If the system is more automated and simplified, it will become easier to learn, thus allowing the main training efforts to focus on mission implementation and not on operation of the system.

Customization

One way to implement flexibility is to enable interface customization in which the operator can fit the interface to the mission and environment specifications. The customization principle can sometimes conflict with the need for a short, effective training process, since the system can change. This may also demand some relearning activities. Despite this, the ability to customize the user interface can be more beneficial for active operators who need to use detailed information during combat and in a "dirty" environment.

Display Size

As UAV systems evolve and are being applied in many new combat environments, one should examine in depth the size limits of the control panels. In our previous works (Minkov et al., 2007; Oron-Gilad et al, forthcoming) empirical results showed that the effect of display size is task dependent. Furthermore, we identified the need for a hand-held control device to control and supervise the UAV system; however, the effectiveness of portable devices needs to be examined further. The importance of hand-held devices is mostly seen in infantry troops where the entire soldier's gear is mounted on the body. In such situations, even if the use of a handling mount is sometimes feasible, it is never comfortable and thus demands another mounting solution.

Extension of Theoretical Models

Oron-Gilad et al. (2005) mentioned that the top-down perspective requires us to be very explicit about the goals, intentions, requirements, aspirations, and limitations of the technology at hand. Our findings here indicate that the technology that is being utilized for operating remote unmanned systems has a significant influence on the system efficiency and perceived appropriateness from the bottom-up perspective as well. We have seen that the majority of issues raised by operators were related to the technology, and a smaller portion of issues was related to methodology and operation strategy. The military's necessity for moment-to-moment information drives the development of technologies that will increase information-gathering capacity. The top-down perspective emphasizes that UAVs are simply one of a suite of "vehicles" or technologies to achieve desired aims.

In contrast, the bottom-up approach confronts the limits and constraints of present-day and near-term operational systems. Thus, when strategists and higher-ranking decision makers (whose view is from the top-down perspective) see technology as the answer, the end point of that general strategy is a situation in which technology becomes the central problem. Therefore, in order to reinforce the importance of technology, and to increase awareness of its limitations and how the end users in the field are affected, we propose to add a fourth dimension to the basic three-dimensional model presented in Figure 12.1. This fourth dimension will be directly related to the level of technology and/or to the automation limitations (as shown in Figure 12.5).

Quick Reference Guide

Common design guidelines for both active and passive operator systems:

- *Modularity and flexibility*. The interface should fit and/or adapt to different mission properties. The operator should be able to choose the most suitable configuration for a specific mission.
- *Automation*. The operators should focus on the main task, which is usually video interpretation and guiding troops. Automation tools and flexible LOA selection can free operators from the technical operation of the system when necessary.
- *Training*. For simple tasks, operating UAV systems is feasible even when based on a short training period. Training efforts should focus more on mission implementation and not merely on the operation of the system.
- *Customization*. Customization can sometimes contradict the need for short, effective training. Despite this, the ability to customize the user interface is beneficial to active operators.
- *Display size*. The effect of display size is task dependent.
- *Hand-held control device*. We identified a need for a hand-held control device mainly for infantry troops.

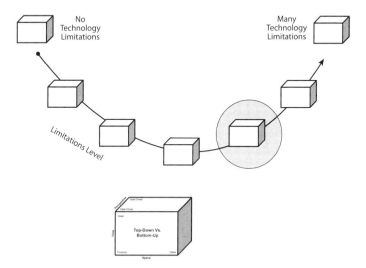

Figure 12.5   A fourth dimension of technology limitations incorporated into the three-dimensional model of operational perspective

References

Ellis, C.A., Gibbs, S.J., & Rein, G.L. (1991). Groupware: some issues and experiences, *Communications of the ACM*, 34(1), 39–58.

Fahlstrom, P. & Gleason, T. (1998). *Introduction to UAV Systems* (2nd edn). Columbia, MD: UAV Systems, Inc.

Hancock, P.A. & Szalma, J.L. (2003). Operator stress and display design. *Ergonomics in Design*, 11(2), 13–18.

Minkov, P. & Oron-Gilad, T. (2007). The effect of display size on performance of operational tasks with UAVs, *Proceedings of the Human Factors and Ergonomics Society 51st Annual Meeting*, Baltimore, MD.

Oron-Gilad T., Chen, J.Y.C., & Hancock P.A. (2005). Unmanned aerial vehicles from the top-down and the bottom-up. In N.J. Cooke, H. Pringle, H. Pedersen, and O. Connor (eds), *Advances in Human Performance and Cognitive Engineering Research, Vol. 7: Human Factors of Remotely Operated Vehicles* (pp. 37–48). San Diego, CA: Elsevier.

Oron-Gilad, T., Minkov, P., & Goshen, E. (forthcoming). The effect of display type on performance of a military urban navigation task: A field study.

Riebeling, S. (2004).Unmanned aerial vehicles do dull, dirty, dangerous work. United States Army public affairs website: http://www.army.mil/info/institution/publicaffairs.

Tiltan Systems Engineering (2008). http://tiltan-se.co.il/page.asp?cat=178&lang=2&type=2 (accessed 14 May 2008).

Wilson, J.R. (2002). UAVs and the human factor. *AIAA: Aerospace America Online*.

Yanco, H.A. & Drury, J.L. (2002). A taxonomy for human-robot interaction, *AAAI Fall Symposium on Human-Robot Interaction*, AAAI Technical Report FS-02-03, pp. 111–19, November.

Chapter 13

Unmanned Aerial Vehicles: Enhancing Video Display Utility with Synthetic Vision Technology

Gloria Calhoun and Mark Draper[1]

U.S. Air Force Research Laboratory

Introduction

Unmanned aerial vehicles (UAVs) have made a definitive contribution to military operations, and their role as a war-fighting tool is expected to increase dramatically. With teleoperation, compared to manned missions, operators are constrained with respect to the visibility of the environment in which the vehicle operates. UAV operators must rely on video imagery transmitted from one or more video cameras mounted on the air platform (Draper et al., 2004). In many cases, this video camera is mounted on a gimbaled turret and thus can be rotated by a remotely situated sensor operator in order to view various points of interest in the UAV's surrounding environment. The display of this video is a key source of information for maintaining operator situation awareness. UAV pilots use this imagery to verify clear path for taxi/runway operations, scan for other air traffic in the area, and identify landmarks and potential obstructions. Sensor operators use the imagery to conduct a wide variety of intelligence, surveillance, and reconnaissance activities as well as to directly support combat operations. However, video imagery quality (and, by extension, operator situation awareness) is often compromised by narrow camera field-of-view, datalink degradations, poor environmental conditions (e.g., dawn/dusk/night, adverse weather, variable clouds), bandwidth limitations, or a highly cluttered visual scene (e.g., in urban areas or mountainous terrain). If video interpretation can be enhanced and made more robust under a variety of situations, UAV mission effectiveness would likely increase substantially.

Synthetic vision systems can potentially enhance video interpretation and the level of UAV operator situation awareness, and, consequently, improve decision making. With this technology, spatially relevant information constructed from databases (e.g., terrain, maps, photo-imagery, pre-mission plan, etc.) as well as from

---

1   Antonio Ayala, Brian Guilfoos, Austen Lefebvre, Brian Mullins, Jeremy Nelson, Heath Ruff and Nick Wright contributed significantly to the empirical work presented here.

numerous information updates via networked communication can be represented as computer symbology and simulated imagery that is overlaid conformal onto the dynamic camera video imagery and presented to operators (see Figure 13.1). For example, a synthetic runway outline can overlap an actual runway in an image. The overlaid symbology is conformal in that it appears to coexist in the real environment, responding to camera movements in the same way as its real-world correlate (i.e., the actual runway). The addition of the computer-generated symbol helps highlight the location of the runway by effectively increasing the signal-to-noise ratio of the imagery, allowing operators to more quickly locate, identify, and act on this critical information. Transient objects that are not present in the geographic database can also be presented, based on information available through intelligence sources and networked communications. In this manner, synthetic vision overlay technology is expected to improve UAV operator situation awareness by highlighting, in real time, key information elements of interest directly on the camera video image (such as threat locations, expected location of targets, landmarks, no-fly zones, runways, and position of friendly forces).

A synthetic vision system can also provide information that does not have a correlate in the actual sensor imagery and hence cannot be seen by the naked eye. In this case, rather than *highlight* existing objects in the imagery, the system *augments* the imagery with additional information that has known geographic coordinates. For instance, ground-based threat lethality envelopes can be represented so that pilots do not fly too near threats, and underground tunnel networks can be visually represented as if the pilot could see through the ground. In addition, the state of a system (e.g., based on radar emissions or machine-to-machine datalink activity)

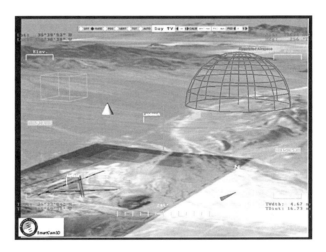

Figure 13.1   Synthetic vision symbology added to simulated UAV gimbal video imagery (symbology marking threat, landmarks, areas of interest and runway)

can be indicated visually. By representing non-visual task-related information in a graphical way, the operator would be able to gain enhanced situation awareness.

Because the symbology created from the synthetic vision system is based on digital data, it is not subject to the limitations of visibility inherent with electro-optical imagery. While darkness, terrain occlusion, smoke, fog, icing, and haze all impact this type of video imagery, the synthetically generated scene can remain unobstructed. Thus, this technology may also help maintain an operator's situation awareness during periods of video datalink degradation and poor visibility by essentially "recreating" the real-world scene via computer graphics. The synthetic vision imagery could totally replace the real sensor image if needed, or a blending of the two information sources could occur by varying the transparency of the synthetic vision imagery. This concept has been evaluated for manned aircraft (Theunissen et al., 2004).

A synthetic vision system can additionally serve a key role in supporting distributive collaborative communication in a net-centric battlespace environment. In this case, the synthetic vision system is applied both as a display and as a control, enabling a net-centric team member to mark a specific spatially referenced point of interest on the camera display, causing matching informative synthetic symbology to appear on the displays of other geographically separated stations in the warfare network. For example, team interaction can be facilitated with networked friendly forces graphically sharing information on their past and present positions as well as planned paths and action points. The friendly forces could also pool their knowledge of neutral and hostile forces to help maintain battlespace awareness.

Whether the synthetic vision system is used to highlight objects in the video, augment the video with additional information, or mitigate negative effects when the video is lost or degraded, the computer-generated information is hypothesized to benefit both the UAV pilot and sensor operator. Pilots controlling UAV flight would be informed by computer-generated symbology denoting no-fly zones, runways, and indication of a safe path through threat regions. Sensor operators performing surveillance and other tasks would benefit from synthetic symbology highlighting ground landmarks, targets, and troop locations, especially when the environment is cluttered. However, the tremendous variety of UAV platforms, level of sophistication, and mission applications suggests that synthetic vision technology may vary widely in its benefits across these UAV characteristics as well as operator task (flight management versus sensor management). This chapter, therefore, will concentrate on how synthetic vision technology may benefit a UAV sensor operator controlling a turreted camera, via teleoperation, of a small to medium altitude UAV.

The expected benefits of synthetic vision systems will certainly be of increasing importance in future envisioned scenarios where operators are tasked to supervise more than one UAV simultaneously in network-centric operations (Hunn, 2006). For instance, synthetic vision systems will be particularly valuable for helping the operator rapidly gain situation awareness of each UAV's respective mission area during scenarios which require an operator to switch attention between multiple UAVs and their associated camera views.

When multiple UAVs are under supervisory control, the amount of information that can be displayed also multiplies. It remains to be determined for multi-UAV control, as well as single-UAV control, what information elements available from the synthetic vision system should *automatically* be presented as opposed to being presented only when requested by the operator. Likewise, the degree to which operators can control features (e.g., opaqueness) of the computer-generated symbology needs to be determined so that mission effectiveness increases without imposing additional workload. This and numerous other human factors issues need to be considered in order to realize the anticipated benefits of UAV synthetic vision systems.

Human Factors Issues with UAV Synthetic Vision Systems

There are multiple human factors questions related to the presentation of individual synthetic symbology elements (Calhoun et al., 2005). For instance, for each line segment, icon, and label used, what is the ideal shape, color, brightness, contrast, size, thickness, style, etc? Should these characteristics vary with changing characteristics of the underlying visual imagery? How should the transparency of the symbology be manipulated such that both the video imagery and the synthetic vision symbology are simultaneously visible? When a terrain overlay is desired, what method should be employed—simple "gridded overlay" (rectangular grids of known size to facilitate depth perception); terrain texturing (e.g., colors correspond to different absolute terrain elevations), or photo-realistic terrain imagery (i.e., from satellite imagery data)?

Issues also surround the blending of synthetic and real images. Specifically, will the symbology elements just be superimposed over the sensor image, like a heads-up display (HUD) from a piloted aircraft, or will elements also be "scene-linked" (referenced to the world) such that they undergo the same visual transformation as real objects depicted in the imagery? An example of the latter is a virtual "billboard" growing larger as the vehicle approaches, just as a real-world object would. Scene-linking symbology in the real image may reinforce other motion cues and benefit visual attention and information retrieval by helping group the synthetic information and real sensor information into one perceptual group. This is supported by object-based models of attention that postulate that complex scenes are parsed into groups of objects, with attention focused on only one object at a time, with object groups defined by contours, color, etc. (McCann and Foyle, 1994). The choice of presentation method will likely be specific to the information element to be displayed. One driving factor is the degree to which the UAV display is anticipated to be overloaded with information, or cluttered.

Information clutter is, in fact, one potentially negative result from the application of a synthetic vision system into UAV operations. While overlaying computer-generated information on a sensor image can help minimize scanning and the effort required to access and monitor information elements, having the additional clutter may also inhibit the processing of the fine detail in the underlying sensor imagery because of the inhibitory effects of overlay clutter. Cognitive tunneling can occur when the operator becomes focused on an element of the synthetic vision symbology (or objects to which attention is directed by the synthetic symbology) to such an extent that other important objects or events in the sensor imagery are not attended to (Yeh and Wickens, 2001). In the case of UAVs, this may result in the operator not detecting unexpected, high-interest targets. Thus, the design of the synthetic vision symbology needs to account for the potentially negative effects of information clutter and cognitive tunneling by only displaying visual elements that will benefit the operator's situation awareness and performance, and employing design features that minimize clutter effects and confusion.

There are numerous other techniques for "view management" which maintain visual constraints on the projections of objects on the display (Bell et al., 2001). By making adjustments in the manner in which synthetic vision symbology is presented, problems with different synthetic elements occluding each other can be minimized, as well as a synthetic element occluding a key element in the real sensor image. Likewise, an intelligent system can ensure that "distant text" does not become illegible, and labels are automatically reoriented and repositioned based on the operator's viewpoint with respect to the object. Besides managing the view to optimize the visibility of the synthetic vision symbology, an intelligent system can highlight newly appearing synthetic elements that are critical for operator attention. Identifying useful coding methods to indicate the criticality, urgency, timeliness, and reliability of information depicted by elements is another issue that needs research.

Regardless of the symbology set overlaid on the video display and the automated view management tools available, operators should be provided with the capability to manually declutter the synthetic vision symbology as needed. "Global" levels of declutter may be optimal in certain situations, whereby the operator can systematically select and deselect general classes of information. Another approach would allow operators to deselect individual symbology elements to allow for maximum flexibility in information presentation for specific situations (for instance, those that might be adjacent to a target that the operator needs to have an unobstructed view of). Providing the operator with the capability to declutter information, as well as allowing them to control other features of the computer-generated symbology and its application, raises the research issue of how these additional control functions should be integrated or added to the operator's conventional controllers. Speech-based control may prove to be a useful interface, helping the operator maintain a heads-up view of the camera display (Draper et al., 2003).

Once a camera view augmentation with synthetic vision technology has been designed, its usability needs to be evaluated with a sensor image that replicates the anticipated scene complexity, sensor view specifics, environmental factors, etc. that can occur in UAV operational applications.

Past Human Factors Research

Presentation of a synthetic visual overlay in a conformal manner with sensor imagery has been demonstrated (in the laboratory) to reduce scanning time; reduce the need to mentally integrate spatial information from disparate sources; and facilitate attentional management (Wickens and Long, 1995). Application research has primarily focused on how synthetic overlays can aid piloting tasks during manned flights. In fact, there is a large body of research showing the utility of synthetic vision systems for both military and civilian manned flight applications. For instance, flight guidance overlay symbology has been found to be valuable for reduced visibility conditions, especially during landings (Prinzel et al., 2002). The functionality demonstrated with synthetic vision for manned aircraft systems can potentially be adapted to UAV control tasks where there is commonality in information requirements between manned and unmanned flight (Tadema et al., 2006).

Synthetic overlay research for UAV applications is scant in comparison to research for manned applications. The majority of efforts have involved developing concept demonstrations and acquiring subjective data from subject-matter experts. For instance, laboratory research in the Netherlands has addressed the feasibility of using synthetic vision technology for UAV operator support. Early work at TNO Human Factors Research Institute investigated synthetic vision technology with regard to teleoperation of remote UAV cameras, including the impact of system lag in the video camera response to control movements (van Erp and van Breda, 1999; van Erp, 2000). More recent research utilized existing synthetic vision software components as part of a UAV mission management system to see if computer-generated imagery can increase situation awareness, both for the current situation as well as understanding the effect of changes on future situations in the mission (Goossens et al., 2004; Tadema and Theunissen, 2003). In a U.S. Army Research Laboratory (ARL) effort, synthetic vision overlay concepts are being employed in the development of displays to support multi-UAV control (Hunn, 2005). Results, to date, suggest that the ability to combine real-time video imagery with virtual databases will enhance situation awareness and improve crew coordination.

The U.S. Air Force Research Laboratory's (AFRL) Human Effectiveness Directorate also has conducted research focused on the utility of synthetic vision systems tailored to UAV operations, as part of their long-term program to address human factors associated with UAV control stations (e.g., Calhoun and Draper, 2006). The present chapter will provide an overview of three AFRL ongoing research areas exploring how synthetic vision systems can improve UAV operators' situation awareness by enhancing the utility of video imagery presented in ground control stations. First, a high-fidelity UAV simulation evaluation of synthetic symbology overlaid on a camera display for several task types will be summarized. Next, research focusing on a picture-in-picture (PIP) concept, where video imagery is surrounded by a border of synthetic generated terrain imagery to expand the operator's field-of-view, will be described. Lastly, recent efforts to use synthetic symbology in a novel task transition aid to help build operator's situation awareness when switching between the camera views of different UAVs will be introduced. All three summarized research areas focused on a single class of UAV application, that of a relatively slow UAV flying at approximately 10,000 feet above ground level. Details on the methodology and results for individual experiments are available in the cited references.

Utility of Synthetic Vision Overlay for Four UAV Task Types

Research Objective

In order to assess the utility of synthetic vision information for UAV sensor operations, computer-generated information was overlaid on a (simulated) live camera display in a high-fidelity UAV simulator, and participants' performance was evaluated on four representative sensor operator tasks (Draper et al., 2006). Participants' tasks included controlling the camera to locate specific ground landmarks in the 360-degree area surrounding the loitering UAV; marking multiple ground targets with synthetic symbology; tracing a synthetically highlighted ground convoy route with the UAV camera boresight; and recording text from synthetic overlaid symbology. This evaluation also explored the impact of the update rate of the UAV telemetry data that is required for proper registration of the synthetic vision symbology/imagery. Since UAVs can have separate communication data links for video imagery and basic system telemetry state data (such as variables related to aircraft and sensor position), it was logical to evaluate conditions where the video imagery and synthetic overlay imagery update at significantly different rates (with the higher bandwidth datalink reserved for the video imagery transmission). Seven different telemetry update rates were examined (0.5, 1, 2, 3, 6, 10, and 24 Hz), representing a spectrum of rates encompassing a variety of UAV applications. The underlying video imagery update rate was held constant at 60 Hz.

Simulation Environment

The UAV sensor operator simulator (Figure 13.2) had head-level and upper 19-inch color displays as well as two 10-inch head-down color displays. During the experiment, participants focused on the head-level display presenting simulated video imagery from a UAV's gimbaled camera, along with HUD sensor symbology. The video imagery was generated from a realistic database model of southern Nevada, and the synthetic vision overlay was implemented using the SmartCam3D System by Rapid Imaging Software, Inc. For the present experiment, the upper display was turned off—participants could only use a flip book (containing visual and text descriptions of each landmark) and the head-level camera display to complete tasks. To change the camera's field-of-view, participants manipulated the right and left joysticks to control the camera's orientation and zoom, respectively. Participants pressed a button on the joystick for the designation steps in each task. The center keyboard was used for the data entry task. Except for the data entry task (which did not involve manual control of the camera), each trial for the other sensor operator tasks was preceded by a 10-second time period in which participants moved the camera around in an "Adaptation Area" in order to become familiar with the appearance/behavior of the synthetic symbology under the "characteristics" (term used with participants, referring to update rate) in effect for that particular trial. The simulation was hosted on seven Pentium PCs.

Figure 13.2   UAV sensor operator simulator

Research Findings

The results demonstrated the potential of synthetic symbology for improving situation awareness, reducing workload, and improving the designation of points of interest at nearly all the update rates evaluated. Participants' performance for all four task types remained fairly constant across the wide range of telemetry update rates. That said, there were instances where performance was better at higher update rates compared to lower update rates, such as target marking and large area search tasks. Since these two tasks are representative of many current UAV sensor operator's activities, these results indicate that the telemetry update rate should be considered carefully in UAV system design. For the task type that required retrieval of specific alphanumeric information from synthetic symbology, the results suggest that the stability of the symbology may be more important than accurate registration of the synthetic symbology with the real world. For this type of task, performance was not negatively impacted by a low update rate or high update rate; rather it was negatively impacted by middle values (2–4 Hz). Update rate had the least effect on the task that required tracing of a synthetically highlighted ground route, because the camera zoom factor was high and minimal camera rotation was required.

Given that participants' subjective ratings were consistent across task types (symbology at high updates rates was generally less degrading/distracting and served as an aid to task performance), and performance tended to improve at the higher rates, the conclusion can be made that the highest telemetry rate supported by the UAV system should be employed when applying a synthetic vision system. One specific recommendation drawn from this data is that telemetry update rates of less than 2 Hz should not be utilized, even for relatively slowly moving UAVs such as the one simulated in this experiment. The results from this experiment also suggest that update rates of 6 and 10 Hz are often adequate, and that performance enhancements may not be realized by spending resources to achieve update rates higher than 10 Hz for this UAV type and application.

Future Research Direction

The above recommendations are only preliminary, as more research is needed to better understand the relationships between synthetic vision overlay technology, task type, UAV/sensor system characteristics, time delay, and registration error. Note that results will also vary with certain UAV performance characteristics such as airspeed, operating altitude, etc. It is also important to consider that there are several points in the overall system that can contribute to both time delay and registration error. For instance, the quality of the UAV positional data is subject to quantization error, random delays, and basic measurement error, besides problems introduced by the telemetry system. If registration errors are systematic rather than variable, operators might be able to adapt. Indeed, that is one research question: How much registration error and time delay is tolerable for a UAV application before task performance degrades substantially? It may also be possible to provide a manual intervention whereby the operator can dynamically recalibrate the correspondence of the synthetic and real world or employ prediction algorithms to overcome the limitations of imprecise and tardy data input to the synthetic vision system.

Further evaluation should also employ a more demanding task environment. The small performance differences between update rates in the reported study may simply reflect human adaptability, bringing additional cognitive resources to bear to ensure that the task is achieved in spite of a subpar update rate. A few subjects mentioned that when the symbology was degraded, they had to compare the pattern of the synthetic symbology with the pattern of target images to discern which had been marked. Another participant commented that when the symbology lagged or was choppy, more concentration was required. Since the time periods of these trials were relatively short, the extra effort to maintain a high performance level did not have to be sustained for a long duration. It may be that larger differences across update rates would have been evident with more numerous trials, longer sessions, and/or more realistic missions where a variety of tasks were being time shared. It also may be that performance decrements would have occurred with less salient targets and/or a more cluttered scene.

Utility of Synthetic Vision Augmentation to Expand Field-of-View

Research Objective

Operators of teleoperated UAVs are prone to poor situation awareness of the remote environment because of the impoverished representations from video feeds which can omit essential cues for building an effective mental model of the environment (Tittle et al., 2002). Augmenting the real-time video presentation may improve operators' situation awareness (Drury et al., 2006). Research is underway to explore the picture-in-picture (PIP) concept where real-world video imagery is condensed such that it occupies a portion of the display area and is "framed" with a synthetic generated imagery border (Figure 13.3). This synthetic imagery was aligned and registered with the UAV video imagery such that it resulted in a virtual expansion of the available sensor field-of-view (well beyond the physical limits of the camera). This potentially mitigates the UAV camera's narrow field-of-view, the basis of the "soda straw" analogy or "keyhole effect," which is a primary complaint of current UAV sensor operators.

Figure 13.3 Picture-in-picture (PIP) concept, with real video imagery surrounded by synthetic generated terrain imagery/symbology (affords virtual expansion of the available sensor field-of-view well beyond the physical limits of the camera)

Three separate studies in this research area have taken place at AFRL. They explore the performance benefits of the PIP concept in respect to synthetic terrain overlay type, update rate of UAV positioning data, PIP size, and registration error. All involved a sensor operator performing large-area searches, using a simulated UAV gimbal camera, for specific ground landmarks in a 360-degree region around the UAV. Virtual symbology ("flags") marked landmarks of interest as well as other objects within the environment.

Simulation Environment

The simulation environment was the same as that described earlier, except that the synthetic vision symbology update rate was not manipulated—data on the (simulated) UAV and camera position in space were updated at 10 Hz.

Research Findings

PIP was *not* found to improve landmark searches in AFRL's first study (Calhoun et al., 2006). To test whether the results reflected the multi-colored, heavily annotated aviation map that was used as the synthetic terrain overlay in the virtual surround (making perception of synthetic overlay symbology difficult), a follow-on study utilized a minimalist synthetic terrain elevation overlay instead, which relied on a few colors to depict relative terrain elevation (see Figure 13.3; Draper et al., 2006). This study, in contrast, showed a definite advantage for PIP—average time to designate landmarks was almost twice as fast with PIP, and participants commented that finding landmarks was faster and involved less workload with PIP.

Research was then conducted to evaluate whether the performance advantage observed with the PIP augmentation is dependent on the size of PIP and the accuracy with which the synthetic flags point to key landmarks within the video imagery, i.e., registration error (Calhoun et al., 2007). Participants' performance was compared across three conditions: no PIP; 50 percent PIP (actual UAV video imagery compressed to 50 percent of its original display size); and 33 percent (UAV video imagery compressed to 33 percent of its original display size). Therefore, 33 percent PIP display utilized a larger synthetic imagery border than a 50 percent PIP display. The effective horizontal × vertical field-of-view, in degrees, of the three PIP conditions was approximately 22 × 16, 44 × 32, and 66 × 48, respectively (note that the field-of-view of the actual UAV video imagery never changes; however, the added field-of-view is produced solely through the addition of synthetic imagery in the surround).

Registration error was also manipulated: the virtual flag symbology either accurately overlaid (i.e., pointed to) respective physical landmarks or it was offset by approximately 165 meters. This factor reflected the variable accuracy in aligning a synthetic world and the real world due to variations in update rate and position-sensing accuracy of the aircraft and camera systems.

The results confirmed that performance on a search task improves significantly when a PIP display is used, particularly with the more compressed video imagery, reducing average designation time by 60 percent. Moreover, subjective data indicated that the participants preferred a PIP display for performing the search and designate task. Participants felt their search was more efficient, and the overall task easier to complete, with PIP. Also, participants were less likely to reinspect previously searched areas with the PIP conditions. The negative impact of registration error between the virtual flags and landmarks was mitigated with the PIP capability enabled. Without PIP, average task completion time was over 50 seconds longer, with high registration error compared to low error (Figure 13.4).

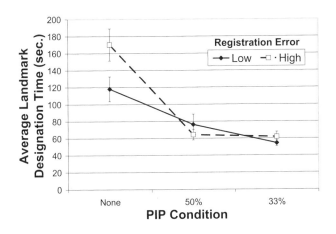

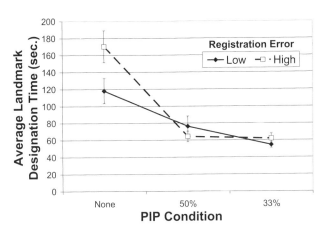

Figure 13.4   Average landmark designation time with each PIP condition as a function of registration error level (error bars are ± standard error of the mean)

In comparing performance between the two PIP sizes, there was little difference in the data. Subjective data were also similar for the two PIP sizes evaluated. However, comments on questions addressing search strategy indicated that the ideal PIP size depended on which step of the task was being completed. Landmark search was rated as faster and easier with the 33 percent PIP, but positioning the cursor over the landmark for designation was easier with the 50 percent PIP (especially when the registration error level was high). Given the previous, along with the finding that most participants recommended that both PIP sizes be made available to operators, it is recommended that more than one size be implemented in applying a PIP augmentation of the camera display for UAV ground station design.

Future Research Directions

Further evaluation of PIP display size is recommended as a function of clutter level, registration error, and for other UAV task applications. Moreover, the optimal number of PIP sizes available to the UAV operator needs to be determined, as well as whether the operator's PIP size selection should be continuous or discrete, the latter involving selection between pre-established sizes. A possibility exists to couple PIP size to control setting: a smaller synthetic border (or perhaps no PIP) would be presented when the operator zooms in the camera view. Likewise, a larger synthetic border would automatically default when the operator zooms out past a certain threshold, indicating a wide area search. Another research issue is whether to fix the sensor imagery to the center of the display or allow the operator to adjust its location within the available display space.

Utility of Synthetic Vision Augmentation as a Camera View Transition Aid

Research Objective

Multi-UAV control applications will require the operator to switch attention between UAVs, each potentially involving very different scenario environments (terrain, threat environment, mission objectives, weather, emergencies, etc.) and task requirements. Not only is there a potential for the negative effects associated with task interruptions and the mental effort required in "context acquisition" after the switch (Olsen and Goodrich, 2003), there is also the potential for negative transfer of context to occur, such that the specific information and tasking involved in the previous mission might delay or degrade the operator's ability to effectively perform tasks in the new mission. For instance, if the operator has a mental model of friendly forces being south of the target in the first mission, will the operator inappropriately apply this mental model to the new mission? A transition aid that employs synthetic vision technology and is designed to enhance an operator's situation awareness when switching between missions in a multi-UAV control environment is now under evaluation. Instead of discretely switching from the camera view on one UAV to the camera view of another, the algorithms driving the transition automatically provide a display format that transitions between the source and new camera views in a semi-continuous manner. With this transition concept, the camera imagery seamlessly fades into a synthetic imagery correlate of the real video image and then uses a "fly out, fly in" metaphor over several seconds, finishing with the transition back from synthetic to real video imagery at the new camera viewpoint. During transition, points of interest are continuously highlighted with overlaid, geo-registered, computer-generated symbology. More specifically, the transition concept provides a three-dimensional perspective of synthetic ground imagery from varying altitudes as the operator switches from an egocentric view (determined by the camera's orientation/viewpoint on the current UAV) to an exocentric view (a global view not tied to any one UAV; a "bird's-eye view"), and then back to an egocentric view (determined by the camera's orientation on the newly selected UAV). The visual momentum provided by a semi-continuous viewpoint transition may aid spatial awareness by helping the operator retain contextual relationships with respect to the overall situation (Azuma et al., 1996). For multi-UAV control, a transition aid that enhances an operator's situation awareness when switching between two camera views (whether viewing the same object/scene or a different one) would be beneficial.

The objective of a recent study was to evaluate whether the transition aid would enhance participants' overall situation awareness and improve performance on a target designation task after switching to a new UAV/camera view (Draper et al., 2008). For this particular experiment, the view smoothly transitioned, retreating over 1.5 seconds upward to an altitude well above the starting camera focal point for the first UAV/mission area. Next, the viewpoint discretely switched to the new location well above the second UAV/mission area, and then descended and rotated, over 5 seconds, to the actual view of the camera located on this UAV. During this transition, the view depicted in the transition consisted of synthetic imagery of the area, with points of interest highlighted with symbology overlays. The evaluation also assessed whether the transition had any negative effects on participants' completion of multiple secondary mission-related tasks.

Simulation Environment

The experiment utilized the Vigilant Spirit multi-UAV operator control station testbed (Figure 13.5). This testbed, developed by AFRL engineers, consisted of two 24-inch monitors, a keyboard, and a right-hand joystick and mouse. The left monitor presented both a global tactical situation display (TSD), showing the path of four UAVs performing missions over different urban areas, and a local TSD (a close-in view, fixed on the center of the loiter pattern of the UAV currently selected). The right monitor presented the view from the gimbaled camera of the specific UAV selected with overlaid synthetic symbology showing

HUD symbology, enemy (red) and friendly (blue) forces, and the Forward Air Control (FAC) center. The video imagery was depicted with the MetaVR Virtual Reality Scene Generator Version 5.3. To the right of the camera window were four UAV thumbnails, the selection of which changed other windows (the local TSD, camera view, and local UAV chat window) to formats specific to that UAV. The other windows on the two monitors (summary panels for each UAV, communications task matrix, health/status matrix, and other chat windows) were used for secondary mission-related tasks. The mouse was used to select windows and complete secondary tasks. Manipulation of the joystick, in two degrees of freedom, controlled the camera orientation in azimuth and elevation. The large center "hat switch" controlled camera zoom, and the center upper button was used to designate locations indicated by the crosshairs.

Research Findings

Results were mixed regarding the effectiveness of this initial design of a transition aid. Participants' ratings on questionnaires indicated that they had more situation awareness in trials with the transition aid, compared to trials without it. However, they failed to perform better in trials with the transition on a probe administered during the mission that was designed to measure context-specific situation awareness. While the transition was not found to hinder performance on secondary tasks, it also did not impact performance on the key task—the average time to locate/designate targets was only slightly faster when the transition was utilized. However, the transition assisted the target designation task in terms of camera movement efficiency. Participants' initial camera movement was more accurate when the transition aid was present. Additionally, the average camera path length

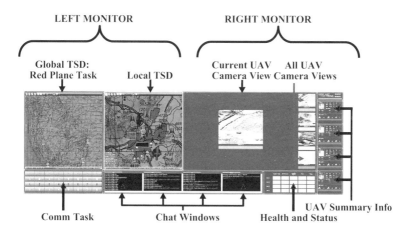

Figure 13.5   Modified Vigilant Spirit multi-UAV operator control station testbed

(length of optimal path between camera initial position and target location divided by length of the camera's actual path) during task completion was less when the transition aid was present when the mission scenario involved changing from a camera view used for a surveillance task to a camera view used for a dynamic target search/designate task (Figure 13.6). The camera path length efficiency measure was worse when both camera views, before and after the transition, were used for a dynamic target search/designate task. Without the transition aid, there was little difference in average path length between the two types of scenario changes.

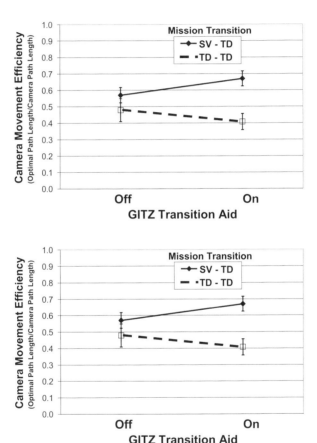

Figure 13.6 **Camera movement efficiency measure as a function of whether the GITZ transition aid was present for two types of mission transitions: surveillance to target designation (SV-TD) and target designation to another target designation (TD-TD). Error bars are ± standard error of the mean**

Future Research Directions

Based on the fact that participants rated the transition format favorably in the questionnaires and that some performance measures showed a benefit, further evaluation of this transition aid for multi-UAV control is recommended. Moreover, the participants identified several issues to be considered in the refinement of the transition aid, ranging from the speed of various segments of the transition to whether or not the operator has direct control over transition parameters. For instance, it is recommended that the "fly out" portion of the transition be minimized, yet remain sufficient, to serve as a stimulus for operators to dissociate themselves from the mission environment of the first UAV. Second, the time spent in an elevated global view over the newly selected UAV location, as well as the "fly in" portion of the transition, should be slightly lengthened, to help operators better acquire a mental map of the new mission area (e.g., threats and terrain feature) for a different UAV. Follow-on research should also examine the nature of missions involved when switching from one UAV/camera view to another and the users' strategy. It has been shown that task switching time is influenced by the stimulus-response mapping characteristics of the two tasks (Rubinstein et al., 2001). This finding, together with the results from the present study showing performance differences as a function of mission scenario type (whether transitioning from a static surveillance mission or another dynamic target search mission), suggest that the design of the transition aid should also consider cognitive processes in task switching.

Summary

Operators of teleoperated robots, such as UAVs, must rely on information from one or more limited field-of-view cameras mounted on the vehicle. Additionally, the video relayed can be compromised by time delays, limited update rates, datalink degradations, and environmental factors. Maintenance of operator situation awareness is challenged by this impoverished yet critical source of information regarding the remote area surrounding the UAV. Integration of synthetic vision systems such that the real-time video stream is augmented with computer-generated symbology can compensate for the undesirable consequences of teleoperation and enhance operator situation awareness. The research described in this chapter demonstrated the utility of augmenting the video display for numerous UAV operator tasks. For the research areas described, performance on sensor management tasks was examined. The results pertaining to update rate and picture-in-picture (PIP) presentation are expected to apply to piloting tasks as well. The potential value of this technology for multi-UAV control applications was also introduced, using synthetic vision to create a dynamic display format to serve as an operator aid for context switching when transitioning attention between UAVs and their associated camera views. For this research area, the results specifically inform applications of single operator supervisory control of multiple UAVs where manual camera control is still occasionally required. However, it is anticipated that a transition aid would also be useful for control paradigms involving multiple operators in a multi-UAV control environment. The generalization of the results presented herein to other UAV operator tasking, missions, and platforms remains to be determined. Additionally, there are numerous other human factors issues pertaining to the design, implementation, and integration of a synthetic vision system in UAV applications that need to be addressed. Follow-on evaluations are required to determine how best to augment real camera imagery with computer-generated information to ensure that UAV operators receive the maximum benefit of this technology.

Chapter Outline

- UAV operators' situation awareness and performance can be compromised by video imagery quality.
- Synthetic vision systems can enhance UAV operations by highlighting, in real time, key information elements of interest directly on the camera video image.
- UAV operator simulation evaluations demonstrated:

  - Computer-generated symbology overlaid on video imagery improved performance on several types of sensor operator tasks.
  - Computer-generated symbology surrounding the video imagery (picture-in-picture) to virtually expand the field-of-view increased operator situation awareness and improved task performance.
  - A computer-generated view that creates a dynamic display format can serve as an operator aid for context switching when transitioning attention between UAVs and their associated camera views.

- It is important to consider human factors issues when augmenting video displays with computer-generated symbology:

  - Symbology design, information clutter, information view management, system performance, operational environment, etc. need to be considered.
  - Use the highest telemetry rate supported by system (>4 Hz desired).

References

Azuma, R., Daily, M., and Krozel, J. (1996). Advanced human-computer interfaces for air traffic management and simulation. In *AIAA Flight Simulation Technologies Conference* (pp. 656–66). Reston, VA: American Institute of Aeronautics and Astronautics.

Bell, B., Feiner, S., and Hollerer, T. (2001, November). View management for virtual and augmented reality, *ACM Symposium on User Interface Software and Technology*, pp. 101–10.

Calhoun, G.L. and Draper, M.H. (2006). Multi-sensory interfaces for remotely operated vehicles. In: N. Cooke, H. Pringle, H. Pedersen, and O. Connor (eds), *Human Factors of Remotely Operated Vehicles* (pp. 149–63). New York: Elsevier.

Calhoun, G.L., Draper, M.H., Abernathy, M.F., Delgado, F., and Patzek, M. (2005). Synthetic vision system for improving unmanned aerial vehicle operator situation awareness. In *Proceedings of SPIE, Enhanced and Synthetic Vision*, vol. 5802, pp. 219–30.

Calhoun, G.L., Draper, M.H., Nelson, J.T., and Ruff, H.A. (2006). Advanced display concepts for UAV sensor operations: landmark cues and picture-in-picture. In *Proceedings of the Human Factors and Ergonomics Society*, pp. 121–5.

Calhoun, G.L., Ruff, H.A, Lefebvre, A., Draper, M.H., and Ayala, A. (2007). "Picture-in-picture" augmentation of uav workstation video display. In *Proceedings of the Human Factors and Ergonomics Society*, pp. 70–74.

Draper, M.H., Calhoun, G.L., Nelson, J., Lefebvre, A., and Ruff, H. (2006, October). Synthetic vision overlay concepts for uninhabited aerial vehicle operations: Evaluation of update rate on four operator tasks. Paper presented at the NATO-HTM-135 Symposium on Human Factors of Uninhabited Military Vehicles as Force Multiplier, Biarritz, France.

Draper, M., Calhoun, G., Ruff, H., Mullins, B., Lefebvre, A., Ayala, A., and Wright, N. (2008, September). Transition aid for changing camera views in UAV operations. In *Proceedings of the First Conference on Humans Operating Unmanned Systems (HUMOUS)*.

Draper, M.H., Calhoun, G.L., Williamson, D., Ruff, H.A., and Barry, T. (2003). Manual versus speech input for unmanned aerial vehicle control station operations. In *Proceedings of the Human Factors and Ergonomics Society*, pp. 109–13.

Draper, M.H., Nelson, W.T., Abernathy, M.F., and Calhoun, G.L. (2004). Synthetic vision overlay for improving UAV operations. In *Proceedings of the Association for Unmanned Vehicle Systems International (AUVSI)*.

Drury, J.L., Richer, J., Rackliffe, N., and Goodrich, M. (2006). Comparing situation awareness for two unmanned aerial vehicle human interface approaches. http://www.mitre.org/work/tech_papers/tech_papers_06/06_0692/06_0692.pdf, accessed 21 March 2008.

Goossens, A.A.H.E., Koeners, G.J.M., and Theunissen, E. (2004). Development and evaluation of level 3 situation awareness support functions for a UAV operator station. In *Proceedings of the 23rd Digital Avionics Systems Conference*, Salt Lake City, UT, pp. 12.D.4-1–12.D.4-8.

Hunn, B.P. (2005). The human challenges of command and control with multiple unmanned aerial vehicles. In *Proceedings of the Human Factors and Ergonomics Society*, pp. 20–24.

Hunn, B.P. (2006). Video imagery's role in network centric, multiple unmanned aerial vehicle (UAV) operations. In N. Cooke, H. Pringle, H. Pedersen, and O. Connor (eds), *Human factors of remotely operated vehicles* (pp. 179–91). New York: Elsevier.

McCann, R.S. and Foyle, D.C. (1994). Superimposed symbology: attentional problems and design solutions. In *SAE Transactions: Journal of Aerospace*, vol. 103, pp. 2009–16.

Prinzel, L.J., Kramer, L.J., Comstock, J.R., Bailey, R.E., Hughes, M.F., and Parrish R.V. (2002). NASA synthetic vision EGE flight test. In *Proceedings of the Human Factors and Ergonomics Society*, pp. 135–9.

Rubinstein, J.S., Meyer, D.E., and Evans, J.E. (2001). Executive control of cognition processes in task switching. *Journal of Experimental Psychology*, 27(4), 763–97.

Olsen, D.R. and Goodrich, M.A. (2003). Metrics for evaluating human-robot interactions. In *Proceedings of PERMIS 2003*. Retrieved 28 March 2008 from http://faculty.cs.byu.edu/~mike/mikeg/papers/OlsenGoodrichPERMIS2003.pdf.

Tadema, J., Koeners, J., and Theunissen, E. (2006, May). Synthetic vision to augment sensor-based vision for remotely piloted vehicles. *Proceedings of SPIE*, vol. 6226, 62260D, Orlando, FL.

Tadema, J. and Theunissen, E. (2003). Feasibility of using synthetic vision technology for UAV operator support. In *Proceedings of the 22nd Digital Avionics Systems Conference*, Indianapolis, IN, pp. 8B1.1–8B1.13.

Theunissen, E., Roefs, F.D., Koeners, G.J.M., Rademaker, R.M., and Etherington, T.J. (2004). Integration of imaging sensor data into a synthetic vision display. In *Proceedings of the 23rd Digital Avionics Systems Conference*, pp. 4A1.1–4A1.10.

Tittle, J., Roesler, A., and Woods, D.D. (2002). The remote perception problem. In *Proceedings of the Human Factors and Ergonomics Society*, pp. 260–64.

van Erp, J.B.F. (2000). Controlling unmanned vehicles: The human factors solution. *RTO Meeting Proceedings 44 (RTO-MP-44)*, pp. B8.1–B8.12.

van Erp, J.B.F. and van Breda, L. (1999). Human factors issues and advanced interface design in maritime unmanned aerial vehicles: A project overview. TNO-report TM-99-A004. Soesterberg, the Netherlands: TNO Human Factors Research Institute.

Wickens, C.D. and Long, J. (1995). Object versus space-based models of visual attention: Implications for the design of head-up displays. *Journal of Experimental Psychology: Applied*, 1(3), 179–93.

Yeh, M. and Wickens, C.D. (2001). Display signaling in augmented reality: Effects of cue reliability and image realism on attention allocation and trust calibration. *Human Factors*, 43(3), 355–65.

PART IV
UGV Research

Chapter 14

Telepresence Control of Unmanned Systems

Chris Jansen and Jan B F. van Erp

Nederlandse Organisatie voor Toegepast Natuurwetenschappelijk Onderzoek

Robots in Non-routine Military Operations

Industrial robots outperform human operators because they are faster, more accurate, and cheaper in the long run. A good example is the pioneering application of welding robots in the automotive industry. Their success can be attributed to the high level of autonomy afforded by the routine character of the perceptual-motor tasks to be performed. Procedures are standardized, and the performance criteria well defined and often unidimensional. The research fields related to robotics have also given attention to more complex, cognitive tasks. See Husbands et al. (2008) for an historical overview of this quest of understanding the human mind and the attempts to replicate it in artificial intelligence (AI). A successful example is playing chess. While chess has been considered as one of the most intelligence-demanding activities, computers nowadays defeat human experts. Whether chess-winning computers therefore have intelligence has been a topic of debate. On the one hand, the game of chess is simple: There are just a few easy to understand, well-defined rules that should be followed, and the 8-by-8 theater is equally simple. In that sense, one might argue that the chess task is as routine as welding: procedures and rules are clear, as is the performance criterion for victory. On the other hand, chess is complex and highly intellectually challenging because the interaction of these simple rules generates many possible strategies and options for each move. In other words, the *problem space* (Newell and Simon, 1972), generated from the finite set of relevant objects and actors, their properties, and their relations, is complex; and, therefore, the task is complex as well. AI has proven to be successful on this complex task because of the well-defined problem space.

   Task performance can be challenging due to a complex problem space (as in chess), but also due to an ill-defined problem space. A problem space is ill defined when there is uncertainty about what objects and actors are involved in the situation at hand, and what their properties, goals, and interrelations are. Klein (1998) states that AI is less suited for tasks involving ill-defined problem spaces. The reason is that AI is very good at *selecting* a good solution among available options, and that this selection process requires a well-defined problem space. Human experts, on the other hand, are good at *generating* new solutions by seeing opportunities and linking new information with seemingly unrelated former experience in similar situations. This ability to generate new solutions is crucial for performing non-routine tasks.

In the military domain, one can think of several tasks that are to a large extent routine, for example land mine detection and compound security. For these kinds of applications (semi-)autonomous robots are currently being developed and deployed. Other military operations are most often not routine at all, but rather full of unanticipated events. For example, one cannot know in advance exactly what to expect in a reconnaissance operation, by definition. It remains a challenge for robots to deal with such dynamic situations that require capacities as dealing with uncertain and incomplete information and intuitive decision making.

In this chapter we focus on employing robots or unmanned systems in these non-routine applications. Even though human performance would most often be superior in doing the task, military unmanned systems are employed to increase safety for military personnel. The main objective for these military robots therefore is to project the human operator's capabilities into distant and often dangerous environments without exposing the operator to the threats that are present at the remote environment. One of the most important performance criteria in human-robot interaction (HRI) in military operations is how high-level cognition (by either an autonomous robot or a teleoperating human) can be optimized at the remote location. The technological challenge in this respect is to design robot systems that perform as well as a human operator performing the task locally. We believe that such a design should benefit from advances in increased autonomy and AI as well as advances in user-interface design and wireless communication technology.

Irrespective of the level of robot autonomy, human operator involvement is important for several reasons (Ianni, et al., 2002). First, all tasks, including those performed by a robot, are part of an overall military operation. So, even when a robot operates autonomously, a human being should be involved for monitoring progress and mission outcome. Also, this autonomous or remotely operated robot will likely collaborate with other soldiers. Concerning such human-robot teaming for accomplishing a common goal, the robot can have several roles, as Sycara and Lewis (2004) identified for software agents. A robot could assist an individual human team member, be a fully fledged teammate itself, or have a supervising/ supporting role to enhance team performance (e.g., team member task allocation, facilitating team communication). A second reason for human involvement in robot autonomy concerns the ethical and political considerations needed in military decision making. Third, as mentioned above, human intervention can enlarge the operational capabilities, particularly in inherently non-routine tasks. The human operator outperforms autonomy if things do not go as planned, which is often the case. The enemy is unpredictable, situations differ from those sketched in a briefing, subsystems fail, or the robot is confronted with sudden threats. In these cases human control, information processing, and decision making qualities as well as awareness of situation and system state are likely to be indispensable to complete the mission. The question is, then, how can this human involvement be best accomplished? Parasuraman et al. (2000) proposed a framework for evaluating the interaction between human operators and semi-automated systems. They identified four levels of information processing in performing the task to which the interaction applies: (1) information acquisition; (2) information analyses; (3) decision and action selection; (4) action implementation. The human-automation interaction is then described by the degree of automation for each of these levels. Analyzing the performance consequences results in the optimal choice for automation degree for each of the information processing levels. The more routine the task, the better AI is equipped to facilitate autonomous robot, and the less the operator needs to be in actual control. The operator can be "back" in the loop, mainly supervising how the robot performs the task. In case the operator needs to intervene, the predictable task characteristics enable him to quickly and easily regain situation awareness (SA), almost from scratch. However, non-routine tasks are generally too complex and unpredictable for deployment of autonomous robots. Furthermore, if a supervising operator would need to intervene, he could not easily and instantaneously build up SA. So, how should remote control of (semi-autonomous) robots be designed in order to optimize performance in non-routine military situations? We argue that the operator should be more directly involved in task performance to build and maintain full SA and to deal with complexities and unexpected events at the moment of occurrence. This calls for an interface that has the operator in the front of the control loop. A promising possibility to establish this is *telepresence*. In this chapter, we will first explain why telepresence human-robot interfaces (HRIs) are promising. Then we will illustrate the performance benefit of telepresence with respect to traditional teleoperation and semi-automation as observed in a reconnaissance task. We will conclude with a discussion on the complex relation between telepresence and task performance.

Telepresence Human-Robot Interfaces

Early HRIs consisted mainly of visual displays containing the sensor image and rudimentary status and planning information. These robots were mostly operated manually with joysticks. In general, two operators were involved: one was controlling the vehicle, the other the sensors. These systems were not easy to operate, mainly because of limitations in the datalink and of difficulties in building up situation awareness (SA) by the operators (Riley et al., 2006). As a result, operators experienced high visual and mental workload, seriously hampering their performance. We consider SA as a critical aspect in controlling robots. We follow Endsley's classic definition (Endsley, 1988; Endsley & Kiris, 1995), which states that situation awareness is the perception of the elements in the environment within a volume of time and space, the comprehension of their meaning, and the projection of their status in the near future. Endsley integrated SA in the perception–decision–action loop and identified three levels of SA. Level 1 refers to the perception of the elements in the environment and their relationship to other points of reference (i.e. internal model); level 2 goes beyond simply being aware of the elements and includes an understanding of the significance of the elements; and level 3 refers to the ability to project the future status of the elements in the environment. Especially for ill-defined, non-routine tasks, the higher SA levels are crucial.

Technological improvements can potentially solve the problems mentioned above: datalinks have become broader and more robust, and vehicles, sensors, and interfaces more intelligent and more advanced. HRIs are being further improved by adding, for example, synthetic visual and audio information, tactile cues (see Haas & van Erp, Chapter 15 this volume). The good sensory feedback provided by these interfaces positions the operator more efficiently in front of the control loop. Taking this good sensory feedback a step further, HRIs could be designed with which the operator is perceptually immersed in the remote location. "Immersion" refers to the interface (and not to a psychological construct like presence) and the extent to which the interface provides sensory cues of the remote environment and blocks out or isolates the user from the real world (Biocca, 1992; Kalawsky, 1993; Sadowski & Stanney, 2002; Witmer & Singer, 1998). Please note that immersion requires good sensory feedback, but that good sensory feedback does not necessarily lead to immersion. Central to this HRI design philosophy is to enable the operator to optimally use his perceptual, cognitive, and psychomotor systems, thereby making the operator's perception and action as similar as possible to those when present at the remote location. In other words, the interface allows the operator to be telepresent.

The idea of telepresence interfaces is not new. A limited database search showed that the first papers with telepresence in the title were published in the early sixties (Caldwell & Reddy, 1960). Thereafter, the term seemed to be forgotten for about 20 years until it re-emerged in the early eighties. The number of published papers with telepresence in the title reached a steady ten per year in the mid-nineties and has risen to more than 20 each year in the last four years. Interestingly, despite the widespread use of the term "telepresence" there is no single, widely accepted definition. Most definitions share the reference to the quality of the *user interface* or to an *experience* that involves displacement of oneself into a remote environment; that is, a more technological approach or a more psychological approach, respectively (Draper et al., 1998). We prefer the psychological approach and reserve the term immersion to refer to the (interface) technology. Wikipedia uses a more technological approach (retrieved December 2007): "Telepresence refers to a set of technologies which allow a person to feel as if they were present, to give the appearance that they were present, or to have an effect, at a location other than their true location." This technological approach refers to the design and efficiency of the displays and controls, and projecting the operator's *capabilities* only (assuming that good sensory feedback automatically results in telepresence). Representative of the psychological approach are Sanchez-Vives & Slater (2005). They believe that the concept of "presence" refers to the phenomenon of behaving and feeling as if we are in the virtual world created by computer displays. (Please note that, according to Hendrix & Barfield (1995), virtual presence and telepresence are the same from a psychological point of view.) The important difference with the psychological approach is that the psychological approach refers to user characteristics, user experiences, and mental state (Akin et al., 1983), and is focused on projecting the operator's *consciousness* (Fontaine, 1992) rather than just his capabilities. In the psychological approach the user interface is often mentioned, but rather as a means than as a goal (see also Draper et al., 1998).

The main reason for the (regained) interest in telepresence is the expected benefit for operator performance. Someone experiencing transparent telepresence would be able to behave, and receive stimuli, as though he were at the remote site (TTRG, 2007). Van Erp and colleagues (2006) hypothesized that telepresence interfaces allow operators to employ their powerful psycho-motor capabilities for many tasks that would otherwise put heavy demands on limited cognitive resources. This shift from the cognitive to the psycho-motor level was expected to reduce workload and increase performance and situational awareness (SA). In the next section, we will present several theoretical approaches to telepresence. After this, we will present new data gathered with a robotic system designed to elicit telepresence, shedding some light on the complex relation between telepresence and task performance. We will limit ourselves to a situation where one operator is in full (manual or semi-autonomous) control of a single robot vehicle. Issues such as team performance and shared situational awareness, multi-robot interaction, and other important human-robot interaction issues are not dealt with in this chapter.

Draper et al. (1998) wrote the first exhaustive paper on explanatory approaches to telepresence. They postulated a model based on dividing attention between the local and the remote environment and the presence of task relevant and irrelevant information in both the local and the remote environment. The amount of telepresence is related to the amount of attention devoted to task-relevant or task irrelevant information in the remote environment. However, task performance is related to the amount of attention devoted to task relevant information, whether available in the local or the remote environment. In other words, processing task relevant, local information is advantageous for performance but may decrease the telepresence experience. Local, task irrelevant information includes local tactile and auditory cues. A similar reasoning can be made regarding SA. Operators may have local SA, remote SA, or a combination of both. Telepresence may be considered the maximization of SA in the remote environment (remote SA) accompanied by loss of SA for the local environment.

Even though Draper's model is successful in describing telepresence situations, Ma and Kaber (2006) conclude that there remains a limited understanding of the factors involved in presence and their relation to performance. Biocca (2001) tried to shed some light on the true nature of presence by suggesting that the philosophy of presence might be most fruitfully approached via the philosophy of mind. It is hypothesized that presence has a link to the mind-body problem and the issue of mediated embodiment (described as the fuzzy boundary between the body and technological extensions of the body). Interestingly, two recent papers in *Science* used telepresence-like technology to evoke out-of-body experiences (OBE). Lenggenhager et al. (2007) used a head-mounted display (HMD) presenting live images of a camera placed 2 meters behind the observer. OBE-like experiences were reported when the observer was subjected to tactile stimulation (stroking) that was in synchrony with stroking of their "mediated body" in the camera image. In this situation, tactile stimulation was used to convince the observers that the body they were looking at was actually their own, and thus that their eyes were dissociated from their body. Ehrsson (2007) used a similar setup with a camera positioned behind the observer, but with one important difference. Again, tactile stimulation of the observer and visual feedback of the stroking were used, but the visual stroking was located at the position of the cameras. In this situation, tactile cues served to convince the observer that their body was located below the cameras. This setup resembled a telepresence situation. Indications that the observer's "consciousness" was actually displaced to the remote (camera) location stems from the fact that threatening the "virtual" body (that underneath the cameras) with a hammer elicited (physical) anxiety responses in the observer, despite the fact that their physical body was in no danger at all. These experiments seem to indicate that haptic information is important in eliciting an out-of-body experience, confirming results obtained with the rubber hand illusion. When a person is watching a fake hand being stroked and tapped in precise synchrony with his own unseen hand, the person will, within a few minutes of stimulation, start experiencing the fake hand as an actual part of his own body. However, when participants try to move the fake hand, or when there is a small delay between the visual and tactile stimulation, the illusion will diminish or break (Armel and Ramachandran, 2003; Botvinick and Cohen, 1998; Tsakiris and Haggard, 2005). These findings indicate that a telepresence interface may benefit from including haptic or multimodal feedback as long as the feedback is in synchrony with user's action and over sensory modalities.

The magnitude of the response may differ from person to person. Sas & O'Hare (2003) investigated the relationship between presence and four personality factors: absorption, creative imagination, empathy, and willingness to experience presence. They found correlations between presence and each of these personality factors. Persons who were more absorbed, more creative, more empathic, or more willing to be transported to the virtual world experienced a greater sense of presence, indicating that individual differences are important. Apart from these personality factors, Witmer & Singer (1998) identified three other factors involved in telepresence. First, control factors, which refer to the extent to which the user has responsive and intuitive control over the environment. Second, distraction factors, indicating the degree to which the operator is immersed in, and focuses his attention on, the remote location. Third, sensory factors refer to the richness and multimodal consistency of available information and how intuitively it is displayed. Slater and Usoh (1993) postulated two sets of determinants of the sense of telepresence: external factors and internal factors. External factors include display quality, the ability to interact with the environment and of the environment to interact with the user, and clarity of causal relationships between user actions and reactions in the remote environment. The internal factors are sensory modality (visual, auditory and kinesthetic) and view point (first, second, or third position, referring to seeing, hearing, or feeling as if one is an actor, an observer, or not present at all).

Above we broadly outlined the concept of telepresence and how it relates to possible applications to military robots performing non-routine tasks. The following part presents research on telepresence, semi-autonomy, and performance for a non-routine task.

Telepresence and Semi-autonomy in a Reconnaissance Task

One application envisioned for military robots is reconnaissance. We consider such a task to be non-routine because the situation is unknown and unpredictable. As indicated above, we expected that having an operator in the loop would be advantageous for robot and sensor control task, resulting in better reconnaissance. In the experiment described below, we compared two offspring of advanced technology, semi-autonomy and telepresence, as control modes for operating an unmanned ground vehicle in a reconnaissance task. We hypothesize that a full telepresence HRI outperforms both teleoperation and semi-autonomous operation.

Method and Participants

Seventy-two male participants were paid to take part in the experiment. The participants were selected on not having experienced motion sickness in the past. Their ages ranged from 18 to 37 years (on average 22.8 years).

Task

Two operators participated in a single run. One of the participants had to remotely operate an unmanned ground vehicle, as well as the camera with pan-tilt-roll unit positioned on top of the vehicle. The vehicle was placed in a large enclosed area of about 140 m^2, unknown to the participants, with several obstacles and objects. The obstacles were tall enough to block the view on the other side of them. While one participant was exploring the area, another participant was watching a screen presenting the same camera images. Prior to exploration of the area, the participants were shown photographs of 15 objects they had to search for while driving through the area (e.g., a telephone, a fire extinguisher, a large plant). Unknown to the participants, only ten of these 15 were actually present. They were given 15 minutes to navigate through the area, look for the objects, and build up a mental image of the area. Both participants were questioned on what objects they had or had not seen, what the area looked like, and where the objects were positioned.

Apparatus

The unmanned ground vehicle has been custom made at TNO Soesterberg and has been used in other studies as well (van Erp et al, 2006). The vehicle has been specifically designed for telepresence control, using a pan-tilt-roll unit with a stereo camera system mounted on top of it. No force feedback was available; the operator did not directly interact with objects in the environment (e.g., picking up, replacing). For the current study, the vehicle could be controlled in three modes using two control stations.

1. The *Telepresence* control station consisted of a head-tracking HMD (dual image, mono VGA), a steering wheel, and an accelerator pedal. The head-tracker directed the pan-tilt-roll unit, and the HMD displayed the mono sensor images. This setup gives the operator the experience of naturally looking around at the remote location. Vehicle control was facilitated by two virtual vertical lines overlayed over the video images (augmented reality). These virtual vehicle references indicated the width of the vehicle as well as the front of the vehicle (see van Erp & Padmos, 2003).
2. The telepresence control station was also used in a second mode called *Autonomous Navigation*. As in Telepresence mode, the operator could look around using the head-tracking HMD system coupled to the pan-tilt-roll unit with stereo cameras. The only difference with the telepresence control mode is that the participant did not use the steering wheel and accelerator; autonomous navigation was simulated by manually pushing the vehicle along a predefined route by an experimenter. Again, virtual vehicle references were present in the video image.
3. The third control mode using the second control station is referred to as *Teleoperation*. This control mode consisted of two joysticks, for vehicle and sensor control, respectively, and a screen on which the sensor image was presented. The roll component of the pan-tilt-roll unit was not controlled, as is the case in commercially available teleoperation control stations. As for telepresence control, virtual vehicle references were presented.

In all modes, the images of the camera system on the vehicle were also presented on a second screen, where another participant watched alongside (*Passive Observation* mode, see below). The control room was an enclosed 40 m^2 area adjacent to the area to be reconnoitered, where the control stations were set up. This room was also used for training, so the participants could practice maneuvering the vehicle through this area before the experiment started. The control area was also used for instructing the participants and for completing the questionnaires.

Design

The between-subject design consisted of two independent variables. The first variable, Control mode, had three levels and determined the way the vehicle was controlled: Telepresence, Autonomous Navigation, and Teleoperation. The second independent variable, with two levels, was Observation mode: Active Observation (the participant actually operated the sensor unit and, depending on the test condition, the vehicle) and Passive Observation (the participant watched the sensor images presented on a separate screen). The combination of control mode and observation mode led to a between-subject design with six conditions, 12 participants per condition. The main dependent variable was the participants' level of situational awareness of the 15 objects of interest. Immediately after exploring the area, participants were asked to write down which of the 15 objects they were instructed to search for were actually encountered in the experiment area. This Object Recall session was followed by an Object Recognition session in which 21 photographs of objects were presented to the participants. Ten objects were on the list and in the area, five objects were on the list but not in the area, and six objects were not on the list but were in the environment. For each photograph, the participants were asked to indicate whether they had seen the object in the area, and how certain they were of their answer.

Procedure

On each day, four sessions were held, with two participants per session. One participant operated the sensor and the vehicle (only the sensor in Autonomous Navigation mode) and the other participant watched the same sensor image as a passive observer. After initial instruction and filling out the informed consent form, the participant operating the vehicle practiced maneuvering the unmanned ground vehicle in the control mode he would use during the test. After this, both participants took a place behind a table in front of the projection screen on which 15 photographs of objects were shown, which the participants had to search for in the experiment area. Each of the 15 objects was shown six times in random order. While the participants were viewing the objects, a second experimenter put the remotely controlled vehicle in its starting position in the experiment area. The trial started when the operating participant had taken their place in the control station, and the other one behind the monitor. The area was explored for 15 minutes. If in control of the vehicle, the participants could manuever it freely through the experiment area. Their chosen path was plotted on a map by the experimenter. Afterwards, both participants were seated behind the table again to fill out questionnaires.

Results

We investigated recall and recognition performance. Recall performance on objects of interest was measured by asking the participants to make a list of the objects presented on pictures before exploring what were actually present in the environment. A correct response was defined as reporting an object that was shown to the participants beforehand as possibly being present in the environment and that indeed was present; an incorrect response was defined as reporting the presence of such an object whereas it actually was not present. We defined as our main dependent variable the number of correct responses minus the number of incorrect responses, as a percentage of the maximum number of correct responses. Maximum score is 100 per cent (that is, (10-0)/10 *100%); minimum score is -50% (that is (0-5)/10 *100%). Figure 14.1 shows the results on object recall obtained in the six experimental conditions. The analysis of variance (ANOVA) revealed a significant main effect for Control mode ($F(2,66)=3.58$, $p=.033$). The Tukey honestly significant difference (HSD) test showed higher scores for Telepresence (65.8 per cent) than for Autonomous Navigation (55.8 per cent); both did not differ from Teleoperation (59.2 per cent). Observation mode ($F(1,66)=2.59$, $p=.112$) and its interaction with Control mode ($F(2,66)<1$) were both non-significant. The later results indicated that the suggested performance benefit in Active Observation (comparing the left bars of each pair in Figure 14.1) for Telepresence as compared to Autonomous Navigation and Teleoperation is statistically non-significant. To investigate the significant main effect for Control mode in more detail, we analyzed correct and incorrect responses separately. These analyses revealed that the higher score in Telepresence mode was due to more correct responses (Control mode: $F(2,66)=3.165$, $p=.028$; Telepresence 71.3 per cent, Autonomous Navigation 62.1 per cent, Teleoperation 67.9 per cent) rather than fewer incorrect responses (ANOVA revealed no significant effects).

All pictures of objects presented prior to reconnoitering the area were presented again afterwards to measure the recognition performance of the objects of interest. For each presented object, the participants had to indicate the extent to which they were sure the object was either present or absent in the environment. Scoring went from very sure of absence to very sure of presence: -2, -1, 0, 1, 2. Each item score was transformed such that correct responses are represented by positive numbers. For example, when a participant correctly indicated that he was sure that an object was absent, that score of -2 was transformed to 2. Likewise, an incorrect indication of an object's presence (score of 1 or 2) was transformed in -1 or -2. Our dependent variable for object recognition was the average transformed item score, theoretically ranging from -2 (all responses were rated very sure but were incorrect) to 2 (all responses were rated very sure and were correct). Figure 14.2 shows the object recognition results obtained in the six experimental conditions. The ANOVA indicated a significant effect for Observation mode ($F(1,66)=8.68$, $p=.004$), with a 0.21 higher score for Passive (1.26) than for Active Observation (1.05). Although the score in Telepresence mode (1.25) seemed to be higher than for Autonomous Navigation (1.13) and Teleoperation (1.08), the main effect for Control mode was non-significant (F(2,66)= 1.93, p=.15), as was the interaction effect (F<1).

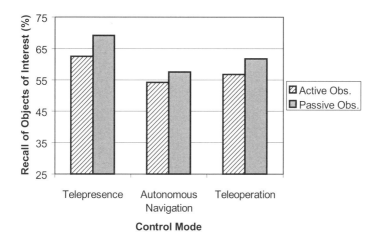

Figure 14.1   Average percentage recall of objects of interest

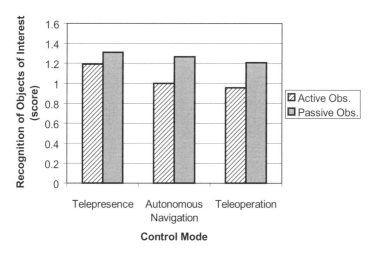

Figure 14.2   Average recognition scores for objects of interest

Conclusion and Discussion

The results show that in this reconnaissance task, considered to be non-routine, telepresence HRI resulted in superior performance. Participants recalled about 10 per cent more objects when in Telepresence mode than when in Autonomous Navigation mode. Note that in both modes the participants exploring the area wore a head-tracking HMD system. Apparently, being able to have control of the vehicle movements also improved reconnaissance performance. This statement is further supported by the observation that performance in Autonomous Navigation mode, using the same the user-friendly, head-tracking HMD system, was no better than the fully joystick-controlled performance, for both sensor and vehicle, Teleoperation mode. The superiority of Telepresence appeared to be present in object recognition as well in Active Observation, but this finding did not reach statistical significance.

A finding that turned out to be significant is that recognition performance was higher when the participant only had to passively look at the sensor images, without having to control sensor and vehicle. This finding corresponds with the idea introduced earlier that being more telepresent does not necessarily guarantee better performance. One can presume that the active operator (wearing an HMD) is more immersed in the remote environment than the participant passively observing the sensor images on a monitor. Apparently, these images on the monitor provided the passive observer with sufficient information for recall and recognition of the objects of interest. The real benefit of telepresence could have been in better spatial SA. We investigated this aspect by asking participants to indicate on a map the locations of the objects. Again, passive observers outperformed active observers operating the robot. Also, we presented the participants with pictures taken from the perspective of the robot of several positions in the remote environment; the participants had to indicate the direction of object locations. The results showed no performance differences between experimental conditions. Thus, our study did not reveal superior performance for telepresence on spatial situation awareness. So should this finding lead to the conclusion that robots should be automated even for non-routine tasks such as reconnaissance? We believe not: performance on recall is worst in Autonomous Navigation mode, while recall is crucial in reconnaissance.

A possible account for better performance in Passive Observation is that participants in Active Observation mode only had a minimal amount of practice before starting the test. These 10 minutes of training were enough for getting acquainted with controlling vehicle and sensors, but not at the level of expert operation. Therefore, too many perceptual and cognitive resources may have been needed in operating the sensor and vehicle, leaving less attention for spotting the objects in the environment. Indeed, we observed a drive–stop–watch pattern in exploration behavior in Telepresence and Teleoperation mode, which suggests that vehicle operation and reconnoitering of the area are difficult to perform simultaneously. Interesting in this respect is the difference between the two conditions. On average, participants in Teleoperation stopped 19.8 times in the

15-minute exploration, while those in Telepresence stopped only 16.6 times. We therefore hypothesize that participants were better able to integrate observing and driving around because operating the vehicle and driving around required less cognitive resources in Telepresence mode. We expect that with more practice in controlling vehicle and sensors, the performance benefit of Telepresence control of both sensor and vehicle will be larger. Even though we observed better performance for Passive Observation than for Active Observation, Sharples et al. (2008) report that passive observers are more susceptible to motion sickness.

Telepresence and Task Performance

Earlier Draper et al. (1998) and Deml (2006) hypothesized that telepresence may or may not enhance performance. Our data showed that the relation between telepresence and performance is indeed not straightforward. Therefore, we introduce a model, outlined in Figure 14.3, to describe telepresence, performance, and workload in telepresence HRI. The model incorporates ideas presented by other researchers mentioned previously.

Any HRI can be described by operator actions providing input to the robot's actuators, and robot's sensors providing perceptions of actuator actions as well as other cues present in the remote environment to the operator. This basic loop is represented by the dark-gray marked blocks and the black arrows. Interfaces providing good sensory feedback enable the operator to be in front of the loop. Telepresence requires in addition immersiveness (Witmer & Singer, 1998), and an attentional focus on the remote environment (Draper, 1998). Therefore, we expand the loop in two ways for describing telepresence and performance; by including proprioception, and by including involvement of cognitive processes.

The first extension takes into account that the operator's actions will not only be fed into the HRI but also directly to the operator through proprioception (in Figure 14.3 depicted by arrow 1, from "action" to "perception"). This proprioception results in an interface that provides good perceptual feedback profitable for operators in the loop tasks. An example is head-slaved camera control where the proprioceptive system gives high-quality and unambiguous information on viewing direction and camera motion (Massey et al., 2006). Note that these actions are actually performed in the operator environment and may therefore also affect the local environment in which the operator is located. All this directly and indirectly perceived information for the basic loop plus proprioception contains task relevant and irrelevant cues from both remote and local environment. With this setup we can describe an important difference between maximum telepresence and performance. For performance, the relevant remote and local cues are crucial. The irrelevant cues should be ignored by the HRI because they will only increase workload without contributing to task performance. For optimal telepresence, however, all remote information is important, both task relevant and task irrelevant cues; all local cues, even when task relevant(!), should be blocked by the HRI. The fewer local cues reach the operator, the more he or she will concentrate on the remote environment or be immersed.

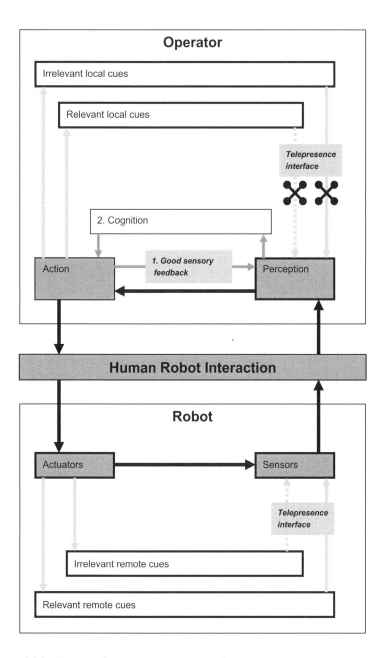

Figure 14.3   Model of telepresence and performance

The second extension of the basic model concerns cognitive processing of information (in Figure 14.3 arrow 2, from "perception" to "cognition" to "action") in addition to perceptual processing (arrow from "perception" directly to "action"). Cognitive resources are needed to analyze and interpret the perceived information, leading to decision-based actions. For example, memory and attention are needed to determine which route to take. This links our model to Endsley's SA model (see also the section on telepresence human-robot interfaces). We distinguish the allocation of cognitive resources to robot control and to actual task completion. In our experiment mentioned above the actual task was reconnaissance, not robot control. We only used a robot because the area was, hypothetically, too dangerous to go in ourselves. Perceptual processing refers to the basic model in Figure 14.3: sensations directly evoke actions, without involving cognition. Examples include maintaining our balance, braking when a child suddenly crosses the road, or other reflexive or highly trained tasks. Such shortcuts resemble the skill-based level of Rasmussen's classical model (Rasmussen, 1982, 1983; Vicente & Rasmussen, 1988), defined as "well-learned sensory motor performance in continuous manual control tasks in stationary conditions." A bit more cognition is involved in Rasmussen's rule-based level, consisting of so-called "if … then" rules. For example, when you see a stop sign while driving, you decelerate. These kinds of processes do not involve conscious decision making, but require a low-level interpretation of the visual information as a stop sign (which is not needed when diving down when a baseball is coming right at you). Bypassing cognition seems trivial for situations like keeping our balance and lane keeping while compensating for side wind. However, when the sensation is mediated by a display, the design of the display is the critical factor in whether this will be possible or not (van Erp & Werkhoven, 2006). We believe that traditional teleoperated systems are hard to operate because the presented information cannot lead to action without cognitive involvement. For example, the operator sees an image from the robot camera displayed on the screen in front of him; for building up correct spatial situation awareness, he has to figure out first what the camera viewing direction is (cognitive processing). Comparing this with the provided proprioceptive information referred to earlier, head-slaved camera control systems instantaneously provide unambiguous information on viewing direction (perceptual processing, no cognition involved). With this second extension of the basic loop, the model states that a good HRI (e.g., head-slaved camera system) affords robot control that is primarily based on perceptual processing, thereby saving the operator's cognitive resources for performing the actual task, like reconnaissance. Or, in terms of Endley's SA model, a good HRI ensures good Level 1 SA (i.e., location and speed of objects and their relations), which is beneficial for Level 2 and 3 SA (see van Erp, 1999). Bypassing cognition in our model also matches the internal factors as mentioned by Slater & Usoh (1993) such as clarity of causal relationships between user actions and reactions in the virtual world. The more unfamiliar an operator is with the interface (due to inferior design or lack of required training), the more cognitive processing is required for robot control and building Level 1 SA, and the fewer cognitive resources are available for the actual task to be performed. We believe that telepresence HRI leads to improved performance when it succeeds in saving as many cognitive resources as possible for performing the actual task.

How can we relate this to our experimental results? In our setting, there were no task relevant cues in the local environment where the operators were located. The operators wearing the head-tracking camera system (in Telepresence mode and Autonomous Navigation mode) were less exposed to task irrelevant local cues and thus had an advantage over operators in Teleoperation mode. But these Teleoperation operators, as well as the Telepresence operators, had proprioceptive feedback (model extension 1) on their own inputs on vehicle control in addition to sensor control, giving them an advantage to Autonomous Navigation operators. However, our results suggested that the advantage for Teleoperation was apparently less than for Telepresence because more cognitive processing (model extension 2) can be expected for keeping track of viewing direction and vehicle displacement. In sum, Telepresence benefits from both advantages, and thus results in superior performance.

We conclude that telepresence HRI *can* result in improved performance, but does not necessarily do so. A performance drawback of telepresence is that its technology is currently suboptimal. For example, HMD displays are inferior to state-of-the-art desktop screens (e.g., on resolution, image refresh rates, field-of-view). Also, most technological requirements, such as minimum delay and large communication bandwidth, are similar to that for any kind of teleoperation, but much more stringent for telepresence. For example, a video signal delay of 200 ms might be awkward to deal with in teleoperation, but is unacceptable in telepresence due to induced operator motion sickness. Sharples et al. (2008) indeed report higher motion sickness ratings for HMD interfaces as compared to desktop and projection systems. Even when these issues are resolved, we believe that the possible surplus value of telepresence HRI is likely to be limited to the following conditions. First, the task to be performed is complex and non-routine. A lot of cognitive processing is needed for interpreting rather incomplete and uncertain information, and for intuitive decision making. The operator needs to perceive as much information as possible in the remote environment because task relevance of information cannot be determined in advance. In these situations, one would like to have a sixth sense. Second, telepresence will boost performance when operator workload is expected to be high. A telepresence HRI will preserve cognitive resources for performing the actual task since the robot is mainly controlled using perceptual processing. Third, telepresence suits mission tasks that allow full immersion in the remote environment. In other words, the local environment of the operator does not contain information cues that have to be integrated with remote information cues. Fourth, the remotely operated robot needs to collaborate and interact with humans in the remote environment. Telepresence facilitates such remote shared situation awareness. Fifth, telepresence should be applied in situations in which operator's emotions such as combat stress, anxiety, and ethical judgments play an important role in successful task performance. We expected these emotions to be more present with telepresence HRIs. If none of these situations applies, multimodal displays (Haas & van Erp, this volume) supporting the basic loop of the model can lead to as good or even better performance.

Key Points

- Robots are often deployed when military tasks are too dangerous to be locally performed by human operators. However, a robot's artificial intelligence cannot always perform as well as a human operator would have done.
- Non-routine tasks are generally ill-defined, with great uncertainty concerning the events that may occur. Therefore, human operators should be more involved in robot control in non-routine than in routine tasks.
- Human operators need to have good situation awareness in order to adequately deal with this uncertainty, and in order to generate solutions for complex, ill-defined problems that occur.
- Telepresence interfaces are promising in providing this good situation awareness because the operators can hear and look around as though they were present in the remote area.
- We observed in a reconnaissance task that telepresence did indeed improve an operator's performance as compared with traditional teleoperation and robot semi-autonomy.
- Telepresence will specifically improve performance when the task is complex and non-routine; when operator workload is high; when full operator immersion is required; when the robot needs to collaborate with other humans in the remote theatre; or when human emotions need to be involved in mission-relevant decision making.

References

Akin, D.L., Minsky, M.L., Thiel, E.D., & Kurtzman, C.R. (1983). *Space Applications of Automation, Robotics and Machine Intelligence Systems (ARAMIS)—Phase II*, vols 1–3 NASA Contractor Reports no. 3734–6, for Contract NAS8-34381. Huntsville, AL: NASA, Marshall Space Flight Center.

Armel, K.C. & Ramachandran, V.S. (2003). Projecting sensations to external objects: Evidence from skin conductance response. *Proceedings of the Royal Society of London B*, 270, pp. 1499–506.

Biocca, F. (1992). Virtual reality technology: A tutorial. *Journal of Communication*, 42(4), 23–72.

Biocca, F. (2001). Inserting the presence of mind into a philosophy of presence: A response to Sheridan and Mantovani and Riva. *Presence: Teleoperators and Virtual Environments*, 10(5), 546–56.

Botvinick, M. & Cohen, J. (1998). Rubber hands "feel" touch that eyes see. *Nature*, 391, pp. 756–57.

Caldwell, D.G. & Reddy, K. (1960). Sensory requirements and performance assessment of tele-presence controller robots. *Proceedings of IEEE International Conference on Robotics and Automation*, pp. 1370–75.

Deml, B. (2006). Human factor issues on the design of telepresence systems. *Presence: Teleoperators and Virtual Environments*, 16(5), 471–87.

Draper, J.V., Kaber, D.B., & Usher, J.M. (1998). Telepresence. *Human Factors*, 40(3), 354–75.

Endsley, M.R. (1988). Design and evaluation for situation awareness enhancement. *Proceedings of the Human Factors Society 32nd Annual Meeting*, pp. 97–101.

Endsley, M.R. & Kiris, E.O. (1995). The out-of-the-loop performance problem and level of control of automation. *Human Factors*, 37(2), 381–94.

Ehrsson, H.H. (2007). The experimental induction of out-of-body experiences. *Science*, 317(5841), 1048.

Fontaine, G. (1992). The experience of a sense of presence in intercultural and international encounters. *Presence*, 1, 482–90.

Haas, E. & van Erp, J.B.F. (2010). Multimodal research for human-robot interactions. This volume.

Hendrix, C. & Barfield, W. (1995). Presence in virtual environments as a function of visual and auditory cues. *Proceedings of the Virtual Reality Annual International Symposium*, pp. 74–82. Piscataway, NJ: IEEE Computer Society.

Husbands, P., Holland, O., & Wheeler, M. (2008). *The Mechanical Mind in History*. Cambridge, MA, London: MIT Press.

Ianni, J.D., Repperger, D., Baker, R.W., & Williams, R.L. (2002). Human interfaces for robotic satellite servicing. *Proceedings of SPIE—The International Society for Optical Engineering*, 4632, pp. 95–103.

Kalawsky, R.S. (1993). *The Science of Virtual Reality and Virtual Environments*. Wokingham, UK: Addison-Wesley.

Klein, G. (1998). *Sources of Power: How People Make Decisions*. Cambridge, MA: MIT Press.

Lenggenhager, B., Tadi, T., Metzinger, T., & Blanke, O. (2007). Video ergo sum: Manipulating bodily self-consciousness. *Science*, 317(5841), 1096–9.

Ma, R. & Kaber, D.B. (2006). Presence, workload and performance effects of synthetic environment design factors. *International Journal of Human Computer Studies*, 64(6), 541–52.

Massey, K., Chatten, J., & Lindoerfer, D. (2006). Human-robotic interface for controlling an armed ground vehicle. *Proceedings of SPIE—The International Society for Optical Engineering*, 6230.

Newell, A. & Simon, H.A. (1972). *Human Problem Solving*. Englewood Cliffs, NJ: Prentice Hall.

Parasuraman, R., Sheridan, T. B., & Wickens, C.D. (2000). A model for types and levels of human interaction with automation. *IEEE Transactions on Systems, Man, and Cybernetics, Part A: Systems and Humans*, 30, 286–97.

Rasmussen, J. (1982). Human errors: A taxonomy for describing human malfunction in industrial installations. *Journal of Occupational Accidents*, 4, 311–33.

Rasmussen, J. (1983). Skills, rules, and knowledge; Signals, signs, and symbols, and other distinctions in human performance models. *IEEE Transactions on Systems, Man, and Cybernetics, Vol. SMC-13*, 3.

Riley, J.M., Murphy, R.R., & Endsley, M.R. (2006). Situation awareness in the control of unmanned ground vehicles. *Advances in Human Performance and Cognitive Engineering Research*, 7, 359–71.

Sadowski, W. & Stanney, K.M., (2002). Measuring and managing presence in virtual environments. In K.M. Stanney, *Handbook of Virtual Environments: Design, Implementation, and Applications*, pp. 791–806. Mahwah, NJ: Lawrence Erlbaum Associates.

Sanchez-Vives, M.V. & Slater, M. (2005). From presence to consciousness through virtual reality. *Nature Reviews Neuroscience*, 6(4), 332–9.

Sas, C. & O'Hare, G.M.P. (2003). Presence equation: An Investigation into cognitive factors underlying presence. *Presence: Teleoperators and Virtual Environments*, 12(5), 523–37.

Sharples, S., Cobb, S., Moody, A. & Wilson, J.R. (2008). Virtual reality induced symptoms and effects (VRISE): Comparison of head mounted display (HMD), desktop and projection display systems. *Displays*, 29(2), 58–69.

Slater, M. & Usoh, M. (1993). Presence in immersive virtual environments. *IEEE Annual Virtual Reality International Symposium*. Piscataway, NJ, pp. 90–96.

Sycara, K. & Lewis, M. (2004). Integrating intelligent agents into human teams. In E. Salas & S.M. Fiore (eds), *Team Cognition: Understanding the Factors that Drive Process and Performance* (pp. 203–32). Washington, DC: American Psychological Association.

Tsakiris M. & Haggard, P. (2005). The rubber hand illusion revisited: Visuotactile integration and self-attribution. *Journal of Experimental Psychology: Human Perception and Performance*, 31, 80–91.

TTRG (Transparent Telepresence Research Group) website. Retrieved December 2007 from http://telepresence.dmem.strath.ac.uk/telepresence.htm.

van Erp, J.B.F. (1999). Situation awareness: theory, metrics and user support. In R.N. Pikaar (ed.), *The Digital Human*. Utrecht: Dutch Ergonomics Society.

van Erp, J.B.F., Duistermaat, M., Jansen, C., Groen. E., & Hoedemaeker, M. (2006). Tele-presence: bringing the operator back in the loop. *NATO RTO Workshop on Virtual Media for Military Applications*. RTO-MP-HFM-136, 9-1–9-18.

van Erp, J.B.F. & Padmos, P. (2003). Image parameters for driving with indirect viewing systems. *Ergonomics*, 46(15), 1471–99.

van Erp, J.B.F. & Werkhoven, P.J. (2006). Validation of principles for tactile navigation displays. In: *Proceedings of the Human Factors and Ergonomics Society 50th Annual Meeting*. Santa Monica, CA, pp. 1687–91.

Vicente, K. J. & Rasmussen, J. (1988). On applying the skills, rules, knowledge framework to interface design. *Proceedings of the Human Factors Society 32nd Annual Meeting*. Santa Monica, CA, pp. 254–58.

Witmer, B.G. & Singer, M.J. (1998). Measuring presence in virtual environments: A presence questionnaire. *Presence: Teleoperators and Virtual Environments*, 7(3), 225–40.

Chapter 15

Multimodal Research for Human Robot Interactions

Ellen C. Haas
U.S. Army Research Laboratory

Jan B.F. van Erp
Nederlandse Organisatie voor Toegepast Natuurwetenschappelijk Onderzoek

Abstract

This chapter describes augmented visual displays and audio and tactile cues used in human-robotic interface (HRI) displays. The first section provides background information on multimodal interfaces. We then describe ground control station interfaces and the developments in augmented visual, auditory, tactile, and multimodal interfaces. The third section deals with specific aspects of operators in moving vehicles, and then we focus on the specific issues that arise when robots are controlled by a dismounted soldier. Conclusions, recommendations, and future directions are described in the final section. As shown in this chapter, integrated visual displays, as well as audio and tactile cues, have great potential in the human-robotic interface.

Multimodal Research for Human Robot Interactions: Background

According to a recent North American Treaty Organization (NATO) report (NATO, 2007), human interface issues associated with individual uninhabited military vehicle (UMV) control station design include providing appropriate situational awareness and effective information presentation and control strategies. As can be seen in the preceding chapters, displays for human-robot interaction (HRI) require some degree of user immersion as well as intuitive, effective human-computer interaction. Early robotic control systems, including teleoperation systems in which the user controlled robots from a distance, mainly used visual feedback from a camera mounted on the robot. From this work, it became clear that this limited feedback resulted in high visual and mental workload and seriously hampered operator performance. As a consequence, researchers started to augment the visual feedback by adding synthetic information to the camera images and by introducing more abstract information in status and tactical displays. Researchers demonstrated that visual displays could be supplemented by audio cues to cue visual attention, convey a variety of complex information, and increase awareness of the surroundings, especially when the visual channel is heavily loaded (Shilling & Shinn-Cunningham, 2002).

Developments in digitization over the past decade made it possible to present audio cues in different spatial locations. In spatial audio displays, also known as 3D audio, a listener using earphones or strategically placed loudspeakers perceives spatialized sounds which appear to originate at different azimuths, elevations, and distances from locations outside the head. Because each sound is presented in a different spatial location, 3D audio displays permit sounds to be presented in locations that are meaningful to the listener. Spatial audio cues have also been shown to increase situational awareness in target search of unmanned aerial vehicle (UAV) displays (Simpson et al., 2004) and of narrow field-of-view visual displays (Haas et al., 2005). Providing spatial auditory display cues can enhance human performance in tasks such as controlling multiple robots (Chong et al., 2002).

Within the last several years, researchers found that haptic cues are very useful in robotic displays applications because they can also provide feedback in demanding environments. Haptic displays generate skin-based as well as proprioceptive (body position, orientation, and movement) information. Tactile displays refer to a type of haptic display that uses pressure or vibration stimulators that interact with the skin (Gemperle et al., 2001). As with auditory displays, tactile displays can be used by themselves or as supplements to visual displays because they can present information without taxing the visual system. Relatively simple displays consisting of one or a small number of tactile elements can be successfully applied as alarms or attention allocation displays. Sklar and Sarter (1999) demonstrated that the use of tactile cues for indicating unexpected changes in status are more effective than visual cues. Tactile displays have also been used to provide a directional cue for warning information (e.g., Fitch et al., 2007).

A second successful tactile display application area is in orientation and navigation tasks. These applications are largely based on relatively complex (3D) displays consisting of a linear array or matrix of actuators covering the observer's torso (Cholewiak & Collins, 2000; van Erp, 2005; van Erp & Verschoor, 2004). These torso displays have been successfully applied to numerous tasks and environments, including in super-agile aircraft (van Erp et al., 2007); helicopters (van Erp et al., 2003); high-speed powerboats (Dobbins & Samway, 2002); and waypoint navigation (van Erp et al., 2005).

Van Erp (2006) mentions two applications relevant to HRI. First, tactile displays can improve the quality and safety of operator control of robotic vehicles. Second, tactile displays can mimic relevant cues arising from the vehicle environment that are available for onboard vehicle operators, but absent for remote control operators. In this chapter, we will mainly focus on the first application.

Over the years, robotic control has changed in terms of scale and environment. The last decade shows a trend towards smaller scale. From large multi-million dollar vehicles controlled via satellite communication at brigade level by multiple operators over large distances and from large control stations, we have moved towards cheap UMVs for dismounted soldiers at group level, hand-launched with only line-of-sight communication and with limited automation and endurance. This transition brings along new challenges with respect to the control interface, probably requiring a different approach than miniaturization of the ground control station interfaces.

Chapter Organization

This chapter is organized to reflect the transition in display scale and environment described above, by describing augmented visual displays, as well as audio and tactile cues used in unimodal (one modality) and multimodal (two or more modalities) displays for different robotic control environments. The following section deals with the more thoroughly developed interfaces in ground control stations. We then focus on the specific aspects of operators in moving vehicles, i.e., operators who are no longer working from a fixed control station, but onboard a helicopter or moving armoured vehicle. Next we concentrate on the specific issues that arise when robots are controlled by a dismounted soldier. Conclusions and recommendations are listed in the final section.

Because research in multimodal displays is relatively new, this chapter will include descriptions of applications in which research exists for only one modality. In cases where research does not exist, conclusions regarding data of related research will be used to draw sensible conclusions on the efficient, effective design and utilization of multimodal displays.

Ground-Based Control Stations

This section will present some of the extensive research performed in the area of ground control station displays for the HRI. Many researchers have explored visual, audio, and tactile modalities, as well as combinations of the three, and much of this research involves teleoperation. This section is only intended to give a concise overview of the developments relevant for HRI designs in the areas of augmented visual displays, auditory displays, tactile displays, and multimodal displays. A thorough overview can be found in a recent NATO report (Chapter 6 NATO, 2007).

Augmented Visual Displays

In the early days of UMVs, high operator workload, limited situational awareness, and suboptimal human performance (especially in controlling remote sensors) led to the wish to augment system visual displays. For instance, to overcome problems associated with a limited field of view, one of the ideas was to implement head-slaved sensor systems combined with head-mounted displays (HMDs). In a sense, these are multimodal devices since they employ the operator's proprioceptive system (including the receptors in the neck muscles) to provide feedback on the sensor's viewing direction and motions (e.g., see Bakker et al., 1999; Kappé et al., 1999). Also, HMDs were expected to improve the operator's sense of telepresence and therewith performance (See Chapter 14, this volume, on the complex relation between telepresence and performance). Some studies showed improved performance, mainly in comparison with a fixed sensor system (Oving & van Erp, 2001). However, not all studies showed beneficial effects of head-slaved systems (de Vries, 2001), while factors such as time delay in the control loop may result in symptoms of motion sickness (Morphew et al., 2004) or increased workload (Wildzunas et al., 1996). Motion sickness is a relevant issue for operators, especially when the visual information is the major source of sensory stimulation, as in the situation where the operator uses an HMD to view the sensor images of the remote environment. For example, Draper et al. (2001) showed that motion sickness was reported much more for either magnification or minification of head-slaved images than for time delays. Several other authors suggested that a head-slaved viewing system that lacks a roll component for deviation from the longitudinal axis (which is often the case) may result in serious motion sickness problems (Oving & Schaap, 2002; Oving & van Erp, 2001, 2002).

A second form of multimodal image augmentation is to add cues that can be used by the peripheral vision system, which specializes in processing motion and orientation cues. Due to factors such as limited field size, low update rate, and zoomed-in images (disturbing the relation between camera motion and optical flow), HMD sensor images lack cues that are normally picked up by peripheral vision. Adding computer-generated information (e.g., distance or position) to the sensor images has been shown to be a useful augmentation (Figure 15.1), for example to improve tracking performance with low update rates and improve SA with zoomed-in images (see van Erp & van Breda, 1999 for an overview). Since 2000, the use of augmented reality has been developed further, but is outside the scope of this chapter.

Auditory Displays

Spatial audio displays have been shown to increase situation awareness in the operation of unmanned robots from ground-based stations (Simpson et al., 2004). Tachi et al. (2003) created a teleoperated humanoid robot with spatial audio and visual displays, with an audio-visual display providing a visual and auditory "view" of the remote environment. The results of this study indicated that presence lent by spatial audio and visual cues created more intuitive robotic control and facilitated the operator's sense of presence

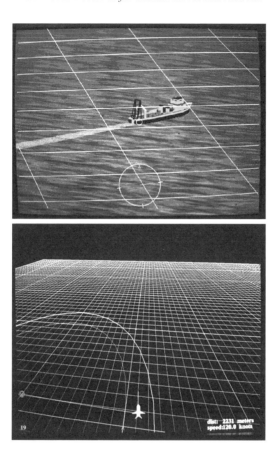

Figure 15.1   **Two examples of image augmentation with cues that tap into the peripheral vision system. The top panel shows a grid overlay over the (simulated) camera image that responds to operator input at high update rates and no delay, and therewith makes tracking performance independent of the update rate and delay of the camera image itself. The lower panel shows a more ecological approach to display, for instance stand-off distance and flying and viewing direction**

Tactile Displays

Tactile alerts may be advantageous in noisy task environments requiring long periods of vigilance, where both audio and visual channels are taxed. Calhoun et al. (2004) explored tactile displays in teleoperation of U.S. Air Force UAVs, and found that tactile displays significantly improve detection of faults in UAV control tasks, and can serve as an effective cueing mechanism.

Lathan and Tracey (2002) showed that when vibrotactile feedback is used in addition to video feedback in a teleoperation task, there were fewer collisions with walls and obstacles (approximately 1 vs. 2.3 per trial). Barnes and Counsell (2003) used haptic feedback to display a robot's autonomous collision avoidance behaviour. Korteling and Van Emmerik (1998) showed that haptic feedback in a joystick has the same profitable effect in a camera control task from a moving UAV as stabilizing the platform. Additionally, in UAV camera control, a position feedback system—moving joystick giving proprioceptive (positional) feedback about relative movement of UAV—increased target-tracking (Korteling & Van der Borg, 1997).

Multimodal Displays

Researchers have compared different modalities to each other by exploring the use of visual, audio and tactile displays separately or in different combinations. Lathan & Tracey (2002) explored remote operator teleoperation with different combinations of visual, auditory, and vibrotactile feedback, and found that providing different modalities alone or in combination did not significantly affect task completion time in a telerobotic navigation task, but that task errors were significantly lower with tactile displays. Gunn et al. (2005) used multimodal displays to communicate threats in a UAV control task, where threat signals led observers to perform a subsequent manual target acquisition. They found that visual, spatial audio, and haptic cues used separately enhanced target acquisition performance over no cueing in the target acquisition phase of the task, and did so to a similar level.

Other researchers explored the simultaneous use of multiple modalities. One application is web-based teleoperation systems, in which remote objects such as robotic arms or mobile robots are controlled through an Internet network by means of a browser (Marin et al., 2005). Baier et al. (2000) found improved task performance for combined haptic and active stereo vision in a teleaction task using Internet protocol communication. Chou & Wang (2001) designed a multimodal interface for Internet-based teleoperation, in which live video images, audio, and force information were organized and presented in a predictive display. They found that presenting multimodal information reduced operator mental workload. Multimodal feedback has also been used in virtual fixtures. In this application, audio, visual, and tactile information is used as perceptual overlays to provide guidance information regarding direction of motion or forbidden regions of the user's workspace (Abbott et al., 2005). An example is given in Figure 15.1, lower panel (van Erp & Kappé, 1997). Virtual fixtures have been found to improve operator speed and accuracy, since they can alert the user of changes in the environment and support hand–eye coordination for object manipulation tasks (Massimino & Sheridan, 1992).

Summary

This section discussed large-scale display applications in stationary ground-based environments. Augmented visual displays, auditory displays, and tactile displays can be used separately or in tandem in multimodal displays to enhance operator performance and situation awareness in teleoperation tasks and in the operation of unmanned robots. To provide an integrated view of the design of multimodal displays in HRI applications, the focus of this chapter will now narrow to describe specific aspects of displays for robot control from moving platforms.

Robot Control from Moving Platforms

In military situations, robot control by a "moving" operator will become more common following the trend towards smaller-scale applications. For example, robots may be controlled by an operator inside an armoured vehicle. Introducing moving operators comes with new challenges. Three major challenges of robot control from moving platforms are the usability of visual displays, the risk of spatial disorientation (SD), and the complexity of multiple frames of reference.

Usability of Visual Displays

One important issue when transferring from a fixed ground-control station to a moving platform is the use of visual displays, which are still the major display modality applied in ground-control systems. Tactile and 3D audio technologies could be crucial in supplementing or replacing visual displays in moving vehicle environments, especially those that have restricted space for additional visual displays and/or contain relatively high levels of vibration and jolt. Vehicle vibration and jolt may lead to visual performance decrements. Researchers found that vehicle vibration adversely affects visual performance (Ishitake et al., 1998) and visual search (Griffin & Lewis, 1978), making cues in other modalities valuable. Several researchers explored the extent to which audio and tactile cues provide effective directional information in localization and visual search tasks performed in moving ground vehicles. Krausman & White (2008) conducted a study examining the detection of tactile signals by users seated on a ride motion simulator. The simulator, located at the U.S. Army Test Center, produced movement dynamics simulating a U.S. Army high-mobility, multipurpose wheeled vehicle (HMMWV) and a Bradley Fighting Vehicle travelling over secondary roads and cross-country terrain at 10–20 mph. This study also compared the usefulness of different types of belt-worn tactors (pancake motor vs. plunger motor) on these simulated terrains. The results indicated that tactor type and position were significant for both vehicles on cross-country terrain, but not on gravel. Users wearing plunger-motor tactors correctly detected significantly more tactile signals than did those wearing pancake-motor tactors, and there were significantly fewer correct tactor detections for tactors mounted at the center of the user's back than at any other location.

Carlander & Eriksson (2006) investigated the use of audio and tactile displays for localization of directional threat cues in a combat vehicle CV-90 light armored tank. They found that 3D audio cueing generated a larger number of errors and greater response times than displays which used both tactile and audio input. This result was probably due to 3D audio front–back cue confusion, which disappeared when tactile cues were added. However, the terrain over which the CV-90 travelled (a large, paved parking lot) was not representative of the full range of terrains over which military vehicles may travel.

Haas & Stachowiak (2007) conducted a study that explored the effect of vehicle motion on the use of separate and redundant spatial audio and tactile displays used as supplements to narrow (30°) field-of-view visual displays in a simulated robotic system target search task. These displays were tested in an HMMWV during engine idle and during vehicle travel over gravel and cross-country terrain. Figure 15.2 shows a photograph of the vehicle used in this study. Results indicated that target search accuracy and response time were unaffected by display modality or movement condition. However, participant responses to workload questionnaires indicated significant modality and terrain differences: perceived workload was significantly lower for combined audio and tactile displays than for each used separately. The researchers hypothesized that the lower workload for the combined audio and tactile display was most likely due to the incorporation of cues from different modalities in an environment with strong noise and vibration distracters.

Haas et al. (2008) studied user cue localization (determining the direction of a cue) in a wide (360°) azimuth range of cue locations during travel in the HMMWV shown in Figure 15.2. Cues were presented with separate and redundant spatial audio and tactile displays. Each participant performed simulated HRI cue localization tasks during engine idle and during vehicle travel over cross-country and gravel terrains. Haas et al. (2008) found that, across all terrain types, tactile displays and combined audio and tactile displays provided significantly greater localization accuracy than did spatial audio displays used alone. Users wearing spatial audio displays reported a higher level of workload than while wearing tactile and audio + tactile displays. Haas et al. (2008) hypothesized that the significantly lower localization accuracy and workload scores for audio displays were most likely due to the use of generic head-related transfer functions and lack of head-trackers. The researchers hypothesized that the success of the multimodal audio and tactile display was most likely due to the incorporation of cues from both modalities in an environment with strong noise and vibration distracters.

Figure 15.2   HMMWV study used in Haas and Stachowiak (2007)

Tactile displays have also been tested in high-speed rigid inflatable boats (RIBs). RIB navigators and drivers are not able to use visual displays or communicate via speech when the RIB is driven at high speeds due to the vehicle vibrations and the high G forces. Both Dobbins & Samways (2002) and van Erp et al. (2005) found that tactile displays were effective on board high-speed boats. Spatial auditory and tactile cues are being considered for aircraft-based displays. Because multimodal display research is fairly new, there is little research involving multimodal cues used in HRI interfaces for aircraft. However, many researchers have explored auditory and tactile cues in aircraft-based applications other than HRI, and a spin-off to the HRI domain is expected (van Erp, 2006). However, considering the experimental results described in this section, it can be concluded that although knowledge on the effect of whole body vibration on tactile and auditory perception is somewhat limited, these displays may have potential in challenging environments with high levels of noise and vibration.

Spatial Disorientation (SD)

Spatial audio and tactile displays have been used in the mitigation of pilot SD in aircraft. For military and civilian pilots, SD is defined as failure to correctly sense the attitude, motion, and/or position of the aircraft with respect to the surface of the Earth, and is a temporary condition caused by the pilot being deprived of an external visual horizon critical to maintaining a correct sense of up and down while flying (Benson, 2003). It has been found to be caused by flight into weather conditions with low or no visibility, night-time flight, instrument failure, and conditions with high pilot visual load, such as combat conditions (Collins & Harrison, 1995). If not appropriately resolved, SD can cause aircraft crashes and fatalities. Thirty percent of U.S. Army military helicopter accidents involved SD as a significant factor (Braithwaite et al., 1998). Resulting costs of SD problems in military and civil aviation have been described by many, including Veronneau & Evans (2004).

Spatial disorientation is only beginning to be recognized as a factor in UAV teleoperation accidents and incidents. UAV SD is defined as a failure of a UAV pilot to correctly sense the attitude, motion, and/or position of the UAV with respect to the surface of the earth (Self et al., 2006). UAV takeoff and landing operations are most vulnerable, especially for pilots with a high visual load. Self et al. noted that UAV control method, visual reference, and operator platform are characteristics most relevant to UAV operator SD. It was also concluded that tactile feedback providing altitude deviation feedback helped UAV pilots to more accurately perform a landing (Aretz et al., 2006). UAV pilot SD is most likely for a manually controlled UAV when operated from a mobile platform (i.e., a ground vehicle or an aircraft), and platform motion type and direction may negatively affect pilot control of UAV manual control (Olson et al., 2006). Olson et al. also noted that pitch axis control was degraded when platform motion was in a different plane of motion from UAV control task. These researchers noted that pitch axis autopilot may mitigate some UAV SD effects. However, research suggests that audio and tactile cues may also mitigate UAV pilot SD, just as they are used to mitigate SD for aircraft pilots.

Multimodal displays have also been developed to mitigate SD. The U.S. Air Force developed the Spatial Orientation Retention Device (SORD) that presents aircraft attitude information to the pilot through multisensory cues, including visual helmet-mounted symbology as well as tactile and audio cues. This system has been prototyped by the U.S. Air Force Research Laboratory at Wright-Patterson Air Force Base, and has been transitioned to a Rotary Wing Brownout Program (Albery 2007). A tactile display developed for military pilots by The Netherlands Organisation for Applied Scientific Research (TNO) can be seen in Figure 15.3.

Multiple Frames of Reference

Spatial disorientation could be thought of as a problem involving one frame of reference – that of the orientation of the moving operator. A related problem that arises for "moving" operators is the complexity of multiple spatial frames of reference (de Vries & Jansen, 2002). Imagine the situation of a co-pilot or gunner (cpg) in an Apache attack helicopter that has control over a UAV and its sensor. This situation requires the cpg to integrate a (north-up) map, the UAV flying direction, the UAV sensor viewing direction, the Apache flying direction, his own viewing direction, and the Apache's sensor and weapon direction. Building an integrated picture from these multiple frames of reference will put a large demand on the cpg's cognitive resources and may result in degraded situation awareness for both earth-fixed and moving objects. To reduce this degradation, the cpg may be supported by automatically aligning different frames. Jansen et al. (2005) advise aligning the UAV sensor image with that of the helicopter. Multimodal 3D audio or tactile displays (van Erp & Verschoor, 2004) may also be a valuable support in this complex situation.

Figure 15.3   Example of a tactile display developed by The Netherlands Organisation for Applied Scientific Research (TNO). The display is designed for military pilots and consists of a matrix of vibrators covering the pilot's torso

Control–display compatibility may become an issue for moving operators having to use multiple frames of reference. For example, for a UAV operator on board a helicopter, the same target may be visible in the UAV's sensor image and the helicopter's sensor image, while they have a viewing angle that may be off by 180 degrees (thus leading to left–right reversals) and constantly changing. Especially for the control of sensor images with a restricted fieldsize control-display compatibility may become an important issue. Misalignment of control and display (Macedo et al., 1998) and/or mismatch of control and object motions in the image (Worringham & Beringer, 1998) will hamper performance. Additional visual (van Erp & Oving, 2002) or multimodal cues may help to reduce these issues.

Summary

The decreased usability of visual displays, the risk of spatial disorientation, and the complexity of multiple frames of reference are three major challenges of robot control from moving platforms. These problems can be mitigated by the use of tactile, spatial audio, and multimodal cues to supplement or replace visual displays. To further provide an integrated view of the multimodal HRI, the focus of this chapter will narrow even further to describe displays for dismounted soldier applications.

Robot Control for Dismounted Soldiers

Further scaling down of the HRI ultimately brings us to the dismounted soldier, where there are even fewer possibilities of using visual displays. There has been no research that deals directly with auditory displays used in robotic control tasks for dismounted soldiers. However, researchers have studied audio displays as one of the unimodal alternative display modalities in dismounted military tasks. Kumagai & Tack (2005) explored visual, auditory, and tactile display modalities used separately in support of wayfinding in wooded terrain during day and night conditions. Canadian infantry soldiers navigated routes using the Future Infantry Navigation Device (FIND) system, which offered wayfinding system in visual, auditory, and tactile modalities. The FIND was compared to the current method of wayfinding, which involved using a compass and pace counting. Results indicated that the FIND was superior to the compass method for locating the waypoints accurately, both in bearing and distance.

Results from several U.S. Army studies demonstrated numerous areas in which tactile systems showed potential for presenting directional information during dismounted soldier tasks. These areas include waypoint navigation, communications in stealth operations, presentation of alerts and warnings, and supplementation of visual displays during high workload situations (Redden, 2007).

Krausman & White (2006) explored tactile display pattern recognition of dismounted soldiers performing combat assault maneuvers on an obstacle course. U.S. Army soldiers received six different tactile patterns while standing still and while maneuvering through obstacles (tires, windows) and performing high crawl operations (crawling with the back raised). Results indicated that specific obstacles had a significant effect on the detection and identification of the tactile signals; a significantly lower percentage of tactile signals were detected during the high crawl than during travel over tires and through windows. Over 80 percent of tactile display patterns were correctly identified while soldiers experienced the obstacles. This study demonstrated that tactile displays can be used to communicate several different meaningful patterns to the user, even under challenging conditions.

In a study comparing handheld visual displays with tactile displays during dismounted night operations, Duistermaat et al. (2007) found that soldier navigation speed was higher and that more targets were detected using tactile displays than with visual displays alone. The soldiers commented that the tactile system provided them with hands-free and eyes-free operation, but that it was less suited to build up global situational awareness compared to a map display.

Pettitt et al. (2006) evaluated soldiers' abilities to interpret and respond to tactile and conventional military visual hand signals while traversing an obstacle course. U.S. Army soldiers received tactile and visual hand and arm signals separately as they negotiated an obstacle course while wearing their standard uniforms and body armor. Tactile signals were presented via a belt worn around the waist. Results demonstrated that soldiers were able to receive, interpret, and accurately respond to tactile commands more quickly than to conventional hand and arm signals. Soldiers commented they were better able to focus more attention on negotiating obstacles and on area situation awareness (SA) when receiving tactile signals than when maintaining visual contact with their leaders in order to receive hand and arm signals.

One innovative soldier information system, the telepresence interface, was designed to provide the experience of "being present at a physically remote location" (Encarta, 2007). Jansen (2006) described a portable telepresence control unit for unmanned vehicles designed for optimal situation awareness by perceptual immersion and enhanced situation awareness by video overlay and virtual 3D sounds. This device, seen in Figure 15.4, is designed and built as an analogue to binoculars. The system consists of a binocular-like handheld device consisting of stereo visual and stereo audio displays, a motion tracker, and controls for operating a UMV and its sensors.

Summary

There is very little research on visual or spatial audio displays in robotic control tasks for dismounted soldiers. The applications described in this section demonstrate visual, audio, and tactile modalities used separately (such as in the FIND system display), but not combined into multimodal displays. Researchers explored tactile displays for non-robotic dismounted operations, including waypoint navigation, communications, presentations of warnings, and supplementation of visual

Figure 15.4 Operator with handheld telepresence system resembling a pair of binoculars and consisting of a stereo HMD including augmented reality, 3D audio with virtual acoustics, and head tracking

displays during high workload operations. Although there are not much data available at present, the trends in innovative soldier information systems show the increasing importance of multimodal information presentation. Innovative soldier information technologies are also appearing, to bring the sense of presence to dismounted soldiers who perform robotic control from remote locations.

Discussion, Conclusions, and Future Directions

The research described in this chapter focused on the use of augmented visual displays, as well as auditory and tactile displays used singly or in combination, to reduce operator workload and enhance situation awareness in the HRI interface. The HRI display of the future may integrate audio with tactile cues in multimodal displays used in different levels of robotic control. As demonstrated here, multiple modalities can be advantageous; when used together, audio and tactile signals can supplement each other in demanding surroundings where variable levels of noise and vibration might mask cues in only one modality.

Audio and tactile modalities work well together because they have much in common. Both are useful if the user's visual field is heavily taxed (i.e., in environments that are poorly lit) or if a visual display is not available. Audio and tactile cues are effective for simple, short messages that do not need to be referred to later. Both are useful for mobile or stationary applications, either can be used to call for immediate response, and both can signal events in time and in space.

When used singly, auditory and tactile cues have varying levels of success. Auditory cues have a large range of frequency, temporal, and spatial cues, and can use evocative cues such as icons, earcons, and speech cues to communicate several different dimensions of information (Haas & Edworthy, 2006). Tactile displays are limited to incorporating temporal (rhythm), spatial, and a small range of frequencies to communicate two or three different dimensions of information at most (Brown et al., 2006; Hoggan & Brewster, 2006). Auditory displays are less effective when they have to compete with high levels of noise (Haas & Edworthy, 2006) from sources such as radio communications. Tactile displays have been shown to be very effective at communicating directional information, but can also display information when mounted on sites such as the torso or arm, although resolution is limited to the allowable spacing between tactors (van Erp, 2002).

The research cited in this chapter demonstrated that spatial audio and tactile cues can communicate directional information, but their ability to do this is constrained by technology. Sensors are needed to communicate directional information from the environment to the display system. Some type of head- or body-tracking apparatus may be needed to communicate in which direction an event occurs in relation to the head or body position of the user. Spatial audio displays may also be limited by how well the mathematical algorithm that defines sound location (the head-related transfer function) corresponds to the physical attributes of the listener (Begault et al., 2001).

There are several practical considerations to using multimodal displays in the battlefield. Researchers and designers should be aware of existing guidelines for designing multimodal displays. These include ISO (1998), van Erp (2002), Sutcliffe (2003), Sarter (2006), and ISO/DIS (2007). General guidelines state that multimodal displays should incorporate manageable information loading by using signals that provide only information that is necessary. As noted in Haas & Stachowiak (2007), signals should also be consistent, and incorporate redundancy whenever possible. Selection of signal dimensions and encoding should exploit learned or natural relationships as much as possible, and both auditory and tactile signals should be easily discernable from other audio and vibrational events in the environment. In addition, both audio and tactile signals should avoid conflict with previously used signals in terms of meaning or characteristics.

In closing, there are several areas that the authors wish to list as having great potential for productive multimodal HRI research. These areas include reproducing remote audio and tactile perceptual cues; defining effective tactile coding strategies; and testing the effectiveness of tactile multimodal displays in moving platforms, at realistic speeds.

In this chapter, we restricted ourselves to applying tactile and audio cues as general applications in multimodal contexts, but said nothing about the use of audio and haptic feedback from the remote environment. This may be a productive area for future research. As stated in the executive summary of the final report of the NATO UMV taskgroup (NATO, 2007): "Indeed, it could be said that the primary goal of the interface should be to reproduce the perceptual cues that are used by the operators in the real environment to perform the task." This expansion of the mediation of the visual cues could include auditory, haptic, and possibly vestibular cues. For example, tactile cues could provide valuable information on robotic ground vehicle performance and environment, where the condition of the road can better be felt than seen. Not much work has been carried out in this area, although Ruff and colleagues (2002) examined the utility of touch displays for alerting the operator to the onset of turbulence. Research should also focus on defining strategies for coding tactile signals for the presentation of multiple dimensions of information. As was previously noted, tactile displays can incorporate temporal (rhythm), spatial, and a small range of frequencies with the potential for communicating at least two different dimensions of information. Tactor location is another possible coding dimension. However, there has been little research in this area, and truly efficient tactile coding strategies have not yet been defined.

As seen in the section describing moving platform displays, tactile cues can be successfully detected and localized in ground vehicles traveling over gravel and cross-country terrain, although the research here described vehicle speeds of only 10 to 15 mph. Future research should explore the effectiveness of tactile displays at a range of speeds relevant to actual vehicle use.

Summary Points

- Augmented visual displays, auditory displays, and tactile displays can be used separately or in tandem in multimodal displays to enhance operator performance and situation awareness in HRI ground control stations, moving platforms, and dismounted operations.
- The decreased usability of visual displays, the risk of spatial disorientation, and the complexity of multiple frames of reference are three major challenges of robot control from moving platforms. These problems can be mitigated by the use of unimodal and multimodal cues to supplement or replace visual displays.
- There is very little research on visual or spatial audio displays in robotic control tasks for dismounted soldiers. Research has shown that tactile displays can be used for nonrobotic dismounted operations including waypoint navigation, communications, presentations of warnings, and supplementation of visual displays during high workload operations.
- Areas in which there is great potential for future research include reproducing remote audio and tactile perceptual cues; defining effective tactile coding strategies; and testing the effectiveness of multimodal displays in moving platforms, at realistic speeds.

References

Abbott, J.J., Marayong, P., & Okamura, A.M. (2005). Haptic virtual fixtures for robot-assisted manipulation. *Proceedings of the 12th International Symposium of Robotics Research*. Retrieved from http://www.haptics.me.jhu.edu/publications/isrr05-abbott.pdf.

Albery, W.B. (2007). Multisensory cueing for enhancing orientation information during flight. *Aviation Space Environmental Medicine*, May, 78 (5 Suppl) B 186–90.

Aretz, D., Andre, T., Self, B., & Brenaman, C. (2006). Effect of tactile feedback on unmanned aerial vehicle landings. *Proceedings of the Interservice/Industry Training, Simulation, and Education Conference (I/ITSEC)*, Orlando, FL.

Baier, H., Buss, M., Freyberger, F., & Schmidt, G. (2000). Benefits of combined active stereo vision and haptic telepresence. *Proceedings of the International Conference on Intelligent Robots and Systems (IROS)*, Munich, Germany, 702–7.

Bakker, N.H., Werkhoven, P.J. & Passenier, P.O. (1999). The effects of proprioceptive and visual feedback on geographical orientation in virtual environments. *Presence*, 8(1), 36–53.

Barnes, D. & Counsell, M. (2003). Haptic communication for mobile robot operations. *Industrial Robot: An International Journal*, 30(6), 552–63.

Begault, D.R., Wenzel, E.M., & Anderson, M.R., (2001). Direct comparison of the impact of head tracking, reverberation, and individualized head-related transfer functions on the spatial perception of a virtual sound source. *Journal of the Audio Engineering Society*, 49, 904–15.

Benson, A.J. (2003). Technical evaluation report. Spatial disorientation in military vehicles: Causes, consequences and cures. *RTO Meeting Proceedings 86 (pp. T1–T7)*. Neuilly-sur-Seine, France: NATO Research and Technology Organization.

Braithwaite, M.G., Durnford, S.J., Crowley, J.S., Rosado, N.R., & Albano, J.P. (1998). Spatial disorientation in US Army rotary-wing operations. *Aviation Space and Environmental Medicine*, 69(11), 1031–7.

Brown, L.M., Brewster, S.A., & Purchase, H.C. (2006). Multidimensional tactons for non-visual information presentation in mobile devices. *MobileHCI*, 231–8.

Calhoun, G., Fontejon, J., Draper, M., Ruff, H., & Guilfoos, B. (2004). Tactile versus aural redundant alert cues for UAV control applications. *Proceedings of the Human Factors and Ergonomics Society 48th Annual Meeting*, 137–41.

Carlander, O. & Eriksson, L. (2006). Uni- and bi-modal threat cueing with vibrotactile and 3D audio technologies in a combat vehicle. *Proceedings of the HFES 50th Annual Meeting*, 1552–6.

Cholewiak, R.W. & Collins, A.A. (2000). The generation of vibrotactile patterns on a linear array: Influences of body site, time, and presentation mode, *Perception and Psychophysics*, 62, 1220–35.

Chong, N.Y., Kotoku, T., Ohba, K., Sasaki, H., Komoriya, K., & Tanie, K. (2002). Multioperator teleoperation of multirobot systems with time delay. II— Testbed description. *Presence: Teleoperators and Virtual Environments*, 11(3), 292–303.

Chou W. & Wang, T. The design of multimodal human-machine interface for teleoperation. *Proceedings of the 2001 IEEE International Conference on Systems, Man, and Cybernetics*, 5, 3187–92.

Collins, H.L. & Harrison, G. (1995). Spatial disorientation episodes among F-15-C pilots during Operation Desert Storm. *Journal of Vestibular Research*, 5(6), 405–10.

De Vries, S.C. (2001). Head-slaved control versus joystick control of a remote camera. *TNO-Report TM-01-B008*, Soesterberg, the Netherlands: TNO Human Factors Research.

De Vries, S.C. & Jansen, C. (2002). Situational awareness of UAV operators onboard moving platforms. In: S. Chatty, J. Hansman, & G. Boy (eds), *Proceedings of the International Conference on Human-Computer Interaction in Aeronautics, HCI-Aero 2002*. Cambridge, MA, 23–25 October, pp. 144–7.

Dobbins, T. & Samway, S. (2002). The use of tactile navigation cues in high-speed craft operations. *Proceedings of the RINA Conference on High Speed Craft: Technology and Operation*. London: Royal Institution of Naval Architects, pp. 13–20.

Draper, M.H., Viire, E.S., Furness, T.A., & Gawron, V.J. (2001). Effects of image scale and system time delay on simulator sickness within head-coupled virtual environments. *Human Factors*, 43(1), 129–46.

Duistermaat, M., Elliott, L.R., van Erp, J.B.F., & Redden, E.S. (2007). Tactile land navigation for dismounted soldiers. In D. de Waard, G.R.J. Hockey, P.Nickel, and K.A. Brookhuis (eds), *Human Factors Issues in Complex System Performance*, Maastricht, the Netherlands: Shaker Publishing, pp. 43–53.

Encarta World English Dictionary [North American Edition] (2007). Redmond, WA: Microsoft Corporation.

Fitch, G.M., Kiefer, R. J., Hankey, J.M. & Kleiner, B.M. (2007). Toward developing an approach for alerting drivers to the direction of a crash threat. *Human Factors*, 49(4) 710–20.

Gemperle, F., Ota, N., & Siewiorek, D. (2001). Design of a wearable tactile display. *Proceedings of the 5th IEEE International Symposium on Wearable Computers*, 5–12.

Griffin, M.J. & Lewis, C.H. (1978). A review of the effects of vibration on visual acuity and continuous manual control, part I: Visual acuity. *Journal of Sound and Vibration*, 56, 383–413.

Gunn, D.V., Warm, J.S., Nelson, W.T., Bolia, R.S., Schumsky, D.A., & Corcoran, K.J. (2005). Target acquisition with UAVs: Vigilance displays and advanced cuing interfaces. *Human Factors*, 47, 488–97.

Haas, E.C. & Edworthy, J. (2006), An introduction to auditory warnings and alarms. In M.S. Wogalter (ed.), *Handbook of Warning*s. London: Lawrence Erlbaum Associates, pp. 189–220.

Haas, E.C., Pillalamarri, R.S., Stachowiak, C.C., and Lattin, M.A. (2005). Auditory cues to assist visual search in robotic system operator control unit displays. ARL-TR-3632. Aberdeen Proving Ground, MD: U.S. Army Research Laboratory.

Haas, E.C. & Stachowiak, C. (2007). Multimodal displays to enhance human-robot interaction on the move. *Proceedings of the Performance Metrics for Intelligent Systems Workshop*. Gaithersburg, MD: National Institute of Standards and Technology.

Haas, E.C., Stachowiak, C., White, T., Feng, T., & Pillalamarri, K. (2008). Enhancing human-robot interaction with multimodal cues while on the move. Aberdeen Proving Ground, MD: U.S. Army Research Laboratory.

Hoggan, E. & Brewster, S. (2006). Crossmodal spatial location: Initial experiments. *NordiCHI 2660: Changing Roles*, 469–72.

Ishitake, T., Ando, H., Miyazaki, Y., & Matoba, F. (1998). Changes of visual performance induced by exposure to whole-body vibration. *Kurume Medical Journal*, 45(1), 59–62.

ISO (International Organization for Standardization) (1998). *ISO 14915: Multimedia User Interface Design Software Ergonomic Requirements, Part 3: Media Combination and Selection.*

ISO/DIS (2007). Ergonomics of human-system interaction: Guidance on tactile and haptic interactions. *Draft International Standard 9241-920.*

Jansen, C. (2006). Telepresence Binoculars (TBI): a technology demonstrator for a telepresence control unit of unmanned vehicles. *Proceedings of the NATO HFM Symposium Human Factors of Uninhabited Military Vehicles as Force Multipliers.* Biarritz, France. RTO-MP-HFM-135; Neuilly-sur-Seine Cedex, France: NATO RTO.

Jansen, C., de Vries, S.C., & Duistermaat, M. (2005). Presenting images from an unmanned aerial vehicle in an attack helicopter cockpit. *Report DV3 2005-A16.* Soesterberg, the Netherlands: TNO Human Factors Research.

Kappé, B., van Erp, J., & Korteling, J.E. (1999). Effects of head-slaved and peripheral displays on lane-keeping performance and spatial orientation. *Human Factors*, 41(3), 453–66.

Korteling, J.E. & van der Borg, W. (1997). Partial camera automation in an unmanned air vehicle. *IEEE Transactions on Systems, Man, and Cybernetics, Part A: Systems and Humans*, 27(2), 256–62.

Korteling, J.E. & van Emmerik, M.L. (1998). Continuous haptic information in target tracking from a moving platform. *Human Factors*, 40(2), 198–208.

Krausman, A. & White, T. (2006). *Tactile Displays and Detectability of Vibrotactile Patterns as Combat Assault Maneuvers are Being Performed* (ARL-TR-3998). Aberdeen Proving Ground, MD: U.S. Army Research Laboratory.

Krausman, A. & White, T. (2008). *Detection and Localization of Vibrotactile Signals in Moving Vehicles* (Technical Report). Aberdeen Proving Ground, MD: U.S. Army Research Laboratory.

Kumagai, J.K. & Tack, D.W. (2005). Alternative display modalities in support of wayfinding (DRDC-Toronto-CR-2005-093). Defence R&D Canada, Toronto, 1 July.

Lathan, C. & Tracey, M. (2002). The effects of operator spatial perception and sensory feedback on human-robot teleoperator performance. *Presence: Teleoperation and Virtual Environments*, 11(4), 368–77.

Macedo, J.A., Kaber, D.B., Endsley, M.R., Powanusorn, P., & Myung, S. (1998). The effect of automated compensation for incongruent axes on teleoperator performance. *Human Factors*, 40(4) 541–53.

Marin, R., Sanz, P.J., Nebot, P., & Wirz, R. (2005). A multimodal interface to control a robot arm via the web: A case study on remote programming. *IEEE Transactions on Industrial Electronics*, 52(6), 1506–20.

Massimino, M. & Sheridan, T. (1992). Sensory substitution of force feedback for the human-machine interface in space teleoperation, *43rd Congress of the International Astronautical Federation, World Space Congress, IAF/iAA-92-0246.*

Morphew, M.E., Shively, J.R., and Casey, D. (2004). Helmet mounted displays for unmanned aerial vehicle control. In C.E. Rash & C.E. Reese (eds), *Proceedings of SPIE, Vol. 5442. Helmet- and Head-Mounted Displays IX: Technologies and Applications*, 93–103.

NATO (2007). Uninhabited military vehicles (UMVs): Human factors issues in augmenting the force. *NATO Research & Technology Organisation Report AC/323(HFM-078)TP/69.* Neuilly-sur-Seine Cedex, France: NATO RTO.

Olson, W.A., DeLauer, E.H., & Fale, C.A. (2006). Spatial disorientation in uninhabited aerial vehicles – A preliminary empirical study. In N.J. Cooke (ed.), *Human Factors of UAVs Workshop*, Retrieved from http://www.cerici.org/pjcts&ws/workshops/uavws06.html.

Oving, A.B. & Schaap, E. (2002). *Motion Sickness When Driving with a Head-slaved Camera System* (TM-02-A046). Soesterberg, the Netherlands: TNO Human Factors Research.

Oving, A.B. & van Erp, J.B.F. (2001). *Armoured Vehicle Driving with a Headslaved Indirect Viewing System: Field Experiments* (Report TM-01-A008). Soesterberg, the Netherlands: TNO Human Factors Research.

Oving, A.B. & van Erp, J.B.F. (2002). *Head Movement When Driving with a Headslaved Camera System* (Report TM-02-A045). Soesterberg, the Netherlands: TNO Human Factors Research.

Pettitt, R.A.; Redden, E.S.; & Carstens, C.C.A. (2006). *Comparison of Hand and Arm Signals to a Covert Tactile Communication System in a Dynamic Environment* (ARL-TR-3838). Aberdeen Proving Ground, MD: U.S. Army Research Laboratory.

Redden, E.S. (2007). Findings from a multi-year investigation of tactile and multi-modal displays. *Proceedings of the Human Factors and Ergonomics 51st Society Annual Meeting*, 1697–700.

Ruff, H.A., Narayanan, S., & Draper, M.H. (2002). Human interaction with levels of automation and decision aid fidelity in the supervisory control of multiple simulated unmanned aerial vehicles. *Presence*, 11, 335–51.

Sarter, N.B. (2006). Multimodal information presentation: Design guidance and research challenges. In A.M. Bisantz (ed.), *Cognitive Engineering Insights for Human Performance and Decision Making, International Journal of Industrial Ergonomics*, 439–45.

Self, B.P., Ercoline, W.R., Olson, W.A., & Tvarynanas, A.P (2006). Spatial disorientation in uninhabited aerial vehicles. In N.J. Cooke (ed.), *Advances in Human Performance and Cognitive Engineering Research*, 7, pp. 133–46.

Shilling, R.D. & Shinn-Cunningham, B. (2002). Virtual auditory displays. In K. Stanney (ed.), *Handbook of Virtual Environments* (pp. 65–92). Mahwah, NJ: Lawrence Erlbaum Associates.

Simpson, B.D., Bolia, R.S., & Draper, M.H. (2004). Spatial auditory display concepts supporting situation awareness for operators of unmanned aerial vehicles. In D.A. Vincenzi, M. Mouloua, & P.A. Hancock (eds), *Human Performance, Situation Awareness, and Automation: Current Research and Trends* (pp. 61–5). Mahwah, NJ: Lawrence Erlbaum Associates.

Sklar, A.E. & Sarter, N.B. (1999). Good vibrations: tactile feedback in support of attention allocation and human-automation coordination in event-driven domains. *Human Factors*, 41(4), 543–52.

Sutcliffe, A. (2003). *Multimedia and Virtual Reality: Designing Multisensory User Interfaces*. Mahwah, NJ: Lawrence Erlbaum Associates.

Tachi, S., Komoriya, S., Sawada, K., Nishiyama, T., Itoko, T., & Kobayashi, M. (2003). Telexistence cockpit for humanoid robot control, *Advanced Robotics*, 17(3) 199–217.

Van Erp, J.B.F. (2002). Guidelines for the use of vibro-tactile displays in human computer interaction, *Proceedings of Eurohaptics*, Edinburgh, 18–22.

Van Erp, J.B.F. (2005). Presenting directions with a vibro-tactile torso display. *Ergonomics*, 48, 302–13.

Van Erp, J.B.F. (2006). Tactile displays: Spin-off from the military cockpit to uninhabited vehicles. *Proceedings of the NATO HFM Symposium Human Factors of Uninhabited Military Vehicles as Force Multipliers* (RTO-MP-HFM-135). Biarritz, France.

Van Erp, J.B.F., Eriksson, L., Levin, B., Carlander, O., Veltman, J.E., & Vos, W.K. (2007). Tactile cueing effects on performance in simulated aerial combat with high acceleration. *Aviation, Space and Environmental Medicine*, 78, 1128–34.

Van Erp, J.B.F. & Kappé, B. (1997). Ecological display design for the control of unmanned airframes. In M.J. Smith, G. Salvendy, & R.J. Koubek (eds), *Advances in Human Factors/Ergonomics, 21B Design of Computing Systems: Social and Ergonomics Considerations*, (pp. 267–70). Amsterdam: Elsevier.

Van Erp, J.B.F. & Oving, A.B. (2002). Control performance with three translational degrees of freedom. *Human Factors*, 44(1) 144–55.

Van Erp, J.B.F. & van Breda, L. (1999). *Human Factors Issues and Advanced Interface Design in Maritime Unmanned Aerial Vehicles: A Project Overview 1995–1998* (Report TM-99-A004). Soesterberg, the Netherlands: TNO Human Factors Research.

Van Erp, J.B.F. & van den Dobbelsteen, J.J. (1998). *Head Slaved Camera Control, Time Delays, and Situation Awareness in UAV Operation* (Report 1998 A 075). Soesterberg, the Netherlands: TNO Human Factors Research.

Van Erp, J.B.F., van Veen, H.A.H.C., Jansen, C., & Dobbins, T. (2005). Waypoint navigation with a vibrotactile waist belt. *ACM Transactions on Applied Perception*, 2(2), 106–17.

Van Erp, J.B.F., Veltman, J.A. & van Veen, H.A.H.C. (2003). A tactile cockpit instrument to support altitude control. *Proceedings of the Human Factors and Ergonomics Society 47th Annual Meeting*, 114–18.

Veronneau S.J. & Evans, R.H. (2004). Spatial disorientation mishap classification. In Zacharan (ed.), *Spatial Disorientation in Aviation*, *Progress in Astronautics and Aeronautics* (pp. 197–241). Reston, VA: American Institute of Aeronautics and Astronautics, Inc.

Wildzunas, R.M., Barron, T.L., & Wiley, R.W. (1996). Visual display delay effects on pilot performance. *Aviation, Space, and Environmental Medicine*, 67(3), 214–21.

Worringham, C.J. & Beringer, D.B. (1998). Directional stimulus-response compatibility: A test of three alternative principles. *Ergonomics*, 41(6), 864–80.

Chapter 16

Robotics Operator Performance in a Multi-Tasking Environment

Jessie Y.C. Chen[1]

U.S. Army Research Laboratory, Human Research and Engineering Directorate

Abstract

A series of four experiments was conducted in a simulated military mounted crewstation environment to examine the workload and performance of the robotics operator in either a single-tasking or a multi-tasking environment. In the multi-tasking environment, the operator also had to concurrently perform gunnery and communication tasks. In both tasking environments, the robotics tasks involved managing a semi-autonomous ground robot or teleoperating a ground robot to conduct reconnaissance tasks. Results showed that, in the single-tasking environment, the operator's performance was significantly worse when the robot had to be teleoperated than when it was semi-autonomous. In the multi-tasking environment, in contrary, the operator's robotics task was significantly worse when the robot was semi-autonomous, implying the operator's increasing reliance on automation when under heavy workload of multiple tasks. The operator's robotics task performance improved when the concurrent gunnery task was automated. However, when the automation was not reliable (i.e., miss-prone or false alarm-prone), the operator's robotics task performance degraded, and the severity of the degradation was affected by both the type of automation unreliability and the operator's attentional control. More specifically, for high attentional control operators, false alarm-prone alerts were more detrimental than miss-prone alerts. For low attentional control operators, conversely, miss-prone automation was

---

1   This study was funded by the U.S. Army's Robotics Collaboration Army Technology Objective (ATO). The author wishes to thank Mr. Michael J. Barnes (ARL-HRED) for his support and guidance throughout the process of this research project. The author also wishes to acknowledge the following individuals for their contributions to the research projects: Dr. Paula Durlach (ARI), Col. Mike Sanders, Maj. Jared Sloan (USMA), Maj. Carla Joyner (USMA), Ms. Laticia Bowens (University of Central Florida), and Mr. Peter Terrence (University of Central Florida). The author would also like to thank Mr. Henry Marshall (RDECOM-STTC) and Dr. Jon Bornstein (ARL) for providing simulation equipment to our research efforts. We would also like to acknowledge LTC Mike Sanders for his contributions to our project.

more harmful than false alarm-prone automation. Finally, our data indicated that the operators' preference of cueing display was related to their spatial ability. Low spatial ability operators preferred visual cueing over tactile cueing, and high spatial ability operators favored tactile cueing over visual cueing.

Robotics Operator Performance in a Multi-Tasking Environment

The goal of this research was to examine the workload and performance of the robotics operator in a simulated military reconnaissance environment. Two types of tasking environments were simulated, one with only the robotics task and the other with multiple tasks. In the multi-tasking environment, the operator also had to concurrently perform gunnery and communication tasks. Past research has shown that a robotics operator's situational awareness (SA) tended to be lower when teleoperation was required than when the robot's level of automation was higher (Dixon et al., 2003; Luck et al., 2006). These findings suggested that the attention on (manual) robotic control might have distracted the operators from focusing on the operator's main task (e.g., target detection). However, not much research has been conducted to examine robotics operator performance in a multi-tasking environment where the robotics is only one of the tasks the operator has to perform. Studies have shown that operators may develop over-reliance on the automatic system when under heavy workload, and this complacency may negatively affect their task performance (Dzindolet et al., 2001; Parasuraman et al., 1993; Thomas & Wickens, 2004; Young & Stanton, 2007a). Therefore, it is likely that in multi-tasking environments where a robotics operator has to concurrently perform other tasks, robotics task performance might be worse if the robot has a higher level of automation but imperfect reliability.

We simulated a generic mounted crewstation environment and conducted a series of four experiments to examine the workload and performance of the robotics operator in either a single-tasking or a multi-tasking environment. In the first experiment, we examined the ways human operators interacted with a simulated semi-autonomous unmanned ground vehicle (UGV), a semi-autonomous unmanned aerial vehicle (UAV), and a teleoperated UGV. Robotics operators performed parallel route reconnaissance missions with each platform alone, and also with all three platforms simultaneously. In the second experiment, we examined the workload and performance of the combined position of gunner and robotics operator. The experimental conditions included a gunner baseline condition and concurrent task conditions where participants simultaneously performed gunnery tasks and one of the following tasks: monitor an autonomous UGV via the video feed,;manage a semi-autonomous UGV; and teleoperate a UGV. Participants also simultaneously performed a communication task. In the third experiment, we investigated the performance of the combined position of gunner and robotics operator, and how aided target recognition (AiTR) capabilities (delivered either through tactile or tactile with visual cueing) for the gunnery task might benefit the concurrent (robotics and communication) tasks. In the fourth experiment, we manipulated the reliability of the AiTR (either false alarm-prone or miss-prone) and examined its effects on the operator's performance of the automated (gunnery) task and the concurrent tasks.

Additionally, we examined how individual difference factors such as spatial ability (SpA) and perceived attentional control (PAC) were related to the operator's task performance. The understanding of the robotics operator's performance and workload is key to successful operation of robotic assets, which are an important part of the U.S. Army's current force and will be an essential part in the Army's future force (Kamsickas, 2003). The findings of these four experiments could provide useful data for the Army's future system design community, and the results on individual differences may have important implications for personnel selection for the Army's future forces.

Experiment 1

In this experiment, the operator's task was to conduct route reconnaissance missions using one or three robots to detect enemy targets along a designated route. There were four experimental conditions: semi-autonomous UAV (UAV condition), semi-autonomous UGV (UGV condition), teleoperated UGV (teleop condition); and mixed (1 UAV + 1 UGV + 1 teleop). Each participant conducted four missions: three with a different robotic asset each and a final mission with all three robotic assets at their disposal. For the UAV and UGV conditions, operators assigned a set of waypoints, and then the robot(s) traveled the route automatically unless the operator intervened to alter their behavior. As the robot traveled, the operator manipulated the sensors searching for targets. The semi-autonomous robots were a UAV and a UGV. The third robot was a ground vehicle requiring teleoperation (teleop). In other words, the operator had to remotely drive this vehicle while manipulating its sensors to search for targets at the same time. All vehicles were equipped with camera sensors, which could be panned and zoomed and could send streaming video back to the operator control station.

Method, Participants and Apparatus

A total of 30 college students, 11 females and 19 males (mean age 20), participated in the study. Participants were compensated with $50 or class credit for their participation in the experiment. The experiment was conducted using the Embedded Combined Arms Team Training and Mission Rehearsal (ECATT-MR) testbed at the Simulation and Training Technology Center of the U.S. Army Research, Development and Engineering Command (RDECOM) in Orlando, Florida. The testbed was equipped with a steering wheel and gas and brake pedals for control of the teleoperated vehicle.

To control the robot in a semi-autonomous fashion, the operator assigned a route to the robot by placing waypoints on the situation awareness map. Once the mission began, the operator used the UAV/UGV screen controls to manage the robots. This screen had a hierarchical menu system that allowed the operator to choose which robot to control, which camera to use, and the movement. For example, the UAV could be ordered to hover or the UGV to halt.

Procedure

Participants received training and practice on the tasks they would need to conduct during an initial session that took approximately three hours. Participants returned one week later to complete the experiment. Before the experimental session, participants took the Cube Comparison test (Ekstrom et al., 1976), a test of spatial ability. In the experimental session, each participant conducted four missions, three with one different robotic asset (i.e., UAV, UGV, and teleop), and a final mission with all three robotic assets. The order of presentation of the single-robot conditions was counterbalanced, while the three-robot (mixed) condition was always the last. Thus, participants had a chance to complete a mission with each asset singly before conducting a mission with all three. For each mission, they were given a specific route to travel, with the requirement to detect and lase as many targets as they could find. They were to arrive at the end point within 30 minutes. Each mission occurred across the same terrain map but used a different route and direction of travel. Assignment of specific routes to asset conditions was counterbalanced across participants. Each route was approximately 4 km and consisted of an assembly area, a start point, two checkpoints, and an end point. Participants were instructed to issue a "location report" at each of these spots. Each mission allowed for the detection of 12 targets, which were a mixture of enemy vehicles and dismounted soldiers. Upon detection of a target, participants were to send a contact report and lase the target. When moving, both the UAV and the UGV traveled at 20 kph; the default altitude for the UAV was 100 m. Periodically, the warning signal for "Communications Fault" would light up, and the participants would need to double click the button to reset it. A workload questionnaire, the NASA Task Load Index (NASA-TLX: Hart & Staveland, 1988) was given at the end of each scenario to assess participants' perceived workload.

Results and Discussion

Data show that giving robotics operators additional assets may not be beneficial for enhancing target detection performance. Essentially, participants failed to discover more targets with three robots compared with the UAV alone or the UGV alone. Moreover, participants experienced a higher workload in the mixed-asset condition ($p < .005$), and over half of participants failed to complete the mission in the allotted time. Participants exhibited little natural tendency to coordinate the use of assets. It appears that giving multiple assets to robotics operators may be counterproductive, at least until the capabilities for automated data fusion, target detection, and target recognition are further developed.

These findings are consistent with those of other robotics control studies (Dixon et al., 2003; Rehfeld et al., 2005). Both Dixon et al. (2003) and Rehfeld et al. reported that giving a second robot to the operator(s) failed to enhance the individual's or team's target detection performance and, in some cases, may even degrade the operators' performance. These findings echoed what has been observed in the field (e.g., using robots for search and rescue efforts in Murphy, 2004) that remote perception is still one of the most fundamental challenges for robotics operators. The findings of Dixon et al. (2003), Rehfeld et al., and the current study suggest that, regardless of the types and homogeneity of the robotic platforms, additional assets do not appear to be beneficial for reconnaissance types of tasks, at least when information must be gleaned from streaming video.

Participants did not appear to take advantage of the multiple perspectives available in the mixed-asset condition. This might be due to several reasons. Firstly, three robots appeared to be more than the operators could handle. Secondly, participants were not specifically given instructions on how to coordinate multiple assets. Had more emphasis and training been provided on asset coordination, the results might have been different. Further research will be required to determine the extent to which experience and training can increase span of control. Lastly, the existing user interface did not support effective integration of sensor information from multiple platforms. An improved user interface should benefit the operator's integration of information from different sources. It is worth noting, however, that integrating information from different viewpoints (e.g., ground view and aerial view) can be challenging to the operator, and the potential human performance issues associated with such displays need to be carefully evaluated (Olmos et al., 2000; Thomas & Wickens, 2000). In addition, operators may be susceptible to saliency effects and anchoring heuristic/bias. Salient information on one display may catch most of the operator's attention, and the operator may form an inaccurate judgment because information from the other sources is not properly attended to and integrated (Wickens, 2005).

Participants with higher SpA were found to perform better in both speed and accuracy across platforms ($p < .05$). Their superior performance was especially consistent when they used the UAV. Results also showed that the target detection performance in the teleop condition was the worst compared with other conditions ($p < .05$). It seems for the teleop condition, a focus on maneuvering was at the expense of time spent searching. Operators needed to focus much of their visual attention on driving the robotic asset instead of using it for scanning and looking for targets. These findings are consistent with Luck et al. (2006) and Dixon et al. (2003), that robotics operators demonstrated higher SA when the robot's level of automation was higher. Luck et al. suggested that the attention on (manual) robotic control might have distracted the operators from focusing on the vehicle's location, which was the study's measure of SA. In Dixon et al. (2003), pilots found more targets when their UAV(s) were autonomous than when they were teleoperated.

Experiment 2

In the second experiment, we examined the workload and performance of the combined position of gunner and robotics operator. The experimental conditions included a gunner baseline (gunner baseline condition) and concurrent task conditions where participants simultaneously performed gunnery tasks and one of the following tasks: monitor a UGV via the video feed (monitor condition), manage a semi-autonomous UGV (auto condition), and teleoperate a UGV (teleop condition). In addition to the gunnery and the robotics tasks, the participants also simultaneously performed a communication task.

Method, Participants and Apparatus

Twenty college students, 3 females and 17 males (mean age 22), participated. They were compensated with $8 per hour or with class credit for their participation in the experiment.

The Tactical Control Unit (TCU) developed by the U.S. Army Research Laboratory's Robotics Collaborative Technology Alliance was used for the robotics tasks. The gunnery component was implemented using an additional screen and controls to simulate the out-of-the-window view and firing capabilities (see Figure 16.1). Participants used designated buttons on the joystick to manipulate the views, zoom in and out, and fire rounds.

To simulate communication with fellow crewmembers in the vehicle, cognitive tests were administered concurrently with the experimental sessions. The questions included simple military-related reasoning tests and simple memory tests. Test questions were prerecorded by a male speaker and were delivered by a synthetic speech program, DECTalk®, at the rate of one question every 33 seconds. An Attentional Control Survey, which measures attention focus and shifting, was

Figure 16.1   Tactical Control Unit (TCU), left, and gunnery station (gunner's out-of-the-window view), right

used to evaluate participants' perceived attentional control (PAC) (Derryberry & Reed, 2002). Two spatial tests were administered to assess participants' SpA: the Cube Comparison Test and the Spatial Orientation Test (SOT). The SOT, which is modeled after the cardinal direction test in Gugerty and Brooks (2004), consists of 32 computerized questions. As in Gugerty and Brooks, the SOT measures participants' ability to determine bearing between objects in an environment. Finally, participants' perceived workload was evaluated using the NASA-TLX questionnaire (Hart & Staveland, 1988).

Procedure

After the informed consent process, participants filled out the Attentional Control Survey and were administered the spatial tests. After these tests, participants received training, which lasted approximately 2 hours. Training was self-paced and was delivered by PowerPoint slides showing the elements of the TCU; steps for completing various tasks; several mini-exercises for practicing the steps; and two exercises for performing the robotic control tasks (one for practicing the teleoperation task and one for practicing the auto control tasks). After the tutorial on TCU, participants were trained on the gunnery tasks.

The experimental session took place on a different day but within a week of the training session. Before the experimental session began, participants were given some practice trials and review materials, if necessary, to refresh their memories. After the refresher training, participants then completed one combined exercise in which they performed all three tasks (i.e., gunnery, robotic control, and communication) at the same time. There were four experimental conditions: gunnery only (gunner baseline condition); gunnery + monitoring one autonomous UGV (monitor condition); gunnery + control of one semi-autonomous UGV (auto condition); gunnery + teleoperating one UGV (teleop condition).

Participants' tasks were to use their robotic asset to locate targets (i.e., enemy dismounted soldiers) in the remote environment, and also find targets in their immediate environment. For the gunner baseline condition, the operator performed only gunnery tasks (i.e., target detection and engagement). In the remaining conditions, the operator had to monitor or manage a UGV while performing gunnery tasks. The monitor condition required the operator to monitor the video feed as the UGV traveled and report any targets detected. The auto condition required the operator to monitor the video feed as the UGV traveled; examine still images generated from the reconnaissance scans, enabled by an aided target recognition (AiTR) capability; and detect targets. The teleop condition required the operator to manually manipulate and drive the UGV along a predetermined route using the TCU to detect targets. In the auto and teleop conditions, upon detecting a target, participants needed to place the target on the map, label the target, and then send a spot-report.

Each participant completed four scenarios, corresponding to the four experimental conditions. Each scenario lasted approximately 15 minutes, and the order of experimental conditions was counterbalanced across participants. The manned vehicle was simulated as traveling along a designated route, which was approximately 4.3 km. There were ten hostile and ten neutral targets (i.e., civilians) along the route in each gunnery scenario. Participants were instructed to engage the hostile targets and verbally report spotting the neutral targets. While the participants were performing their gunnery and/or robotic control tasks, they would have to perform the communication (i.e., cognitive) tasks simultaneously. Participants filled out the NASA-TLX after each scenario. There were 2-minute breaks between scenarios.

Results and Discussion

Results showed that gunner's target detection performance degraded significantly when he or she had to concurrently monitor, manage, or teleoperate a UGV ($p < 0.001$, Figure 16.2). The gunners' performance in the three concurrent-task conditions were all significantly different from one another, with the monitor condition being the highest and teleop condition being the lowest ($p < 0.05$). These results suggest that, if it is necessary for the gunner to concurrently access information from the robotic assets, then the robotic tasks should be limited to activities such as monitoring. If excessive manipulation of the Warfighter-machine interface is required, as in the auto and teleop conditions, their gunnery performance will be significantly affected. Participants' SpA (based on their composite spatial test scores) was found to be an accurate predictor of their gunnery performance ($p < 0.005$).

For the robotics tasks (see Figure 16.2), there were significant differences among the monitor, auto, and teleop conditions in human target detection performance ($p < 0.05$), with the auto condition being the lowest (only 53 percent were detected). The inferior performance associated with the semi-autonomous UGV seemed to reflect participants' over-reliance on the AiTR capability, and failure to detect more targets along the route that were not picked up by the AiTR. In contrast, in Experiment 1, participants had the lowest target detection using the teleop. However, in Experiment 1's UGV condition, the AiTR capability was not available and the only task automated was vehicle navigation. Results of the current study are consistent with automation research that operators may develop over-reliance on the automatic system, and that this complacency may negatively affect their task performance (Dzindolet et al. 2001; Parasuraman et al., 1993; Thomas & Wickens, 2000; Young & Stanton, 2007a). None of the individual difference factors were found to significantly correlate with participants' robotics task performance.

Participants' communication task performance degraded when their robotics tasks became more challenging (i.e., auto and teleop conditions) ($p < 0.005$). It is interesting to note that participants appeared to be able to perform their communication task at similar levels in the gunner baseline and monitor conditions. This suggests that, in the monitor condition, participants had sufficient

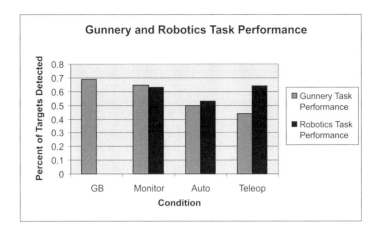

Figure 16.2   Gunnery and robotics task performance

cognitive resources left to perform the communication tasks. Participants with higher PAC performed better on the concurrent communication task ($p < 0.05$), although they performed at a similar level on their gunnery and robotic control tasks as those with lower PAC. These results suggest that participants devoted most of their attention resources to the gunnery and robotic tasks, and only those with higher attention allocation skills could more successfully perform the tertiary communication tasks. Since communication will be a critical part in the future military task environment, these results may have important implications for personnel selection for the Army's future forces.

Participants' perceived workload increased almost linearly in order from the gunner baseline, monitor, auto and to the teleop condition ($p < 0.0001$), and the differences among the four conditions were all statistically significant ($p < 0.05$, see Figure 16.3). These results are consistent with Schipani (2003), who evaluated robotics operator workload in a field setting. Additionally, it appears that participants with higher SpA, although performing better on the tasks, did not perceive the tasks as less demanding. On the other hand, the correlations between participants' PAC and their workload were as expected: those with higher attentional control thought the tasks were less demanding. It is worth noting that only in the most challenging condition, teleop, did the correlation reach significance ($p < 0.05$). In other words, as the tasks became harder, the differences in the levels of perceived workload between those with higher and lower PAC appeared to widen.

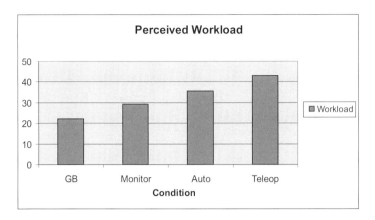

Figure 16.3　Perceived workload

Experiment 3

In this experiment, we examined if and how tactile cueing, which delivered simulated AiTR capabilities (i.e., cues to a potential target position relative to the gunner), enhanced gunner performance in a military multi-tasking environment similar to the one in Experiment 2.

Method, Participants and Apparatus

Twenty college students, 4 females and 16 males (mean age 21), participated. They were compensated with $8 per hour or with class credit for their participation in the experiment.

The TCU and the gunnery station were the same as in Experiment 2. Additionally, to augment target detection in the gunnery component, visual and tactile alerts were used to cue the participant to the direction of a target as determined by the AiTR. The visual alerts were displayed in the lower right area of the screen, and the targets consisted of icons presented around the overhead view diagram of the participant's gunner station on the simulated vehicle. As the gunner rotated the view, the turret portion of the vehicle diagram moved along the eight possible orientations to allow the gunner to place his or her field of view on the cued target. Tactually, target positions relative to the gunner were presented using eight electromechanical transducers known as "tactors," which were arranged equidistantly on an elasticized belt worn around the abdomen just above navel height.

The overall design of the study is a 2×3 repeated-measures design. The independent variables are AiTR type (baseline [BL]—no alerts vs. tactile alerts only [Tac] vs. tactile + visual alerts [TacVis]); and robotics task type (auto vs. teleop).

The reliability level of the alerts was 100 percent. However, only hostile targets were cued, not the neutral targets. The participants had to detect the neutral targets on their own.

Procedure

After the informed consent process, participants filled out the Attentional Control Survey and were administered the spatial tests, Cube Comparison, Hidden Patterns, and SOT (Ekstrom et al., 1976; Gugerty & Brooks, 2004). After these tests, participants received training, which lasted approximately 2 hours. The training procedure was the same as in Experiment 2.

The experimental session took place on a different day but within a week of the training session. Before the experimental session began, participants were given some practice trials and review materials, if necessary, to refresh their memories. After the refresher training, participants changed into one of the laboratory cotton T-shirts in order to standardize how the tactors were applied to the skin. The experimenter arranged the tactors so that they were equidistant around the participant's abdomen. The experimenter then explained the nature of the AiTR system and the corresponding visual or tactile cues that would be provided. Participants then completed one combined exercise in which they performed all three tasks (i.e., gunnery, robotics, and communication tasks) at the same time.

The gunnery and robotics tasks (i.e., auto and teleop) were identical to those in Experiment 2. In total, there were six 15-minute scenarios, corresponding to the six experimental conditions, the order of which was counterbalanced across participants. While the participants were performing their gunnery and robotics control tasks, they simultaneously performed the communication task by answering questions delivered to them via DECtalk®. There were 2-minute breaks between experimental scenarios. Participants filled out the NASA-TLX after they completed each scenario, and also a usability survey on their preferences for presentation of AiTR information at the end of the experimental session. Participants rated their preference on a five-point scale: entirely visual; predominately visual; both visual and tactile; predominately tactile; entirely tactile.

Results and Discussion

Results showed that gunner's target detection performance improved significantly when his or her task was assisted by AiTR—from about 52 percent to about 84 percent ($p < 0.001$). Consistent with the findings of Experiment 2, participants' SpA (based on their composite spatial test scores) was found to be an accurate predictor of their gunnery performance ($p < 0.05$). Those with higher SpA consistently outperformed those with lower SpA throughout the scenarios.

It was also found that the gunners' detection of neutral targets (which was not aided by AiTR) was significantly worse when the gunners had to teleoperate a robotic asset (vs. when the asset was semi-autonomous) or when the gunnery task was aided by AiTR ($p < 0.01$ and $p < 0.001$, respectively). These findings suggest that participants devoted significantly less visual attention to the gunnery station when their robotic asset required teleoperation or when their gunnery task was assisted by AiTR. On average, participants detected 45 percent of the neutral targets when there was no AiTR; they only detected 28 percent when AiTR was present.

For the robotics task, participants detected significantly more hostile targets when teleoperating than when the robot was semi-autonomous ($p < 0.01$). Additionally, when their gunnery task was aided by AiTR, participants detected more targets when teleoperating, but detected fewer targets when their UGV was semi-autonomous—robotics task × AiTR interaction was significant, $p < 0.005$ (see Figure 16.4). These findings are consistent with those of Experiment 2 and reflected operators' over-reliance on the automation available in the auto condition to help them with their robotics task. However, they failed to detect those targets not captured by the robot (although these targets were visible to them).

Results of the current study are consistent with automation research that operators may develop over-reliance on the automatic system, especially when under heavy workload, and this complacency may negatively affect their task performance (Dzindolet et al. 2001; Parasuraman et al., 1993; Thomas & Wickens, 2004; Young & Stanton, 2007a). Our results also showed that participants' teleoperation performance improved significantly when their gunnery task was assisted by AiTR ($p < 0.05$). Therefore, AiTR was not only beneficial for the automated task (i.e., gunnery) but also the concurrent task (i.e., robotics). This finding is consistent with previous research on the effects of automating the primary task on enhancing the concurrent visual tasks (Dixon et al., 2003). Additionally, our data show that AiTR was more beneficial for enhancing the concurrent robotics performance for those with lower SpA than for those with higher SpA ($p < 0.05$). Similarly, AiTR appeared to benefit those with lower PAC more than those with higher PAC

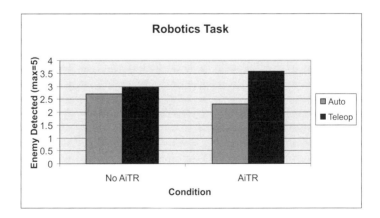

Figure 16.4   Robotics task performance

(difference was marginally significant, $p = 0.052$). When AiTR was available to assist those operators with low SpA and PAC, the performance of their concurrent task was improved to a similar level as those with higher SpA and PAC. These results are consistent with findings previously reported (Young & Stanton, 2007b) and may have important implications for system design and personnel selection for the future military programs.

Participants' communication task performance improved when their gunnery task was aided by AiTR ($p < 0.05$). Again, this result suggests that AiTR not only enhanced the tasks it was designed for, but also benefited concurrent tasks. It also shows that our cognitive communication task was sensitive to the task load manipulations we implemented for the primary task (AiTR vs. no AiTR).

Participants' perceived workload was found to be affected by the type of concurrent robotics task as well as whether their gunnery task was aided by AiTR (Figure 16.5). They had a higher workload level when their gunnery task was unassisted by AiTR or when they teleoperated the robotic asset ($p < 0.05$). These results are consistent with Experiment 2.

According to the usability survey on preference of AiTR display, 65 percent of our participants responded that they either relied predominantly or entirely on the tactile AiTR display. Only 15 percent responded that they either relied predominantly or entirely on the visual AiTR display. AiTR preference was also significantly correlated with SpA and PAC ($p < 0.05$). Perhaps those with higher SpA can more easily make use of the spatial tactile signals in the dual task setting, and therefore have a stronger preference for something that makes the gunner task easier to complete. Individuals with lower SpA, on the other hand, may have not utilized the spatial tactile cues to their full extent, and therefore continued to prefer the visual AiTR display. However, this preference may have caused degraded target detection performance due to more visual attention being devoted to the visual AiTR display than the simulated environment. In contrast, those who were

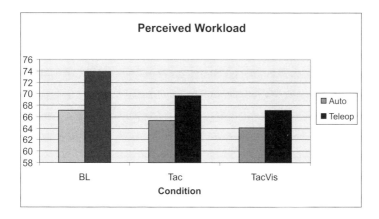

Figure 16.5   Perceived workload (auto vs. teleoperation)

more spatial relied on the directional information of the tactile display to help them with the visually demanding tasks, resulting in more effective performance.

Experiment 4

In this experiment, we followed the paradigm of Experiment 3 and examined if and how tactile/visual cueing with imperfect reliability affected the operator's performance. The AiTR was either miss-prone (MP) or false alarm-prone (FAP).

Method, Participants and Apparatus

Twenty-four college students, 4 females and 20 males (mean age 22), participated. They were compensated $15 per hour or with class credit for their participation in the experiment.

The TCU and the gunnery station were the same as in Experiments 2 and 3. Both visual and tactile cueing were provided to the participants. There were three types of robotics tasks: monitor, auto, and teleop (identical to Experiment 2). The overall design of the study is a 2×3 mixed design. The between-subject variable is AiTR type (FAP vs. MP). The within-subject variable is robotics task type (monitor vs. auto vs. teleop). The reliability levels of the FAP and MP alerts were both 60 percent. The FAP condition consisted of 10 hits (i.e., alerts when there were targets); 8 FAs (i.e., alerts when there were no targets); 0 misses (i.e., no alerts when there were targets); and 2 correct rejections (CRs), i.e., no alerts when there were no targets). The MP condition consisted of 2 hits, 0 FAs, 8 misses, and 10 CRs. Additionally, only hostile targets were cued, not the neutral targets. The participants were instructed to detect the neutral targets on their own. The alerts (both true and false) did not occur when neutral targets appeared in the environment.

Procedure

Participants were randomly assigned to the MP or the FAP group. The preliminary session (surveys and spatial tests) and training session were identical to those of Experiment 3. The experimental session immediately followed the training session. The gunnery and robotics tasks (i.e., monitor, auto, and teleop) were identical to those in Experiment 2. Each participant completed three 15-minute scenarios, corresponding to the three robotics tasking conditions, the order of which was counterbalanced across participants. While the participants were performing their gunnery and robotics tasks, they simultaneously performed the communication task by answering questions delivered to them via DECtalk®. There were two-minute breaks between experimental scenarios. Participants filled out the NASA-TLX after they completed each scenario, and also a usability survey on their preferences for presentation of AiTR information at the end of the experimental session.

Results and Discussion

Results showed that operator's gunnery task performance in detecting hostile targets was significantly better in the monitor condition than in the other two robotics task conditions ($p < 0.05$), consistent with the findings of Experiment 2. In Experiment 2 and the current experiment, the workload associated with the monitor condition was significantly lower than the other robotics conditions. These results suggest that the operator had more visual and mental resources for the gunnery task when the robotics task was simply monitoring the video feed, compared with the other two robotics conditions. Also consistent with Experiments 2 and 3, participants' SpA was found to be an accurate predictor of their gunnery performance ($p < 0.05$). Results also showed that there was a significant interaction between types of unreliable AiTR and participants' PAC ($p < 0.05$). Those with lower PAC had a better performance with the FAP cueing, and those with higher PAC had a better performance with the MP cueing. For those with high PAC, our data are consistent with the notion that operator reliance on and compliance with automation are independent constructs and are separately affected by system misses and false alarms. Based on Figure 16.6 (thick solid line), it is evident that high PAC participants did not comply with alerts in the FAP condition. Since the FAP AiTR had a 0 percent miss rate, a full compliance should result in a detection rate over 80 percent, as reported in Experiment 3. As predicted, Figure 16.6 shows that, in MP conditions, high PAC participants did not rely on the AiTR and detected more targets than were cued. However, an examination of the data for the low PAC participants revealed a complete opposite trend (thin solid line). Specifically, with the FAP condition, low PAC participants showed a strong compliance with the alerts, which resulted in a good performance in target detection (at a similar level as in Experiment 3). With the MP condition, however, low PAC participants evidently over-relied on the automation and, therefore, had a very poor performance.

It was also found that the gunner's detection of neutral targets (which was not aided by AiTR) was significantly worse when he or she had to teleoperate a robotic asset (vs. when the asset was semi-autonomous): $p < 0.05$. This finding is consistent with Experiments 2 and 3 and suggests that participants devoted significantly less visual attention to the gunnery station when their robotic asset required teleoperation. Our data also showed that those with lower PAC performed at about the same level regardless of the AiTR type, while those with higher PAC performed better with the MP cueing (interaction between AiTR type and PAC was significant, $p < 0.05$). This suggests that higher PAC participants devoted more visual attention to the gunnery station (implying a reduced reliance on the automation for the gunnery task) when the AiTR was MP than when the AiTR was FAP. Although we did not measure participants' scanning behaviors, the detection rate of neutral targets on the gunnery station provides an estimate of the amount of operators' visual attention on the automated task environment. Again, the data of high PAC participants seem to support the hypothesis that MP automation reduces

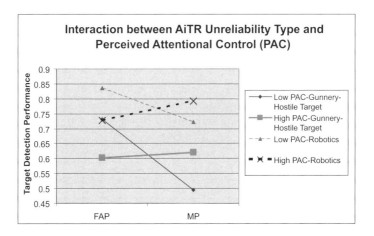

Figure 16.6   Interaction between AiTR unreliability type and PAC

operator reliance. However, the same phenomenon was not observed for the low PAC participants.

For the robotics tasks, the data showed that participants had the best performance when the task was only monitoring the video feed ($p < 0.001$). Moreover, the monitor task performance stayed at the same level regardless of the AiTR types. The data of the teleop performance are consistent with previous studies (Dixon & Wickens, 2006; Wickens et al., 2005) that MP automation degrades concurrent task performance more than FAP ($p < 0.05$). However, the same trend was not observed for the other two less challenging robotics tasks. Therefore, it appears that the adverse effect of MP automation on concurrent tasks is only manifest in more challenging task conditions. Our data also showed that, again, there was a significant interaction between AiTR type and PAC ($p < 0.05$, see Figure 16.6). Consistent with the previous two performance measures (gunnery-hostile and gunnery-neutral), the low PAC participants exhibited a higher level of performance degradation with the MP conditions (thin dotted line). The performance of the high PAC participants (thick dotted line), on the other hand, showed a completely opposite trend. These results suggest that the high PAC participants' reduced compliance with the FAP alerts did not help them with their concurrent task, compared with the MP conditions; in fact, it resulted in a poorer targeting performance. Conversely, their reduced reliance on the MP alerts did not hurt their performance as did the FAP alerts. Overall, the low PAC participants showed the most pronounced adverse effect of MP alerts on concurrent performance. Conversely, the FAP alerts not only helped them with their automated task but also with their concurrent task.

Participants' communication task performance was significantly better when their robotics task was monitor than when it was teleop ($p < 0.05$), consistent with the finding of Experiment 2. Participants' perceived workload was found to be affected by the type of concurrent robotics task ($p < 0.001$). The workload was significantly higher in the teleop condition than in the auto condition, which in turn was significantly higher than in the monitor condition. These results are consistent with Experiment 2.

Our results also show that low SpA participants preferred visual cueing over tactile cueing, and high SpA participants favored tactile cueing over visual cueing ($p < 0.01$). This confirmed a trend that was first observed in Experiment 3. The implication of this finding for user interface design is discussed in the following section.

General Discussion

We simulated a military mounted crewstation environment and conducted a series of four experiments to examine the workload and performance of the robotics operator in either a single-tasking or a multi-tasking environment (Chen & Barnes, 2008). In the multi-tasking environment, the operator also had to perform gunnery and communication tasks. In both tasking environments, the robotics tasks involved monitoring or managing a semi-autonomous ground robot or teleoperating a ground robot to conduct reconnaissance tasks. In terms of robotics task performance, we observed very different patterns when robotics tasks were performed singly or were in a multi-tasking environment. Our results showed that, when robotics was the only task, teleoperation tended to degrade task performance due to operator attention being diverted to the driving (i.e., teleoperation) task. When robotics was only one of the tasks the operator had to perform concurrently, however, a complete reverse pattern was observed. Participants' robotics task performance was *better* when they teleoperated vs. when the robot was semi-autonomous. These results are consistent with automation research that operators may develop over-reliance on the automatic system when under heavy workload (e.g., multi-tasking), and this complacency may negatively affect their task performance (Dzindolet et al. 2001; Parasuraman et al., 1993; Thomas & Wickens, 2000; Young & Stanton, 2007a). Thomas & Wickens (2000) showed that when participants had access to information gathered from automatically panning cameras, they tended to prematurely close the automatic panning feature prior to finishing examining the entire environment. Participants manually panning the cameras, however, had significantly higher target detection performance, which indicated more adequate panning. It is worth noting that these findings, along with the results of our studies, do not necessarily suggest that manual manipulation of sensor devices be used instead of AiTR devices. However, the issue of over-reliance on these automatic capabilities needs to be taken into account when designing the user interface where these features are one of the components.

Results of Experiment 3 demonstrated that automation was not only beneficial to the automated task but also to the concurrent tasks. When the participants' gunnery task was aided by AiTR, their robotics and communication tasks also improved. Our data also show that AiTR was more beneficial for enhancing the concurrent robotics performance for those with lower SpA than for those with higher SpA. Similarly, AiTR appeared to benefit those with lower PAC more than those with higher PAC. When AiTR was available to assist those operators with low SpA and PAC, the performance of their concurrent task was improved to a similar level as those with higher SpA and PAC. These results are consistent with other findings showing that vehicle automation helps reduce the performance gap between experts and novices (Young & Stanton, 2007b). These results may have important implications for system design and personnel selection for the future military programs.

Results of Experiment 4 seem to suggest that, overall, for high PAC participants, FAP alerts were more detrimental than MP alerts. FAP alerts not only affected their automated task but also the concurrent task. This finding is consistent with the conclusion of Dixon et al. (2007) that FAP degraded overall performance more than MP automation. However, it is worth noting that, for low PAC participants, we observed the opposite pattern: MP automation was more harmful than FAP automation. The overall data suggest that low PAC participants had a higher trust in the automation system than did high PAC participants. It is likely that low PAC participants had more difficulty in performing multiple tasks concurrently and had to rely on automation when available. High PAC participants, on the other hand, tended to rely on their own multi-tasking ability to perform the tasks. Our results are consistent with past research (Lee & Moray, 1994; Lee & See, 2004) that self-confidence is a critical factor in mediating the effect of trust (in automation) on reliance (on the automatic system). Lee and Moray found that, when self-confidence exceeded trust, operators tended to use manual control. When trust exceeded self-confidence, automation was used more. Our present data suggest that this relationship between self-confidence and level of reliance is also moderated by operators' PAC.

Our data repeatedly show the superior performance by participants with higher SpA, and these results are consistent with past research (Lathan & Tracey, 2002; Vincow, 1998). Thomas and Wickens (2004) showed that there were individual differences in scanning effectiveness and its associated target detection performance. However, Thomas and Wickens did not examine the characteristics of those participants who had more effective scanning strategies. The findings of our studies indicate that SpA may be an important factor for determining scanning effectiveness. Our findings also support the recommendation by Lathan and Tracey that military missions can benefit from selecting personnel with higher SpA to operate robotic devices. Also, training interventions that could enhance the spatial interpretations required might be of benefit (Rodes et al., 2005).

Our results show that low SpA participants preferred visual cueing over tactile cueing, and high SpA participants favored tactile cueing over visual cueing. Perhaps those with higher SpA can more easily employ the spatial tactile signals in the dual task setting and therefore have a stronger preference for something that makes the gunner task easier to complete. Individuals with lower SpA, in contrast, may have not utilized the spatial tactile cues to their full extent and therefore continued to prefer the visual AiTR display. It is likely that, in our study, those who preferred visual AiTR displays might be more iconic in their mental representations (Kozhevnikov et al., 2002). However, this preference may have caused degraded target detection performance because more visual attention was devoted to the visual AiTR display than to the simulated environment. In contrast, those who were more spatial relied on the directional information of the tactile display to help them with the visually demanding tasks, resulting in a more effective performance. These findings may have important implications for personnel selection, system designs, and training development. One recommendation is that, to better enhance the task performance for low SpA individuals, the visual cueing display should be more integrated with the visual scene. For example, augmented reality (i.e., visual overlays) is a potential technique to embed directional information onto the video (Calhoun & Draper, 2006).

Our results repeatedly showed that participants' communication task performance tended to degrade when the concurrent robotics task became more challenging (e.g., requiring teleoperation). According to Naveh-Benjamin et al. (2000), information encoding processes require more attention than retrieval and are more prone to the effects of competing demands of multitasking. It is, therefore, likely that the information-encoding process of the communication task in our studies was more disrupted by the more challenging teleop task than by other robotics tasks.

Across the four experiments, participants consistently rated the teleop conditions as the highest-workload conditions. These results are consistent with past research such as Schipani (2003), who evaluated robotic operator workload in a field setting. Although many of the ground robotic assets in the U.S. Army's future combat systems will be semi-autonomous, teleoperation will still be an important part of any missions involving robotic assets (e.g., when robots encounter obstacles or other problems). The higher workload associated with teleoperation needs to be taken into account when designing the user interfaces for the robotic assets.

Conclusions

We simulated a military mounted crewstation environment and conducted a series of four experiments to examine the workload and performance of the robotics operator in either a single-tasking or a multi-tasking environment. We have the following conclusions:

- When robotics was the only task, teleoperation tended to degrade task performance due to operator attention being diverted to the driving (i.e., teleoperation) task. When robotics was only one of the tasks the operator had to perform concurrently, however, participants' robotics task performance was *better* when they teleoperated vs. when the robot was semi-autonomous.

- Teleoperation created significantly more workload for the operators than did other robotics tasking conditions that did not require manual operation of the robot.
- Automation was not only beneficial to the automated task but also to the concurrent tasks.
- For participants with better attentional control, false alarm-prone alerts were more detrimental than miss-prone alerts. In contrast, for participants with poor attentional control, miss-prone automation was more harmful than false alarm-prone automation.
- Participants' spatial ability was found to be a reliable predictor of their targeting task performance. Additionally, those with lower spatial ability tended to prefer visual cueing over tactile cueing (in a visually intensive, multi-tasking environment), and those with higher spatial ability tended to favor tactile cueing over visual cueing.

References

Calhoun, G. & Draper, M. (2006). Multi-sensory interfaces for remotely operated vehicles. In N.J. Cooke, H. Pringle, H. Pedersen, & O. Connor (eds), *Human Factors of Remotely Operated Vehicles* (pp. 149–63). Oxford: Elsevier.

Derryberry, D. & Reed, M. (2002). Anxiety-related attentional biases and their regulation by attentional control. *Journal of Abnormal Psychology*, 111, 225–36.

Dixon, S. R. & Wickens, C. D. (2006). Automation reliability in unmanned aerial vehicle control: A reliance-compliance model of automation dependence in high workload. *Human Factors*, 48, 474–86.

Dixon, S. R., Wickens, C. D., & Chang, D. (2003). Comparing quantitative model predictions to experimental data in multiple-UAV flight control. *Proceedings of the Human Factors and Ergonomics Society 47th Annual Meeting* (pp. 104–8). Santa Monica, CA: HFES.

Dixon, S. R., Wickens, C. D., & McCarley, J. S. (2007). On the independence of compliance and reliance: Are automation false alarms worse than misses? *Human Factors*, 49, 564–72.

Dzindolet, M. T., Pierce, L. G., Beck, H. P., Dawe, L. A., & Anderson, B. W. (2001). Predicting misuse and disuse of combat identification systems. *Military Psychology*, 13, 147–64.

Ekstrom, R. B., French, J. W., and Harman, H. H. (1976). *Kit of Factor-referenced Cognitive Tests*. Princeton, NJ: Educational Testing Service.

Gugerty, L. & Brooks, J. (2004). Reference-frame misalignment and cardinal direction judgments: Group differences and strategies. *Journal of Experimental Psychology: Applied*, 10, 75–88.

Hart, S. G. & Staveland, L. E. (1988). Development of NASA TLX (Task Load Index): Results of empirical and theoretical research. In P.A. Hancock & N. Meshkati (eds), *Human Mental Workload* (pp. 139–83). Amsterdam: North-Holland.

Kamsickas, G. (2003). *Future Combat Systems (FCS) Concept and Technology Development (CTD) Phase-Unmanned Combat Demonstration-Final Report* (Tech. Rep. D786-1006102). Seattle, WA: The Boeing Company.

Kozhevnikov, M., Hegarty, M., & Mayer, R. E. (2002). Revising the visualizer-verbalizer dimension: Evidence for two types of visualizers. *Cognition and Instruction*, 20(1), 47–77.

Lathan, C. E. & Tracey, M. (2002). The effects of operator spatial perception and sensory feedback on human-robot teleoperation performance. *Presence*, 11, 368–77.

Lee, J. D. & Moray, N. (1994). Trust, self-confidence, and operators' adaptation to automation. *International Journal of Human-Computer Studies*, 40, 153–84.

Lee, J. D. & See. K. A. (2004). Trust in automation: Designing for appropriate reliance. *Human Factors*, 46, 50–80.

Luck, J. P., Allender, L., & Russell, D. C. (2006). An investigation of real world control of robotic assets under communication latency. *Proceedings of the ACM Conference on Human-Robot Interaction*, Salt Lake City, UT.

Murphy, R. R. (2004). Human-robot interaction in rescue robotics. *IEEE Systems, Man and Cybernetics Part C: Applications and Reviews*, 34, 138–53.

Naveh-Benjamin, M., Craik, F. I. M., Perretta, J. G., & Tonev, S. T. (2000). The effects of divided attention on encoding and retrieval processes: The resiliency of retrieval processes. *The Quarterly Journal of Experimental Psychology*, 53A, 609–25.

Olmos, O., Wickens, C. D., & Chudy, A. (2000). Tactical displays for combat awareness: An examination of dimensionality and frame of reference concepts and the application of cognitive engineering. *International Journal of Aviation Psychology*, 10, 247–71.

Parasuraman, R., Molloy, R., & Singh, I. (1993). Performance consequences of automation-induced "complacency." *International Journal of Aviation Psychology*, 3, 1–23.

Rehfeld, S. A., Jentsch, F. G., Curtis, M., & Fincannon, T. (2005). Collaborative teamwork with unmanned ground vehicles in military missions. *Proceedings of the 11th Annual Human-Computer Interaction International Conference*, Las Vegas, NV.

Rodes, W., Brooks, J., & Gugerty, L. (2005). Using verbal protocol analysis and cognitive modeling to understand strategies used for cardinal direction judgments. *Proceedings of Human Factors of UAVs Workshop*, Mesa, AZ.

Schipani, S. (2003). An evaluation of operator workload during partially-autonomous vehicle operations. *Proceedings of PerMIS 2003*, 16-18 September, Washington, DC. Available online at http://www.isd.mel.nist.gov/research_ areas/research_engineering/Performance_Metrics/PerMIS_2003/Proceedings/ Schipani.pdf (accessed 23 February 2004).

Thomas, L. C. & Wickens, C. D. (2000). *Effects of Display Frames of Reference on Spatial Judgments and Change Detection* (Tech. Rep. ARL-00-14/FED-LAB-00-4). Urbana-Champaign, IL: University of Illinois, Aviation Research Laboratory.

Thomas, L. C. & Wickens, C. D. (2004). Eye-tracking and individual differences in off-normal event detection when flying with a synthetic vision system display. *Proceedings of the Human Factors and Ergonomics Society 48th Annual Meeting* (pp. 223–7).

Vincow, M. A. (1998). *Frame of Reference and Navigation Through Document Visualizations*. Unpublished dissertation, University of Illinois.

Wickens, C. D. (2005). Attentional tunneling and task management. *Proceedings of the 13th International Symposium on Aviation Psychology*.

Wickens, C. D., Dixon, S. R., Goh, J., & Hammer, B. (2005). *Pilot Dependence on Imperfect Diagnostic Automation in Simulated UAV Flights: An Attentional Visual Scanning Analysis* (Tech. Rep. AHFD-05-02/MAAD-05-02). Urbana-Champaign, IL: University of Illinois, Institute of Aviation, Aviation Human Factors Division.

Young, M. S. & Stanton, N. A. (2007a). Back to the future: Brake reaction times for manual and automated vehicles. *Ergonomics*, 50, 46–58.

Young, M. S. & Stanton, N. A. (2007b). What's skill got to do with it? Vehicle automation and driver mental workload. *Ergonomics*, 50, 1324–39.

Chapter 17

A Cognitive Systems Engineering Approach for Human-Robot Interaction: Lessons from an Examination of Temporal Latency

Laurel Allender

U.S. Army Research Laboratory

Examining Human-Robot Interaction

Advanced robotics technology holds promise for extending, enhancing, even replacing humans in the performance of dangerous or tedious tasks. Search and rescue robots can go where humans cannot—or dare not—and become remote eyes and ears for human first responders and rescue teams (Casper & Murphy, 2003). Robots that provide some level of care or supervision in medical contexts are being tested and are intended to enhance the well being of the person being cared for as well as lightening the loads of other care-providers (e.g., Gockley & Matarić, 2006). Robots are even being evaluated in social and educational settings with children to understand the development, or evolution, of naturalistic human-robot interaction (e.g., Tanaka et al., 2006). Of course, robots may also be fun, such as the cocktail-serving robots offered in the 2003 Neiman Marcus Christmas catalog.

For the military, robotic technology is a serious endeavor and is being considered for a wide range of mission-oriented tasks. There have been some significant successes with respect to autonomous mobility, for example the U.S. Army Research Laboratory's Experimental Unmanned Vehicle (XUV) program (Committee on Army Unmanned Ground Vehicle Technology, 2002) and the various competing teams that have participated in the Defense Advanced Research Projects Agency Grand Challenges, (DARPA, 2008). Small robots made headlines in 2002 when they were tested as military search robots in Afghanistan (e.g., Kirsner, 2002). The U.S. Army's future combat systems is proposed to include six types of unmanned air and ground autonomous vehicles for reconnaissance and surveillance, supply and support, and offensive and defensive missions (Future Combat Systems, 2008).

Accompanying these technological advances (both achieved and proposed) has been the growth, or explosion really, of questions about and research into human-robot *interaction* (HRI). HRI goes beyond the physical control device or the video feedback display of the human-to-robot *interface* to include the full gamut of perceptual, attentional, cognitive, collaborative, and social interactions. Indeed, it is asserted that HRI comprises a convergence of characteristics that differentiate it from human-machine interaction generally or even from automation-intensive human-computer interaction (Scholtz, 2003), including the dynamic control, mobility, and autonomy of the robot and changing human-robot roles and responsibilities.

The Context for Human-Robot Interaction

A simplified depiction of HRI is shown in Figure 17.1. We can think of the four elements—the *robotic technology*, the *communication network* and its associated characteristics such as bandwidth; the *controls and displays* that the human uses to interact with the robot; and the *human operator* him- or herself—as typically proceeding linearly in time from left to right, and then back again.

Of course, there are many interaction effects across the four basic elements of HRI. For example, the nature of the robotic technology and its onboard sensors will affect how much information is required or is available to be pushed through the network. A 2D sensor will provide less dense information to be pushed over the network than will full-streaming video. The slower the onboard processer or the smaller the memory, the more frequently a robot will need to "ask for" updates over the network from a central system processor or database or from the operator. Likewise, the technology drives the features and capabilities of the controls and displays and, ultimately, both enables and places a particular "demand" on the operator. A robot that must be teleoperated via a steering wheel or joystick will place a nearly constant motor control demand on the operator, whereas a robot with waypoint path planning will require only periodic inputs and monitoring.

Note, however, that this simple flow is depicted as being embedded in a larger context and it is this context that must be considered to explore HRI fully. A given context brings a host of additional factors that shape the interaction aspect of HRI. There are environmental factors such as terrain and weather; equipment issues such as hardware and software reliability; network and communication hardware factors such as shared bandwidth; information overload and task allocation issues, which may even be exacerbated by the robot interface; and human operator characteristics such as level of training or fatigue. Further, the element labeled

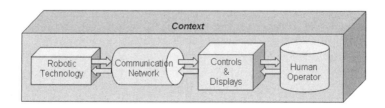

Figure 17.1   A simplified depiction of the basic HRI elements

"human operator" may in fact be a team of stakeholders who may not interact physically with the robotic system, but may place demands on it for information, for example requesting a particular image view or a sensor update. The broader context for the robot-based mission or tasks may well include other, essentially simultaneous tasks. These "other" tasks may be related to the robot-based mission, such as communicating with a teammate or monitoring system status, or they may be part of another goal altogether, such as self-protection. Thus, it must be understood that, even while the robotic system is being controlled directly or managed as a means to accomplish an end, various other goals, missions, and tasks are being coordinated and managed as well. Although there are considerable efforts in the areas of adaptive automation (e.g., Parasuraman et al., 2007), for the most part today, it is the human operator who coordinates, manages, and adapts.

Studying the Incorporation of Technology into Work

Different scientific disciplines have focused on separate phases or aspects of HRI and have each contributed uniquely to creating the context for HRI. Given all the possible interactions among the basic HRI elements and the larger goal-oriented work context, however, a fuller examination of HRI is required. Cognitive systems engineering (CSE) is cited (Endsley et al., 2007) as being at the forefront of current perspectives for studying systems in the context of work and, therefore, a promising framework for understanding HRI. CSE builds on traditional human factors engineering, which historically has been applied to the physical interface, and on the cognitive engineering approach, which added consideration of human operator mental activity, and asserts that the full intersection of people, technology, and work must be considered jointly (Woods & Hollnagel, 2006). The CSE approach embraces examination of the way in which technology use actually unfolds over time.

With virtually any technology, but especially with one as highly automated and autonomous as robotics technology, as people incorporate the technology into the work context, the unexpected can occur. Sarter et al. (1997) used the term "automation surprise." Once the basic operation of the technology is learned, people will begin to employ it to accomplish their goals; however, their method may not be the one the technologists and designers intended. Since the technology adds a "layer" between the person and the task, the existing task or job is not simply automated, but is fundamentally changed. Further, the introduction of technology adds new tasks; for example, the technology itself must be monitored and maintained, and, when it does not function as expected, it must be diagnosed. The less reliable the technology, or the more the way it is used varies from the original concept, the more it needs to be "tended." All too often the result is that the technology requires substantial mental effort and time on behalf of the human operator; consequently, he or she has less mental effort and time available to devote to the other tasks at hand, including the mission the technology was intended to support in the first place. Even the most carefully crafted interface will probably not support all the ways the robotic technology ultimately is used. Thus, automation surprise sets up a reverberation of effects, of complex interactions that can be best examined by adopting a CSE perspective.

Temporal Latency in HRI

The CSE perspective demands that the four basic HRI elements shown in Figure 17.1 be examined in context and, of particular interest here, that context includes a temporal dimension. Indeed, networked robotic systems inherently include some degree of latency, which can range quite broadly. Delays between Earth and a robot on Mars are about 45 minutes and delays between a laptop and a robot in a laboratory around 0.5 seconds (Goodrich et al., 2001). Temporal latency effects can be compounded by limited network bandwidth, as highlighted recently (Osborn, 2008, p. 1):

> Within five years, the U.S. Army may have too little radio spectrum to allow its next-generation, networked force to work as it is being designed to do, in particular as the U.S. Army introduces elements of an ever-more-networked force that moves vast amounts of data from soldier-mounted sensors, aerial and ground robots, manned vehicles and more. "Message latency (delay) affected warfighter effectiveness by as much as 50 percent for a selected scenario," a recent RAND study said.

Thus, latency in displaying robot position or sensor information to the operator or in communicating control actions from the operator to the robot is expected to have critically negative effects on HRI measures of performance and effectiveness. It can cause errors, disrupt work, and possibly lead to unsafe workarounds or to abandoning the technology altogether. A recent NATO report (NATO, 2007) called out temporal latency as a main contributor to uncertainty in HRI and, therefore, to distrust of the system. Latency is described as being attributable to either a delay in human control to robot action, termed the *response delay*, or a delay in the result of that action being fed back to the operator, termed the *information delay*. These two types of latency are similar in some respects to the two halves of the human perception-action (or sensory-motor) cycle. The human system naturally includes latencies due to neural processing time; however, the key difference between the human system and the human-robot is one of scale. The human perception-action cycle has evolved over millions of years such that the latencies are fully calibrated and incorporated into the system. HRI-related latencies may well exceed what the human system can accommodate without some overt adaptation. (For a discussion of the human temporal system in human system interaction generally, see de Pontbriand et al., 2007.)

In the next section, the literature on temporal latency effects in HRI will be reviewed, highlighting the effects on control responses and on information acquisition. Following that, assessment of situation awareness (SA) in HRI will be discussed. This background sets the stage for two experiments exploring temporal latency effects in HRI in a real-world context. The chapter concludes with lessons learned for HRI from the CSE perspective.

Temporal Latency and the Effect on Control

Performance decrements in HRI-related contexts effects due to temporal latency have been identified in research from the U.S. Department of Defense, the National Aeronautics and Space Administration (NASA), universities, research institutions, and industry. These studies involve a wide range of applications: robotic vehicles (DePiero et al., 1990; Adams, 1962); simulated vehicles (Cunningham et al., 2001a; Cunningham et al., 2001b); robotic manipulator arms (Lane et al., 2002); and underwater robots (Bulich et al., 2004). Research across all these domains has identified significant performance decrements due to latency within the control–information loop.

Latency duration   Typical ground robot applications have latency durations ranging from several hundred milliseconds to several seconds, and latencies in this range have been found to significantly affect performance (Bulich et al., 2004; Lane et al., 2002; Cunningham et al., 2001a,; DePiero et al., 1992; McGovern, 1990; Adams, 1962). Of course, the negative impact of latency is a function not only of duration per se, but is also dependent upon the form of latency, the difficulty of the task, and the operator's ability to adapt to the latency, each of which will be discussed in turn.

Latency variability   Latency may have a constant duration, such as with hardware-induced or path-induced delays, but it may also be variable. Lane et al. (2002) found that a short, variable lag was more detrimental than a longer, fixed one. Similarly, Watson et al. (1998) found impaired performance on a grasping and placement task once the standard deviation of latency moved above 82 ms. DePiero et al. (1992) offer this conclusion: "Driving experience using the (*simulator*) system has demonstrated the significance of latency on driver performance. Both a low and a constant value of latency is very important."

Task difficulty   Research has shown that task difficulty can interact with latency effects in robotic control, with task difficulty being characterized by three factors: the required accuracy; the required timeliness; and, in some cases, the required frequency of control. A general rule of thumb that the time to complete a task increases as task difficulty increases has been seen in a wide variety of tasks, including operating remotely controlled vehicles (Adams, 1962; Fox, 1962), unmanned air vehicles, underwater robots, and robotic manipulators (Lane et al., 2002). Furthermore, when difficulty is due to high accuracy requirements, there is often a speed/accuracy tradeoff. Accordingly, the faster one performs the task, the lower the accuracy, and vice versa. When there is not a requirement to be fast during teleoperation, good accuracy can be maintained even with the temporal latency present in the system by making small, brief control movements (Ranadive, 1979).

Control adaptation   Humans have the ability to adapt to less than perfect systems and often adopt control strategies to do so. Using control actions to temper the speed/accuracy tradeoff is one example of adaptation. Another is adopting a "move-and-wait" strategy. The latency duration at which participants generally switch to such a strategy varies. For tasks with low accuracy and timeliness requirements, users are often able to complete the task without adopting a move-and-wait strategy at all (Thurling et al., 2000; Suomela et al., 1999) and even for other more difficult tasks, the latency duration at which users adopt the move-and-wait strategy can vary greatly (Lane et al., 2002; Ranadive, 1979).

Some adaptations to latency are more subtle. A process known as spatial visual-motor adaptation, in which a recalibration of sensory-motor spatial relationships occurs within the human brain, can allow humans to adapt to a constant latency even after only a few minutes of exposure to the latency (Krotkov et al., 1996). This adaptation seems to be an extension of the naturally occurring adaptation to latency in the perception–action cycle, although the adaptation to external latencies can vary considerably depending on a host of factors including: task requirements, uncertainty in the environment, age of the person, etc. (Cunningham et al., 2001a, 2001b).

The research on human adaptation to latency in robot control is mixed. On the one hand, adaptation has been seen in many remote driving and manipulation investigations (Cunningham et al., 2001a, 2001b; Suomela et al., 1999). In these investigations introduction of latency after a baseline familiarization period at first impairs performance, but the baseline performance level is soon regained or even exceeded though additional training or practice in the latency condition. In some cases participants spontaneously reported that towards the end of training the visual feedback seemed simultaneous (Cunningham et al., 2001a). This suggests that the participants were in some way altering the existing visual-motor transformations used in driving. Supporting this conclusion, a strong negative effect on performance was seen again when the latency was subsequently removed.

On the other hand, however, some research found little evidence of compensation for external latency (Cunningham et al., 2001b). Held et al. (1966) found that visual-motor adaptation was unachievable for latency even as small as 0.3 s in a robotic manipulator experiment, and concluded that people dissociate the teleoperator hand movements from those of their own hand at these durations. Accordingly, both the nature of the underlying mechanism and our ability to apply it to robotic control are still sources of considerable debate (Cunningham et al., 2001b). So although well-trained or highly practiced operators may be able to adapt to some degree to a constant temporal latency, precisely how internal, essentially automatic adaptation, and explicit control strategies combine is not fully understood.

Temporal Latency and the Effect on Situation Awareness (SA)

Temporal latency has been shown to have a negative effect on an operator's ability to control or manage a robotic system, but what is the effect on his or her SA while *using* a robotic system? We use the term "SA" here in Endsley's (1995) sense of a state of awareness comprising three levels: perception, understanding, and projection. Even without temporal latency, maintaining situational awareness while using robotic systems is a challenge. Riley et al. (2006) enumerate several HRI SA issues, including robot localization and orientation. Even with high bandwidth and low latency robotic teleoperation systems, operators can get disoriented and lost fairly easily, even with landmarks and a map of the area (McGovern, 1990). Yanco & Drury (2004) estimated that robot operators spend as much as 30 percent of the time acquiring or reacquiring SA, even while performing time-critical tasks.

Level 1 SA: perception   The decrements in driving performance shown with remote driving (e.g., Mitchel et al., 1994) can at least in part be attributed to the differences in visual cues that feed the perceptual level of SA, for example decrements in depth perception (McGovern, 1990; Fong et al., 2001), obstacles detection (Fong et al., 2001), and motion perception (Murphy et al., 1993). Blackburn et al. (2002, p. 9) drew a parallel between teleoperation and driving at night:

> The difficulties of navigation via video are similar to the difficulties many people experience when trying to drive an automobile at night in the country. Valuable contextual information that would normally be available to the person through peripheral vision in daylight is denied in the dark of night, while attention is focused on the illuminated field of view of the headlights.

Level 2 SA: understanding   This altered spatial context affects the way in which Level 1 SA, perception, feeds Level 2 SA, understanding. Indeed, Oron-Gilad et al. (2006) note that perception is an active and continuous process and that SA naturally emerges from that process, whereas remote- or robot-fed perception is passive and relatively discrete and, therefore, does not support the development of SA in the same fashion. When the feed from perception to understanding is further disrupted in the temporal domain by communication and control latency, the disconnect is compounded. Thus, the human now must undertake the cognitive workload of managing disruption in both the spatial and temporal domains. Participants in a study involving control of an undersea robot rated a significant increase in the required level of attention and concentration as latency duration increased (Bulich et al., 2004). Thomas et al. (1997) note that when there was a delay between a "move" command sent to a remote camera and the image returned back to the operator, the operator had to explicitly verify that the intended action occurred, which resulted in some loss of information such as the position of landmarks. This can be a powerful effect, as seen in one teleoperation study (Kay and Thorpe, 1997, p. 1156) where "operators expressed the feeling that the image that they were looking at was taken from a completely different angle than they had expected and had nothing to do with the path they had picked." The disorientation and associated lack of Level 1 situational awareness can severely limit the utility of information provided by the robot when required for Level 2 or Level 3-based tasks (e.g., Darken et al., 2001).

Sellner et al. (2006) employed a video replay method that helped to tease apart Level 1 perception (i.e., viewing time) and Level 2 SA comprehension (i.e., decision–reaction time) in a study of a robot troubleshooting task. They found a time (i.e., amount of information) by accuracy tradeoff when considering only video replay length; however, when considering video replay length *plus* decision time, they observed a U-shaped relationship such that decision time was faster with either 5 or 10 seconds of replay than at either no replay or 20 seconds. This would seem to indicate a non-linear relationship between temporal latency and performance with respect to Level 2 SA.

Methods for studying SA in HRI   In order to more completely understand and measure SA in HRI, variations or extensions of the concept have been proposed. Riley et al. (2004) investigated the relationship between SA and telepresence based on the assumption that "being there" is a fundamental component of building SA. Drury et al. (2003) proposed five categories of SA that captured the various possible pairings among teams of multiple humans and robots as a function of the mission context. Drury et al. (2007) proposed an expansion that emphasized SA with respect to robot location, activity, surroundings, and status as well as overall mission. These frameworks have been applied to after-action ratings of critical HRI incidents (Scholtz et al., 2004) and to comparisons of display types via protocol analysis (Drury et al., 2007), illustrating their value for understanding the various aspects of HRI.

Using Levels of Automation (LOAs) to Manage the Effects of Temporal Latency

It has been proposed that increasing the level of automation locomotion and navigation of robotic systems should not only relieve the operator of workload (e.g., Schipani, 2003; Sellner et al., 2006), but that such autonomy should mitigate latency-induced difficulties. Sheridan (1993, p. 602) states: "There have been many demonstrations that supervisory control not only circumvents the time latency problem but also can speed up certain teleoperations beyond direct manual control even when there is no time latency …" For example, if a robot is able to avoid obstacles on its own, not only is the operator no longer burdened by fine-motor tracking requirements, but he or she should not need to wait on delayed feedback from the robot.

At the same time, there is a risk that, with the robot moving and/or making decisions on its own, the operator may be more prone to losing SA, particularly when dealing with temporal latency. Substantial research has shown negative SA effects due to "out-of-the-loop" issues associated with higher levels of automation in robotic control (Darken et al., 2001; Endsley & Kiris, 1995). In addition, under conditions of temporal latency, robot positions or actions can become disassociated from operator perceptions: Kay and Thorpe (1997) found that operators expressed that the robot had moved incorrectly; and Held et al. (1966) concluded that subjects dissociated movement of a robotic hand from that of their own with durations as low as 0.3 seconds. It seems then, that not only can robot operators lose SA when temporal latency is present, but that loss may lead to a lack of trust in the robotic system, which further exacerbates the problems of maintaining control and of maintaining SA (Jentsch et al., 2008).

Two Latency Experiments

In this section, we summarize the results of two experiments that further explored the HRI context under conditions of bandwidth-induced temporal latency that gave rise to a range of latency durations and latency variability. Various LOAs adapted from Endsley & Kaber (1999) (see Table 17.1) were evaluated with respect to human control of the robot and to use of the information returned from the robot using both performance and SA measures. From a CSE perspective, using this combination of measures should enable a fuller understanding of the HRI context than using one type of measure or the other alone.

Table 17.1   Level of automation (LOAs) adapted from Endsley & Kaber (1999)

		Roles			
Level of Automation	Endsley and Kaber LOA	Monitoring	Generating	Selecting	Implementing
Teleoperation	Manual Control	Human	Human	Human	Human
Guarded Teleoperation	Action Support	Human and Computer	Human	Human	Human and Computer
Obstacle Avoidance/ Waypoint Navigation	Decision Support	Human and Computer	Human and Computer	Human	Computer
Autonomous Waypoint Navigation	Automated Decision Making	Human and Computer	Human and Computer	Computer	Computer

The mission was to navigate a small robot (upper left, Figure 17.2) through an office environment and conduct a search and rescue effort. The robot was equipped with an onboard real-time video and a laser that displayed a minute's worth of position data to the operator (upper right, Figure 17.2). The operator also had a paper map (lower left, Figure 17.2) with points designated for stopping and gathering more detailed reconnaissance information. The courses, while not rubble-strewn, did contain several obstacles (boxes, trash bins, chairs), closed doors, and open doors not shown on the paper map. Latency duration ranged from no latency to short (1 second), medium (2 seconds), and long (4 seconds). Latency variability (in the short and long conditions) was set so that it varied as much as 50 percent around the mean: the short duration latency ranged from 1 to 3 seconds, with an average of 2 seconds; and the long duration latency ranged from 2 to 6 seconds, with an average of 4 seconds.

The four LOAs were:

- T—Teleoperation: joystick control;
- G—Guarded Teleoperation: joystick control with collision avoidance;
- O—Obstacle Avoidance Waypoint Navigation: operator-selected waypoints for simple shortest route movement with obstacle avoidance;
- A—Autonomous Waypoint Navigation: operator-selected waypoints with "best" path planning.

In order to assess the control loop and the information loop, four classes of measures were collected:

- Robot control, e.g., overall mission time (i.e., the time to complete a course) and driving errors (including hitting a wall, backing up).
- Instantaneous SA, e.g., navigation errors including *direction-room errors* (moving in the wrong direction or entering the wrong room; *recon errors* (missing a reconnaissance stop or stopping in the wrong location); and *lost-stop errors* (stopping due to being lost).
- SA position and route recall measures, where errors were deviations between the robot's actual location and route and that recalled and recorded on the map (lower right, Figure 17.2) and were of three types: *position, room-route*, and *instances of driving errors*.
- Ratings of the percentage of attention spent on control and navigation or localization (for a total of 100 percent).

Figure 17.2 Experimental equipment and an example course map and data
Upper left: Pioneer 3DX robot with top-mounted camera and SICK lasers. Upper right: Display showing the robot video display (left) and laser range data (right) with the robot marked by a red dot at the center. Lower: Example course map with desired route and reconnaissance spins on the left and one participant's recalled route on the right.

Results

Temporal latency and control: time and errors The first part of the temporal latency-HRI story is the control leg, told through overall mission time and driving errors, and is summarized here (see Luck et al., 2006 for details). Both latency and variability were found to be significant drivers of control performance. The longer the latency was, the longer the overall mission time. This was true whether the latency was constant, and therefore predictable, or variable and unpredictable. Likewise, the more variability present in the latency, the longer the overall mission time. These two factors—latency duration and latency variability—interacted such that a long and variable latency resulted in the worst control performance overall.

Across the four levels of robot autonomy, from low to high (T, G, O, and A), it was clear that increasing autonomy indeed helped to mitigate the negative control performance effects of latency duration and latency variability. Performance was better at higher LOAs—mission times were shorter and there were fewer driving errors. When latency was introduced, at low LOAs, longer latency durations resulted in longer mission times and driving errors, whereas at higher LOAs longer latency duration did not lengthen mission times and even decreased mission times and errors in some instances. This pattern held whether the latency was of constant or variable duration. In terms of driving errors, increasing the latency duration increased the number of errors in the low LOAs by approximately 33 percent relative to no latency, but the number of driving errors remained low in the high LOAs, regardless of latency duration.

Temporal latency and SA   Mission time and driving errors, of course, are the control-related measures. The information-related measures (navigation, SA recall errors, and concentration) are a little more subtle and offer a different view into the HRI context. For navigation and recall errors, it was generally the case that neither the duration nor variability of the latency had a large effect as compared to the effect observed between having latency and not having latency. These results were unexpected. The same was true of LOA: LOA did not have a large effect on SA, except in the case of full and guarded teleoperation (T and G) where there were more navigation errors in the category of stopped or lost than in the higher LOAs.

Granted, it was the case that overall SA performance was fairly good; however, upon examination of the errors that did occur as a function of the amount of latency, a pattern of results emerged showing the interplay of timing and LOA on SA recall errors. There was a rough U-shaped interaction of latency duration and variability for SA recall errors and for the subset of drive marking errors where constant-short and variable-long latencies yielded more errors than constant-long and variable-short latencies (Figure 17.3). The advantage of a higher LOA was also shown to ameliorate the negative effects of latency to some degree, particularly for SA recall large position marking types of errors (left side, Figure 17.4). Further, there was another interaction effect where long latencies produced more errors for lower LOAs than did short latencies, but short latencies produced more errors for the fully autonomous condition than did long latencies (right side, Figure 17.4). Finally, the concentration measure showed that the higher the LOA, the less concentration was rated as being required for control activities. The high LOA (A) short and medium latency conditions were rated as requiring the least amount of overall concentration, while even with high LOA, the long latency condition was rated as requiring the same amount of overall concentration as no latency at all, although the distribution between concentration for control vs. SA was different (Figure 17.5).

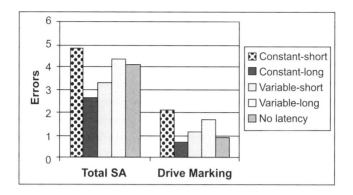

Figure 17.3   **SA marking errors by latency duration and variability highlighting the duration by variability interaction**

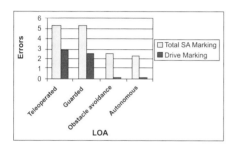
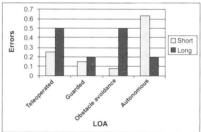

Figure 17.4   **Left: Total SA errors and just drive marking errors by LOA showing the large contribution of drive marking errors at low LOAs to overall SA errors. Right: SA recall large position errors for short and long durations by LOA broken out for short and long latency durations**

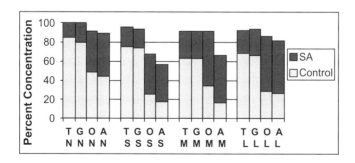

Figure 17.5   **Breakdown of concentration by LOA and latency duration**

Discussion

For the task space used in the experiments reported here, the effects of latency duration and variability on control performance were clear: Increasing latency duration resulted in worse performance; increasing latency variability resulted in worse performance; and long and variable latency resulted in the worst performance of all. As hoped for, increasing the autonomy of the robot served to mitigate those negative effects. When the nearly constant motor control required for full teleoperation was replaced successively by system capabilities, overall control performance improved. Technology fulfilled its intended purpose in relieving operator workload.

The SA measures, however, paint a somewhat more complex picture of operator performance and activity. Latency duration and variability did not significantly affect SA, neither SA related to navigation nor to recall of the path or errors along the path in comparison to no latency. Also, LOA affected only errors of being lost or stopping to get the robot's bearings.

Still, the trends of increased errors that did occur when comparing levels of latency duration and variability along with the levels of LOA indicate that the ability of humans to adapt is not linear with respect to time, to variability, or to the progression of automation support. Increasing LOA helped mitigate real-time SA errors to some degree; however, the SA errors attributed to latency were not made real-time, but were made during recall. Specifically, the SA recall errors were ones of mistaking where the robot had traveled over the course of the mission, mainly mistaking or omitting locations where obstacles were hit or where a portion of the route had to be retraced rather than rooms entered or specific mission-related start and stop points. There were more of these types of errors with constant-short and variable-long latencies than with constant-long and variable-short latencies.

While it could be argued that these drive-marking errors are the least significant type of SA error since they did not impact mission accomplishment per se, nevertheless, it is useful to speculate about the non-linear nature of the SA decrement. This is especially so since the control effects showed a more expected, linear pattern of results. The answer may lie in part with strategies employed. The fact that SA did not diminish, and sometimes improved at long latencies, may be due to either a strategy of slowing down in order to manage the increased control difficulty or to a discrete move-and-wait strategy, both of which allow more "time" for SA. The finding that SA recall for errors in driving was worst for short-constant and for long and variable latencies invites some speculation at the micro-strategy level of adaptation as well. It may be the case that, even in the range of a few seconds, those seconds matter. When the latency is short but constant, the operators may have been kept busy by the strict rhythmicity of the latency; and when it was long and variable, the rhythm may have been just outside the window of natural adaptation, and also kept busy managing the disconnects arising from latency. At the middle ground however, at short-variable and long-constant latencies, it may be the case that the operators could work within the time window and adapt slightly more easily to the latency pace and variation. The ratings of required concentration support this view: the longest latency condition coupled with the highest autonomy had the highest percentage of concentration being devoted to SA activities relative to control.

Conclusions

The cognitive systems engineering (CSE) perspective advocates a broader perspective on understanding people, technology, and work than traditional approaches. The experiments presented here offered a glimpse into the domain of HRI and how the CSE perspective proved essential for exploring human-robot work. Across the performance measures and the whole gamut of SA measures, there were different patterns of results. The CSE perspective brought them together for a richer understanding of HRI, which will ultimately help shape the implementation of robotic systems.

Of course, the CSE perspective also highlights the limits of this research. First, although operators in the experiments were responsible for both control of the robot and for certain SA-related activities, this falls far short of the dynamic and multi-tasking environment that will be encountered in the field. There is a growing literature on multi-tasking in HRI, such as control of more than one robot (e.g., Chadwick, 2006; de Visser & Parasuraman, 2007) and performing tasks not related to the robot (Chen & Joyner, 2009; Sellner et al., 2006), as well as field observation and experimentation (e.g., Casper & Murphy, 2003; Riley & Endsley, 2004; Schipani, 2003). Also, in addition to robotic autonomy, new HRI control and display interfaces are being developed, each of which may provide support for one aspect of the work environment, but not necessarily for other aspects (e.g., Drury et al., 2006, 2007).

Another implication of adopting the CSE perspective is the set of measures that are used. The measures must go beyond time on task, errors, or subjective preference. The constructs and typical measures of mental workload (e.g., Hart, 2006) and SA (Endsley, 1995) are central in CSE, but in and of themselves tell only part of the story. Beyond usability, technology must support cognitive work and new metrics are being suggested to measure how effectively the technology itself supports cognitive work (Hoffman et al., forthcoming). For HRI, frameworks have been proposed to capture mutual human-robot awareness (Drury et al., 2003); SA with tailored subcategories (Drury et al., 2007); and mixed human-robot team processes (Johnson, 2002). New measurement approaches have been proposed that calculate neglect tolerance and interface efficiency (Olsen & Goodrich, 2003) and, similarly, interaction and neglect efficiency (Crandall & Cummings, 2007). But even these measures must be extended to capture fully the level of automation, the speed of robot, the relative difficulty in terms of the required accuracy of control, the required timeliness of control, and the required frequency of control, or to generalize to other robotics applications. Indeed HRI metrics must be meaningful to human factors engineers, cognitive scientists, engineers, and roboticists (see Steinfeld et al., 2006, for a summary).

Thus, there is a growing body of research that explores HRI, and a promising set of measurement approaches to more fully address HRI. The CSE approach helps ensure that researchers consider multiple layers of the HRI story. As in the experiments described in this chapter, the result is a richer understanding of the impact of robotic elements on performance and understanding.

References

Adams, J. L. (1962). An investigation of the effects of time lag due to long transmission distances upon remote control, Phase II. NASA TN-1351.-Phase I. NASA Tech. Note D1211, 1961-Phase II vehicle experiments, Phase III conclusions. NASA Tech. Note D1351.

Blackburn, M.R. Everett, H.R., & Laird, R.T. (2002). After action report to the unmanned ground vehicles/systems Joint Program Office: Center for the Robotic Assisted Search and Rescue (CRASAR) related efforts at the World Trade Center. Technical Document 3141, Space and Naval Warfare Systems Center, San Diego, CA.

Bulich, C., Klein, A., Watson, R., & Kitts C. (2004). Characterization of delay-induced piloting instability for the Triton undersea robot. *Proceedings of the IEEE Aerospace Conference*, Big Sky, MT.

Casper, J. & Murphy, R. R. (2003). Human-robot interactions during the robot-assisted urban search and rescue response at the World Trade Center. *IEEE Transactions on Systems, Man, and Cybernetics, Part B: Cybernetics*, 33(3), pp. 367–85.

Chadwick, R. (2006). Operating multiple semi-autonomous robots: Monitoring, responding, detecting. *Proceedings of the Human Factors and Ergonomics Society 50th Annual Meeting*. San Francisco, CA, pp 329–33.

Chen, J. Y. C. & Joyner, C. T. (2009). Concurrent performance of a gunner's task and robotics operator's tasks in a multitasking environment. *Military Psychology*, 21(1), 98–113.

Committee on Army Unmanned Ground Vehicle Technology, Board on Army Science and Technology (2002). *Technology Development for Army Unmanned Ground Vehicles*. The National Academies Press: Washington, DC.

Crandall, J. W. & Cummings, M. L. (2007). Developing performance metrics for the supervisory control of multiple robots. *Proceedings of the Human Robot Interaction Conference 2007*, Arlington, VA, pp. 33–40.

Cunningham, D. W., Billock, V. A., & Tsou, B. H. (2001a). Sensorimotor adaptation to violations of temporal contiguity. *Psychological Science*, 12(6).

Cunningham, D. W., Chatziastros, A., von der Heyde, M., & Bulthoff, H. (2001b). Driving in the future: temporal visuomotor adaptation and generalization. *Vision*, 1(2), 88–98.

Darken, R. P., Kempster, K., & Peterson, B. (2001). Effects of streaming video quality of service on spatial comprehension in a reconnaissance task. *Proceedings of the Annual Interservice/Industry Training and Education Conference*. Orlando, FL, November.

DARPA (2008). http://www.darpa.mil/grandchallenge, downloaded 11 February 2008.

de Pontbriand, R. J., Allender, L. E., & Doyle III, F. J. (2007). Temporal regulation and control. In P. A. Hancock and J. L. Szalma (eds), *Stress and Performance*. Aldershot, UK: Ashgate.

de Visser, E. J. & Parasuraman, R. (2007). Effects of imperfect automation and task load on human supervision of multiple uninhabited vehicles. *Proceedings of the Human Factors and Ergonomics Society 51st Annual Meeting*, pp. 1081–5.

DePiero, F. W., Noell, T. E., & Gee, T. F. (1992). Remote driving with reduced bandwidth communication. *Proceedings of the Space Operations Application and Research Symposium (SOAR)*, Houston, TX.

Drury, J. L., Keyes, B., & Yanco, H. A. (2007). LASSOing HRI: Analyzing situation awareness in map-centric and video-centric interfaces. *Proceedings of the ACM/IEEE International Conference on Human Robot Interaction*, Arlington, VA, pp. 279–86.

Drury, J. L., Scholtz, J., and &, H. A. (2003). Awareness in human-robot interactions. *Proceedings of the IEEE Conference on Systems, Man, and Cybernetics*, Washington, DC, pp. 912–18.

Drury, J. L., Yanco, H. A., Howell, W., Minten, B., & Casper, J. (2006). Changing shape: Improving situation awareness for a polymorphic robot. *Proceedings of the 2006 Conference on Human-Robot Interaction (HRI 06)*, Salt Lake City, pp. 72–79.

Endsley, M. R. (1995). Toward a theory of situation awareness in dynamic systems. *Human Factors*, 37(1), 32–64.

Endsley, M. R., Hoffman, R., Kaber, D., & Roth, E. (2007). Cognitive engineering and decision making: An overview and future course. *Journal of Cognitive Engineering and Decision Making*, 1(1), 1–21.

Endsley, M. R. & Kaber, D. B. (1999). Level of automation effects on performance, situation awareness and workload in a dynamic control task, *Ergonomics*, 42(3), 462–92.

Endsley, M. R. & Kiris, E. O. (1995). The out-of-the- loop performance problem and level of control in automation. *Human Factors*, 37, 381–94.

Fong, T., Thorpe, C., & Baur, C. (2001). Advanced interfaces for vehicle remote operation: collaborative control, sensor fusion displays, and remote driving tools. *Autonomous Robots*, 11 (ANS-4).

Fox, G. J. (1962). Perceptual motor factors in the remote control of a lunar vehicle, Phase 1: the effects of a communication time delay on steering performance as a function of vehicle speed and course complexity. Grumman Aircraft Report No ADR09-07061.1.

Future Combat Systems (FCS) (2008). http://www.fcs.army.mil/systems, downloaded 11 February 2008.

Gockley, R. & Matarić, M. J. (2006). Encouraging physical therapy compliance with a hands-off mobile robot. *Proceedings of the 2006 Conference on Human-Robot Interaction (HRI '06)*, Salt Lake City, pp. 150–55.

Goodrich, M. A., Olsen, D. R. Jr., Crandall, J. W., & Palmer, T.J. (2001). Experiments in adjustable autonomy, *Proceedings of the IJCAI01 Workshop on Autonomy, Delegation, and Control: Interacting with Autonomous Agent*, Seattle, WA.

Hart, S. G. (2006). NASA-Task Load Index (NASA-TLX): 20 Years Later. *Proceedings of the Human Factors and Ergonomics Society 50th Annual Meeting*, pp. 904–8.

Held, R., Efstathiou, A., & Greene, M. (1966). Adaptation to displaced and delayed visual feedback from the hand. *Journal of Experimental Psychology*, 72, 887–91.

Hoffman, R. R., Marx, M., Hancock, P. A., Amin, R., McDermott, P., & Brents, C. (forthcoming). The metrics problem in the study of cognitive work and a suggestion concerning a class of solutions.

Jentsch, F., Fiore, S., Rehfeld, S., & Salas, E. (2008). Military teams and tele-operation of unmanned ground vehicles: An initial literature review. Technical Report TPL-03-2004, University of Central Florida, Orlando.

Johnson, C. L. (2002). Etiquette within and between large human-robot teams. American Association for Artificial Intelligence Fall Symposium: Etiquette for Human-Computer Work, 15–17 November, North Falmouth, MA.

Kay, J. S. & Thorpe, C. (1997). An examination of the STRIPE vehicle teleoperation system. *Proceedings of the IEEE International Conference on Intelligent Robots and Systems*, vol. 2, pp. 1152–7.

Kirsner, S. (2002). Making robots, with dreams of Henry Ford. *New York Times*. 26 December. http://www.nytimes.com.

Krotkov, E., Simmons, R., Cozman, F., & Koenig, S. (1996). Safeguarded teleoperation for lunar rovers: From human factors to field trials. *Proceedings of the IEEE Planetary Rover Technology and Systems Workshop*, Minneapolis, MN.

Lane, J. C., Carignan, C. R., Sullivan, B. R., Akin, D. L., Hunt, T., & Cohen, R. (2002). Effects of time delay on telerobotic control of neutral buoyancy vehicles. *Proceedings of the IEEE International Conference on Robotics and Automation*. Washington, DC, May (ANS-24).

Luck, J., McDermott, P. L., Allender, L., & Russell, D. (2006). An investigation of real world control of robotic assets under communication latency. *Proceedings of the Human-Robot Interaction Conference*, 2–4 March. Salt Lake City, UT.

McGovern, D. E. (1990). Experiences and results in remote operation of land vehicles. Sandia National Laboratories, Report SAND90-0299-UC-515 (ANS-8).

Mitchel, B. Yeager, E. Suarez, M. Griffin, J., & Seibert, G. (1994). Cognitive strategies in UGV navigation, Cybernet Systems Corporation (CSC 94-233-2-1) for U.S. Army Research Laboratory Human Research and Engineering Directorate contract DAAA-15-93-C-0037.

Murphy, K., Juberts, M., Legowik, S., Nashman, M., Schneiderman, H., Scott, H., & Szabo, S. (1993). Ground vehicle control at NIST: From remote operation to autonomy. Robot Systems Division: National Institute of Standards and Technology (ANS-3).

NATO (2007). RTO Human Factors and Medicine Panel Task Group (HFM-078/TG-017). Uninhabited military vehicles (UMVs): Human factors issues in augmenting the force. RTO Technical Report (RTO-TR-078).

Olsen, D. R. Jr., & Goodrich, M.A. (2003). Metrics for evaluating human-robot interactions. *Proceedings of PERMIS 2003.*

Oron-Gilad, T., Chen, J. Y. C., & Hancock, P. A. (2006). Remotely operated vehicles (ROVs) from the top-down and the bottom-up. In N. J. Cooke, H. L. Pringle, H. K. Pedersen, & O. Connor (eds), *Human Factors of Remotely Operated Vehicles*. New York: Elsevier.

Osborn, K. (2008). U.S. Army faces spectrum crunch, *Defense News*, 7 January, p. 1.

Parasuraman, R., Barnes, M., & Cosenzo, K. (2007). Adaptive automation for human-robot teaming in future command and control systems. *The International C2 Journal*, 1(2), 42–68.

Ranadive, V. (1979). Video resolution, frame rate, and grayscale tradeoffs under limited bandwidth for undersea teleoperation. MS thesis, MIT.

Riley, J. M. & Endsley, M. R. (2004). The hunt for situation awareness: Human-robot interaction in search and rescue. *Proceedings of the Human Factors and Ergonomics Society 48th Annual Meeting.*

Riley, J. M., Kaber, D. B., & Draper, J. V. (2004). Situation awareness and attention allocation measures for quantifying telepresence experiences in teleoperation. *Human Factors and Ergonomics in Manufacturing*, 14(1), 51–67. Published online in *Wiley InterScience* (http://www.interscience.wiley.com).

Riley, J. M., Murphy, R. R., & Endsley, M. R. (2006). Situation awareness in the control of unmanned ground vehicles. In N. J. Cooke, H. L. Pringle, H. K. Pedersen, & O. Connor (eds), *Human Factors of Remotely Operated Vehicles*. New York: Elsevier.

Sarter, N. B., Woods, D. D., & Billings, C.E. (1997). Automation surprises. In G. Salvendy (ed.), *Handbook of Human Factors and Ergonomics* (2nd edn). New York: Wiley, pp. 1926–43.

Schipani, S. P. (2003). An evaluation of operator workload, during partially-autonomous vehicle operations. *Proceedings of Performance Metrics for Intelligent Systems (PerMIS '03)*, NIST special publication 1014. http://www.isd.mel.nist.gov/research_areas/research_engineering/Performance_Metrics/PerMIS_2003/Proceedings/.

Scholtz, J. (2003). Theory and evaluation of human robot interaction. *Proceedings of the 36th Annual Hawaii International Conference on System Sciences (HICSS '03)*, p. 25.1.

Scholtz, J., Young, J., Drury, J. L., & Yanco, H. A. (2004). Evaluation of human-robot awareness in search and rescue. DARPA MARS Program.

Sellner, B. P., Hiatt, L. M., Simmons, R., & Singh, S. (2006). Attaining situation awareness for sliding autonomy. *Proceedings of Human Robot Interaction*, Salt Lake City, UT, pp. 80–87.

Sheridan, T. B. (1993). Space teleoperation through time delay: Review and prognosis. *IEEE Transactions on Robotics and Automation*, 9(5), 592–606.

Steinfeld, A., Fong, T. Kaber, D., Lewis, M., Scholtz, J., Schultz, A., & Goodrich, M. (2006). Common metrics for human-robot interaction. *Proceedings of the Human-Robot Interaction Conference*, 2–4 March. Salt Lake City, UT, pp. 33–40.

Suomela, J., Savela, M., & Halme, A. (1999). Tele-existence techniques of different enhancement degrees in front end loader teleoperation. *Proceedings of the FSR*, Pittsburgh, PA.

Tanaka, F., Movellan, J. R., Fortenberry, B., & Aisaka, K. (2006). Daily HRI evaluation at a classroom environment: Reports from dance interaction experiments. *Proceedings of the 2006 Conference on Human-Robot Interaction (HRI '06)*, Salt Lake City, pp. 3–9.

Thomas, G., Robinson, W. D., & Dow, S. (1997). Improving the visual experience for mobile robotics. *Seventh Annual Iowa Space Grant Proceedings*. Drake University, November, pp. 10–20.

Thurling, A. J., Adams, A. J., Greene, K. A., & Punjani, S. M. (2000). Improving UAV handling qualities using time delay compensation. USAF Test Pilot School, Edwards AFB, CA (ANS2-7).

Watson, B., Walker, N., Ribarsky, W., & Spaulding, V. (1998). Effects of variation in system responsiveness on user performance in virtual environments. *Human Factors*, 40, 403–14.

Woods, D. D. & Hollnagel, E. (2006). *Joint Cognitive Systems: Patterns in Cognitive Systems*. Boca Raton, FL: CRC Press.

Yanco, H. A. and Drury, J. (2004). "Where am I?" Acquiring situation awareness using a remote robotic platform. *Proceedings of the IEEE Conference on Systems, Man and Cybernetics*, October.

Chapter 18

Robotic Control Systems for Dismounted Soldiers

Elizabeth S. Redden and Linda R. Elliott

U.S. Army Research Laboratory

Introduction

In the early twentieth century the term "robot" was first used. The word robot was based upon the Czech word for "forced labor" (Wikipedia, 2008). Robots were considered to be especially desirable for certain work because they would never get tired; they could endure physical conditions that were uncomfortable or even dangerous; they did not get bored; and they would not die. Isaac Asimov first coined the word "robotics" in his science fiction work in the 1940s to describe the technology of robots, and predicted the rise of the robotics industry. The brief history of robotics has resulted in the maturation of technology for increasingly sophisticated robots with enhanced and more diverse capabilities.

In the near future, robots will be capable of fighting together with their human controllers in dismounted battles. The lack of autonomy of these near-term robots makes them more like a tool, and not the fully autonomous androids that populate science fiction novels. These near-term robots are really human-centered automation, which Hancock (1996) describes as a stage in evolution that precedes autonomous machine intention. The most common robots currently in use by the military are small, flat, teleoperated machines that are mounted on miniature tank treads (Grabianowski, 2005). Many are able to tackle almost any terrain, and often have sensors for audio-video surveillance and chemical detection. They are also capable of carrying weapon packages mounted to their main chassis. While varying in size and weight, many of these robots are man-portable, and thus more relevant to dismounted warfighter operations, where warfighters are in the field and away from their vehicles. Even in their current nonfighting roles (e.g., explosive ordnance disposal, mine clearing, etc.), robots are now helping dismounted warfighters, and their numbers and missions are increasing. It is important to consider how to design the robots and their control systems (e.g., displays and controllers) to best support the dismounted warfighter's ability to fight. A dismounted warfighter cannot easily carry the relatively large control system typically found in stationary environments and vehicles. The dismounted warfighter is already overloaded with the weight of weapons, ammunition, water, and equipment. They cannot carry current robotic control systems as well as warfighting equipment over long distances, and for long periods of time. Even relatively small robots, such as the Matilda, PackBot, and TALON, provide teleoperation interfaces that are large and heavy.

Dismounted warfighters need an innovative robotic control system with minimal weight and bulk that allows them to control and task various small robotic platforms. Current robotic interfaces add a great deal of additional weight to the dismounted warfighter. As a result, it is important to document the range of robotic interface sizes that can be used in different environments, and to understand the trade-offs involved when choosing various interfaces for the environments in which they will be used. Our body of research focuses on how best to scale down the size and weight of displays and controls.

Scalability

The trend for interface design is moving away from "one size fits all" toward a scalable family of products with common architecture, but sized depending on the mission. The term "scalability" encompasses the various ways in which devices can be made more effective in terms of size and weight. It also addresses development of intuitive displays of information that enhance transfer of training among various controls and displays, and minimize information overload. The key to ensuring that a system is scalable is to consider its context of use. An example can be made with email access. In the past individuals were only able to use their desktop computers to access email, but they now use personal digital assistants (PDAs), BlackBerries, and cell phones when they are out of their offices. The question for robotic control system design for the dismounted warfighter is: How small and lightweight can we get and still be "good enough?" So far, studies indicate that the impact of size and weight is task-dependent. Control system size may affect performance and workload, but specific effects will depend on such factors as task demands (e.g., visual recognition of large icons versus more demanding visual search, tracking, navigation, teleoperation, or reading demands); the nature of display content (e.g., text, maps, aerial photographs, close camera views); display capabilities (e.g., color, high resolution, and zooming/panning); the environmental clothing being worn by the operator; and the number of operations that must be controlled on the robot.

Several streams of research are under way to determine the degree and manner in which robotic controls and displays can be scaled to be more effective for dismounted warfighters (Barnes et al., 2000, 2005; Chen et al., 2006, 2007; Renfro et al., 2007; Stafford et al., 2007a, 2007b; Stafford, 2008). In this chapter, we focus on control system size and weight, and their impact on operational effectiveness. Warfighters operate in a large range of environments, from the relatively stable and spacious environment of a tactical operations center (TOC), to the cramped environment of a vehicle, to the rugged and challenging environment of the dismounted warfighter. All of these environments have an impact on the size and configuration of a robotic interface. Small displays and input devices can potentially impact the complexity of operation, speed of operation, and the accuracy of input.

One factor affecting scalability of control systems to the environment is screen size. Designing for one "optimum" screen size might seem to be a good idea because it might seem to better control the presentation of the interface. However, if a display is designed that only works on one "fixed" browser window size, it may not work well on others. It might even result in important features disappearing off the edge of the screen. Another factor affecting the scalability of control systems is the number and type of physical controls that must be provided. It stands to reason that as the controller size is reduced, the number of individual controls must also be reduced or the size of the individual controls must be reduced.

The following two sections provide overviews of recent studies conducted at Fort Benning, Georgia, that compared control systems in terms of their screen size and controller features. These studies were conducted using U.S. Army soldiers operating in a dismounted operational environment.

Screen Size Study

The popularity of very large screen displays seems ever increasing. There is perhaps an assumption that larger displays are better for all tasks. Certainly, people enjoy larger screens for entertainment, given the popularity of increasingly larger TV displays. The motivation behind the development for many of the large-screen displays is often to provide an immersive experience (Tan et al., 2003), or a sense of presence for both virtual reality and home theaters. Another rationale for larger screens is that various research efforts have indicated or predicted some decrement in performance and/or workload with smaller displays. For example, larger, high-resolution displays have been associated with improved task performance in information-rich virtual environments (Ni et al., 2006). In a recent comparison of UAV display sizes, Minkov et al. (2007) found that small (2-inch) handheld displays were associated with greater workload than larger ones.

At the same time there is also a growing trend for smaller displays. Full-scale movies are now a popular option for iPods. Explicit directions are provided on handheld global positioning systems (GPS). Functions previously performed using large desktop computers can now be performed using BlackBerries. Studies have focused on how to improve web interaction on these small displays (Jones et al., 1999; de Bruijn et al., 2001). Companies such as Microsoft are focused on packaging many functions into small PDA or cell phone displays (Ross, 2007). These functions include graphic maps, GPS directions, personal organizers, email software, photos, and movies.

Display size can affect performance and workload. Specific effects can depend on the nature of the task, the nature of display content, and the availability of factors such as color, high resolution, and zooming/panning. For example, Wickens (2005) found the effects of display size can depend upon display clutter, resolution, multimodality, and dimensionality (2D versus 3D). Research with pilots (Wickens et al., 2003) found that display size had no impact on overall surveillance, change detection, tracking, or response time when using complex and dynamic aerial map displays with icons. In a subsequent study using a low-fidelity 2D and 3D tracking task, Stelzer and Wickens (2006) found that performance was degraded when urgency was high and displays were smaller. They also compared effects of display size during surveillance and target search tasks using complex and dynamic aerial map display icons. In their study, smaller displays were associated with a higher rate of error in heading and altitude, as well as increased deviations from flight path. However, display size did not affect response time or ability to estimate distance between aircraft, detect changes, or maintain target distance from the lead aircraft.

A recent study using simulations of UAV views predicted performance degradation with smaller displays (Minkov et al., 2007). The task simulated a situation where the dismounted warfighter operated an aerial UAV and compared 15-inch and 7-inch displays with 2-inch displays that were held against the eye to simulate a 50-inch display from a distance of 6.5 feet. The screen was split to provide a map view (with zooming and panning capability) and an aerial video image. Operators were required to navigate the UAV through a route with five waypoints in a relatively urban area. The target detection tasks required the operator to scan and detect targets, such as tanks or armored vehicles. Display size did not affect the number of waypoints navigated; however, operators navigated waypoints more slowly with the 2-inch display. There was no difference between the 7-inch and 15-inch displays. For the target detection task, there was no significant difference between the 7-inch and 15-inch displays; however, operators found significantly fewer targets with the 2-inch display than with the 15-inch display. NASA Task Load Index (TLX) scores indicated the lowest level of workload for the 15-inch display and the highest workload level for the 2-inch display. Questionnaire-based ratings indicated that operators preferred the 7-inch display for navigation, and the 2-inch display for target detection, even though their performance was less effective with the 2-inch display. Researchers concluded that the 7-inch display was most appropriate, and also pointed out that workload would likely be even higher for the 2-inch display if the task duration was longer, as the device was not designed for continuous use, but rather for short periods. In contrast, Stafford and Hancock (Stafford, 2008) administered the Automated Neuropsychological Assessment Metrics (ANAM) test battery—which is comprised of basic tests of perception, reaction time, and working memory—to subjects through a variety of display sizes, and found no significant difference in performance scores. However, the ANAM does not include complex visual search or recognition tasks. Again, we see that the effects of display size depend on particular task demands and device capabilities.

A recent experiment conducted at Fort Benning varied display sizes on a teleoperation driving task and target search typical of dismounted soldiers (Redden et al., 2008). This study examined the effect of four display options that can be currently used for ground robot teleoperation. The task demands for teleoperation of a ground robot entailed using a joystick to maneuver a far robot, while looking at a display from a camera mounted on the front of the robot. The operator in this study had to stay within a prescribed route, and search for visual targets representing potential enemy soldiers and improvised explosive devices (IEDs) that lay on the ground at varying distances from the route. While there was a strong visual search component to the task, the demands differed from that of other display size studies (e.g., tracking very small icons on a cluttered and dynamic map).

The sizes of the displays investigated were limited to those that are currently or soon available to the dismounted warfighter. Thus warfighters using these displays would not have to carry a separate display for robotic control; instead, robotic control could be performed using a display that they already carry, rather than adding additional equipment. The four display sizes were the Force XXI Battle Command Brigade-and-Below (FBCB2) size display used in vehicles (approximately 10.4 inch in diagonal); the Commander's Digital Assistant (CDA) size display (approximately 6.5 inch in diagonal) used by lower echelon commanders; a PDA-size display (approximately 3.5 inch in diagonal) that is being considered for squad members by the Future Force Warrior (FFW) program; and a monocular goggle-mounted display (GMD) that is being considered for squad leaders by the FFW program.

The study was conducted with Officer Candidate School (OCS) students who had experience as enlisted soldiers and who came from varied military occupational specialties (MOSs). In the experiment, soldiers teleoperated a TALON, a lightweight robot designed for missions ranging from reconnaissance to weapons delivery, using its suitcase controller to drive the robot. A two-way radio frequency line provided continuous data and video feedback for precise vehicle positioning. Three of the four different display sizes used to conduct this experiment were presented on an iExplorer to control the number of variables present in the study. The fourth display (the GMD) was mounted to a pair of Oakley glasses. Both the iExplorer and the GMD were plugged into the existing TALON control system so that its radio could be used.

The robotic driving course consisted of an oval-shaped course with four different lanes. Each lane had three legs, and a total length of approximately 300 meters. Leg A, the first leg of each lane, was marked with white 2-inch wide cotton tape (commonly referred to as engineering tape), and required the soldier to drive as quickly as possible to the end of the leg. An obstacle placed on the path was located at the end of the first leg. For Leg B, the operator was required to negotiate around the obstacle by going off the path and then returning to the path by the shortest and easiest route. Hazards such as mockup enemy soldiers, booby traps, improvised explosive devices (IEDs), and mines that could be clearly seen by the driving camera were placed along the rest of the lane (Leg C) between the obstacle and the end of the course. The robotic operators teleoperated the TALON from inside a tent that was placed behind a beam, to block the line of sight between the operators and the course.

It is important to consider the tasks that were examined during this field experiment when interpreting the results. The teleoperation task was confined to a marked course (Legs A and C), and to a very short offroad movement around an obstacle. The target detection task was limited to targets within the immediate vicinity of the robot. Therefore, the performance of interest was (a) camera-driven navigation and (b) camera-driven target recognition and identification.

The bottom-line finding from this experiment was that soldiers were able to perform the teleoperation and detection tasks required by this course with all four devices. All soldiers completed the courses in similar amounts of time (with the exception of the GMD, which was slower on obstacle avoidance) and with similar degrees of error.

Additional examination of the results, the literature, and soldier comments sheds further light on these findings. Although the soldiers performed *significantly* worse with the GMD on only one measure (Leg B completion times), it should be noted that there was a trend for worse performance with the GMD on every performance measure. Soldiers found it to cause significantly more work, effort, and frustration than the 6.5-inch and 3.5-inch displays, and more mental demand than all three of the other displays. It was also the least preferred device, and caused greatly increased physical problems. Soldiers complained of too much visual stimulation because of the strong visual images presented in each eye, and binocular rivalry (a term used when dissimilar images are presented to the two eyes which causes an unstable state of alternating periods of monocular dominance). Overall, dominance often appears as a fragmented mixture of the two eyes' views resulting in a dynamic, patchwork appearance (Lee & Blake, 1999; Mazumder et al., 1997; Meenes, 1999). Most of the soldiers either closed their left eye, or covered it with their hand, in an attempt to reduce the problem.

On the positive side, the GMD used during this experiment was reported to be comfortable to wear (lightweight). Soldiers did express concern that the design did not let them wear their ballistic eye protection, but the manufacturer stated that there were plans to accommodate this problem. There were also no reports of fogging, which is frequently the case with night-vision goggles, and some closed goggles such as the Sun, Wind, and Dust (SWD) goggles.

The smaller display sizes (6.5 inch and 3.5 inch) did not appear to cause any problems for these soldiers during the time periods in which they used them. Most of the soldiers positioned themselves closer to the smaller screens than they did to the 10.4-inch device. This may have caused more accommodation and convergence of their eyes. While both accommodation and convergence have been cited as contributing to eyestrain (Jaschinski-Kruza, 1988; Owens & Wolf-Kelly, 1987), the accommodative and convergent demand required by moving the smaller screen sizes closer to their eyes was not reported to cause discomfort for the periods of time they used the displays. In fact, the soldiers rated the comfort of viewing the 6.5-inch and 3.5-inch displays higher than they did the comfort of viewing the larger display (10.4 inch) and the GMD. This may have been due to their fairly young ages (the majority was under 30), since the resting point of accommodation increases with age (Ankrum, 1996) or the viewing distance required by the smallest display size was not close enough to cause problems. This hypothesis is somewhat supported by a study of three viewing distances, in which Rempel et al. (2007) found that the near distance (24 inches) was associated with significantly less blurred vision, less eye irritation, fewer headaches, and improved convergence recovery than the middle and far distances (33 and 39 inches).

The findings that soldier performance with the smaller display sizes (6.5 inch and 3.5 inch) was equal to performance with the larger display size (10.4 inch) are consistent with several other experiments that found no performance differences created by the effect of display size (Alexander et al., 2005; Stelzer & Wickens, 2006; Stark et al., 2001; Muthard & Wickens, 2004). However, it should be noted that although the soldiers did not perform *significantly* worse with the largest display than they did with the smaller displays, there was a trend for worse performance with the 10.4-inch display than with the 6.5-inch and 3.5-inch displays and soldiers rated it significantly more frustrating than the 3.5-inch display. Alexander et al. (2005) hypothesized that a smaller display could be of benefit to traffic surveillance by pilots, given that fewer eye movements would be required to scan the small display area as compared with a larger display area. Some soldier comments were consistent with this theory, and none were inconsistent. However, Martin-Emerson and Wickens (1997) found little added cost to increasingly longer eye movements. A similar hypothesis is that the smaller screens present information within the subjects' macular vision, which is responsible for the clearest and most distinct vision, and limit the need for the use of peripheral vision and the edges of the screen. Visual input to the macula occupies a substantial portion of the brain's visual capacity, and the smaller display sizes confine the information from the robotic scene to this area of highly concentrated cones. Further support for this hypothesis is provided by Parasuraman (1986), who reported an "edge effect" which he described as an observer's penchant to limit monitoring the edge of displays. Wickens et al. (2003) found that this edge effect resulted in peripheral changes being less likely to be detected, and the possibility of additional effort and neck strain. The larger displays could thus hinder situation awareness (SA) because a larger display area would fall into the peripheral vision of the operator. However, two experiments conducted by Muthard and Wickens (2004) to assess the impact of display size on flight control, airspace surveillance, and goal-directed target search showed mixed support for these hypotheses. Pilots exhibited less flight path error with larger displays, and surveillance and search were unaffected by display size.

Another factor that probably influenced the soldiers' ratings of the largest display was the mismatch between the camera and the display resolutions. This could have had some effect on results. The subjects complained that the system did not have good resolution, although it had the highest resolution of the devices used. The resolution of the 10.4-inch screen was 800×600, but the source resolution of the camera was only 640×480. The match between the camera and display resolution with the 6.5-inch display (640×480 for both) could be the reason they rated the middle-size screen as having the sharpest image. However, the camera used during this field experiment has a resolution that is typical of cameras on small robots, so the resulting mismatch could easily occur in the field.

Given the high dedication to operational realism, some problems that usually occur in the field also occurred in the experiment. In field operations, operator performance is sometimes degraded because of problems with robotic systems. Carlson et al. (2004) reviewed data from 10 studies of 15 different UGVs, and found that reliability of UGV performance in the field tends to be low (i.e., between 6 and 20 hours between failures). The common causes included limited wireless communication range and insufficient bandwidth for video-based feedback. These issues also often adversely impacted the human operator's remote perception and ability to teleoperate. Two problems with connectivity and communication occurred in this study. The first was the frequent lag in video reception, which caused operators to have delayed responses. At other times, the operators' video pictures froze, and the operators continued to reapply the same command over and over when, in fact, the system was responding. Because the operator received no feedback the continued commands resulted in the operator driving the vehicle too far forward or backward. Another problem that was noted during the experiment was glare on the display screen. During this experiment, the soldiers worked from a tent that helped reduce the problem with glare. However, dismounted soldiers will not always be shaded from the sun—glare is a problem that needs to be addressed. While adding some degree of variance overall, the variance arising from these operational and environmental concerns occurred randomly, and was not correlated with any particular device.

In summary, this study on display size indicates promising avenues for development and adaptation of displays for the dismounted warfighter. The soldiers in this study generally performed well with the smaller displays and expressed a preference for the smaller size screen displays. GMDs also represent a promising option in terms of size, weight, lack of glare, and perceived screen size and resolution. However, the soldiers reported more eyestrain and other physical problems with the GMD. These issues and tradeoffs are apparent in this study and need further research and development to deliver enhanced alternatives for the dismounted warfighter.

Control Scalability Study

The previous section described a study on display size that addressed scaling robot control systems by decreasing display size. However, control systems also consist of a controller that must also be scaled for use by dismounted soldiers. This section describes a study that addresses relative characteristics of different controller devices.

Optimal scaling of controllers for dismounted soldiers is challenged by the number and type of controls that can be provided. Control media for the teleoperation of robots can range from joysticks, toggles, keyboards, scroll wheels, buttons, pucks, pedals, touch screens, rockers, switchers, gesture and voice controls, mice, game controllers, and tactile mice, haptic joysticks, and vocal joysticks (Akamatsu & Sato, 1994; Chen et al., 2007; Cockburn & Brewster, 2005; He & Agah, 2001; Malkin et al., 2005; Neumann, 2006). Controls may perform single operations (e.g., a push button); but more complex systems perform multiple operations depending on the mode the system is in, or depending on the position pressed on the control (e.g., a rocker switch). All these input devices have demonstrated the potential for good performance under specific conditions and for specific tasks—none is ideal for all situations. For example, Cosenzo and Stafford (2007) compared a U.S. Army Tank-Automotive Research, Development, and Engineering Center (TARDEC)-designed, software-based joystick as an alternative to a traditional physical joystick. The software-based joystick was presented on the display, and the operators used a stylus or finger to teleoperate a TALON robot. The soldiers were able to perform all functions with the smaller software joystick, and reported it to be more comfortable to use and more precise than the traditional joystick. However, Chronis and Skubic (2000) found a software joystick to be very awkward for navigating a robot on a desired path. The hardware joystick was found to be somewhat more user-friendly. Thus, it is important to investigate control choice on performance under specific conditions and for specific tasks.

When scaling controls for the dismounted warfighter, size, weight, and ease of use are of primary importance. First, we want to identify devices or device characteristics that can be made smaller and lighter. There are two potential ways to reduce the size and weight of a robotic controller, given a set number of functions. It stands to reason that as the controller size is reduced, the size of the individual controls must be reduced or the number of individual controls must be reduced.

The most obvious approach is to reduce the size of the individual controls on the controller while reducing overall controller size. If we simply make existing controllers smaller, the similarity of controls would enhance transfer of training—an advantage. While miniaturizing controls sounds like a good approach for scaling down the size of controllers, there is a limit on how far this approach can be taken before performance is degraded. Logically, one of the primary limits to the size of a control used to manipulate a robot is the size of the smallest part that a human can easily manipulate. Also, soldiers have to be able to operate controls while wearing gloves, without accidentally activating adjacent controls. So, how small can we go before the buttons and spacing are too difficult for operational performance or reduce operational effectiveness?

The second approach is to reduce the number of controls on the controller, which would also allow the size of the overall controller size to be reduced. This can be achieved in two ways. The first way is by using the same control to accomplish multiple tasks. For example, a driving joystick can become a camera pan and tilt controller when the system is placed in another mode of operation. However, multifunction controls can increase control activation time and cognitive complexity. The end result is that such a controller may require many hours of training before a user is competent and comfortable enough to use the system during a real mission, and can hinder performance by requiring increased time and effort. The second way is the use of novel nonphysical controllers such as speech and gesture recognition systems (Chen et al., 2007) and vocal joysticks (Malkin et al., 2005) to reduce the number of physical controls. These technologies have demonstrated great potential for future applications, but are not yet ready for use on the dismounted warfighter's battlefield.

In a recent study performed at Fort Benning to investigate near-term alternatives for reduction of controller size, we conducted an operational investigation of teleoperation control performance using three different robot controller devices (Pettitt et al., 2008). The emphasis was on realism and operational relevance, in order to determine what systems are most suited for more immediate deployment. Twelve soldiers from OCS and three soldiers from Headquarters Company 1st Battalion 11th Infantry Regiment participated. After training on the operation of the iRobot PackBot, each soldier completed training and practice on a driving course, using each controller type. The first controller was the legacy system, and was representative of current small robot controllers. The second controller was chosen to represent the concept of miniaturizing the controls along with the controller. Both the first and second types of controller had a similar number of single-function controls (30 and 26 possible control manipulations). The third controller was chosen to represent the concept of reducing controller size by reducing the number of controls. This controller included several multifunction controls, and provided only 11 different control manipulations. Each controller was capable of performing all of the functions required to complete the robotic course.

The experiment was conducted on a driving course approximately 200 meters in length. The driving lane on the course was approximately 1 meter wide, and clearly marked with white engineer tape on the left and right sides. Along the course, there were seven stations where the operator conducted a specified maneuver or reconnaissance task. Five of the stations are shown in Figure 18.1. The seven stations were: stair ascend/descend; window reconnaissance; target identification/tracking; hill climb; bunker reconnaissance; tunnel reconnaissance; and zigzag negotiation. Soldiers teleoperating the PackBot were located at a remote stationary position inside a tent. The tent served to keep the operator and the controller equipment out of the elements, and to block the operator's line of sight to the robot.

A repeated measures analysis of variance (ANOVA) indicated that there was a significant difference among the mean course completion times for the three controllers. Follow-on pair-wise comparisons were conducted using Holm's Bonferroni correction. The average course completion time for the multifunction controller was significantly slower than the average times for the legacy and miniaturized controllers. Soldiers attributed the time required to switch function modes and the limited ability to perform simultaneous tasks when teleoperating the robot using the multifunction controller as the primary reasons for its slower

Figure 18.1   Maneuver and reconnaissance task stations

completion times. Performing some tasks simultaneously using the multifunction controller, such as driving the robot forward while raising the sensor arm, was not possible due to having to switch modes of the multifunction controls. There were no statistically significant differences between the controllers in terms of off-course errors and driving errors.

When asked to rate the characteristics of each controller, the miniaturized controller was rated highest for size, weight, shape, and comfort of use. However, several soldiers stated that controlling the robot's rate of speed and direction using the miniaturized controller was difficult, due to the sensitivity of its driving controls. Soldiers stated the large size of the legacy controller's controls allowed them to be labeled, and made finding the correct control easier. Performing multiple tasks simultaneously (i.e., raising the control arm while tilting the sensor head) was rated as very difficult or stated to be impossible with the multifunction controller.

Soldiers were able to successfully perform all the driving and reconnaissance tasks presented on the robotic driving course using all three controllers. Unfortunately, as in many field experiments, we had to use controllers that were available, so the differences between controllers included a variety of factors (e.g., different control spacings, number of hands used to activate the controls, control clustering and location, control labeling, etc.). However, we did achieve insight into factors involved when reducing controller size. The results indicated that reducing a controller size by shrinking the size of the individual controls impacted performance less than reducing its size through multifunction controls. The multifunction controller was more difficult to master and use than the controller with reduced control sizes, because switching between functions was effortful, time consuming, and confusing.

The difficulty of using the multifunction controls also increased perceived workload as rated using the NASA TLX. This is a particularly important finding for dismounted operations because Sterling and Perala (2007) found that robotic teleoperators supporting infantry units had higher workloads and stress than teleoperators supporting other types of units.

Summary and Conclusions

The robot warfighter will soon change the military world both tactically and strategically. Tactically, soldiers will be able to deploy robots for a variety of specific tasks and missions. Strategically, commanders may choose different tactics because of these enhanced capabilities. Currently military robots are teleoperated, and look and act somewhat like remote-controlled cars. Even when future robots become semi-autonomous or fully autonomous, teleoperation will still be an important default mode of operation (Chen et al., 2007). Because teleoperation requires human control, robotic controllers and displays will continue to be important components of the robotic system, and must be carried on the battlefield. Dismounted warfighters are already overloaded, and it is critical to ensure that the size and weight of these controllers and displays are minimized, while addressing issues related to capabilities, durability, and ease of use.

It appears that there is no single answer to the question "How small can we reduce a display for robotic teleoperation?" Although the study reported here indicates that a 3.5 inch display is sufficient for remote driving and detection of targets that are of immediate threat to the vehicle (within 10 meters), it may be that other tasks may require larger displays (Redden et al., 2008). Research indicates that size requirements are dependent upon a number of factors, such as task and environmental demands, individual differences (e.g., experience, expertise), and display characteristics. For example, Minkov et al. (2007) found some degradation with smaller camera displays showing an aerial perspective, where points of interest can be relatively small. In contrast, the Redden et al. study found no degradation with a ground-level camera that was relatively close to targets of interest. The study described here focused only on remote driving on a pre-specified course, and did not include the task of remote navigation. The requirement to view a map display will likely increase the size requirement of the display, depending on the degree of clutter and/or the need to view the map simultaneously with the driving (camera) display. Additional studies should be performed that include other teleoperation tasks and capabilities relevant to dismounted warfighters—such as navigation using a map, precise sensor manipulation, the use of other modality displays to supplement the visual display, and/or long distance target identification in a cluttered environment—before the absolute limit in display size reduction can be found.

There is a growing interest in GMD displays by designers seeking to reduce size, weight, and glare. GMDs are small and lightweight while providing a relatively large perceived display area with minimal glare. However, it should be noted that our study revealed some issues and problems. Soldiers performed significantly worse with the GMD on time to complete the obstacle avoidance task, and there was a trend toward worse performance on all measures. In addition, soldiers reported higher levels of effort and frustration, increased physical problems, and indicated it as the least preferred device. Soldiers complained of too much visual stimulation. Most of the soldiers either closed their left eye, or covered it with their hand, in their attempts to reduce the problem. Soldiers also expressed concern that the design did not let them wear their ballistic eye protection. On the positive side, the GMD used during this experiment was reported to be comfortable to wear and lightweight. There were also no reports of fogging. Our study underlines the need for further refinement of goggle-mounted displays to reduce these problems, and also further refinement of other displays to reduce the ever-present problem of daylight glare.

Issues pertaining to capabilities to scroll, zoom, and pan need to be addressed. There is no doubt that these capabilities will moderate effects of display size. The use of overlays will also affect the degree of clutter and operator capability to utilize increased types and amounts of information. These capabilities will also increase complexity of controls and menu navigation. While potentially very useful, features such as zoom, pan, and overlays may cause the operator to lose track of where he or she is. To address this, there is an emerging class of displays called "foveating" displays that help the viewer build and comprehend a large picture while easily attaining more detailed views. One example is the ZoneZoom, developed by a Microsoft researcher, that divides a given view into nine segments. If you want to zoom in on a particular area, tap the key corresponding to that area. It can be further subdivided in nine segments and so on (Ross, 2007).

Control system size issues need to be further pursued. A literature search on this topic revealed that very little has been done in this area. Pettitt et al., 2008 found that miniaturization of controls show more promise than the use of multiple multifunction controls, but more study needs to be done in the area to address type and layout of the controls. Reconfigurable controllers that allow miniaturization and consolidation of controls need to be developed so that experimenters can control extraneous variables while only manipulating selected variables. Innovative ways to reduce the number of physical controls on the controller (e.g., speech and gesture control) have high potential, once capabilities demonstrate high reliability.

Further investigation into ways to incorporate the controllers and displays into items that are already being carried should also take place. Items such as the CDA, PDA, helmet/goggle-mounted devices, and weapon-mounted devices should be considered as platforms for robotic controllers and displays. It is also imperative to investigate the effect of controller and display operation on the warfighter's SA of his environment, and to ensure that SA is not compromised. As unmanned ground vehicle technologies improve, the military expects to incorporate more robots into its forces for use by dismounted warfighters. Though these robots are seen as force multipliers, bulky and complex control system interfaces that demand the full attention of operators can be deadly on patrol or during covert operations. Warfighters need small, lightweight, simple, intuitive ways to control these robotic assets.

References

Akamatsu, M., & Sato, S. (1994). A multi-modal mouse with tactile and force feedback. *International Journal of Human-Computer Studies*, 40, 443–53.

Alexander, A., Wickens, C., & Hardy, T. (2005). Synthetic vision systems. The effects of guidance symbology, display size, and field of view. *Human Factors*, 47, 693–708.

Ankrum, D. R. (1996). Viewing distance at computer workstations (Guidelines for monitor placement). *WorkPlace Ergonomics*, Sept./Oct., 10–12.

Barnes, M., Everett, H. R., & Rudakevych, P. (2005). ThrowBot: Design considerations for a man-portable throwable robot. *Proceedings of SPIE Proc. 5804: Unmanned Ground Vehicle Technology VII*. Retrieved (8 May 2008) from http://www.spawar.navy.mil/robots/pubs/SPIE5804-59.pdf.

Barnes, M. J., Knapp, B. G., Tillman, B. W., Walters, B. A., & Velicki, D. (2000). *Crew Systems Analysis of Unmanned Aerial Vehicle (UAV) Future Job and Tasking Environments* (Tech. Rep. ARL-TR-2081). Aberdeen Proving Ground, MD: U.S. Army Research Laboratory,

Carlson, J., Murphy, R. R., & Nelson, A. (2004). Follow-up analysis of mobile robot failures. *Proceedings of the 2004 IEEE International Conference on Robotics and Automation(ICRA)*, 5, 4987–94.

Chen, J., Haas, E., & Barnes, M. (2007). Human performance issues and user interface design for teleoperated robots. *IEEE Transactions on Systems, Man, and Cybernetics*, 37, 1231–45.

Chen, J., Haas, E., Pillalamarri, K., & Jacobson, C. (2006). *Human-robot Interface: Issues in Operator Performance, Interface Design, and Technologies* (ARL-TR-3834). Aberdeen Proving Ground, MD: U.S. Army Research Laboratory.

Chronis, G. & Skubic, M. (2000). Experiments in programming by demonstration: Training a neural network for navigation behaviors. *Proceedings of the International Symposium on Robotics and Automation*. Monterrey, Mexico. Retrieved (8 May 2008) from http://www.cs.missouri.edu/~skubicm/Papers/.

Cockburn, A. & Brewster, S. (2005). Multimodal feedback for the acquisition of small targets. *Ergonomics*, 48, 1129–50.

Cosenzo, K. A. & Stafford, S. (2007). Usability assessment of displays for dismounted soldier applications (ARL-TR-4326). Aberdeen Proving Ground, MD: U.S. Army Research Laboratory.

de Bruijn O., Spence R., & Yin Chong M. (2001). RSVP browser: Web browsing on small screen devices. London: Department of Electrical and Electronic Engineering, Imperial College of Science, Technology and Medicine. Retrieved (8 May 2008) from http://personal.cis.strath.ac.uk/~mdd/mobilehci01/procs/debruijn_cr.pdf.

Grabianowski, E. (2005). How military robots work. Retrieved (17 January 2008) from http://science.howstuffworks.com/military-robot.htm.

Hancock, P.A. (1996). On convergent technological evolution. *Ergonomics in Design*, 41, 22–9.

He, F. & Agah, A. (2001). Multi-modal human interactions with an intelligent interface utilizing images, sounds, and force feedback. *Journal of Intelligent and Robotic Systems*, 32, 171–90.

Jaschinski-Kruza, W. (1988). Visual strain during VDU work: The effect of viewing distance and dark focus. *Ergonomics*, 31, 1449–65.

Jones, M., Marsden, G., Mohd-Nasir, N., Boone, K., & Buchanan, G. (1999). Improving web interaction on small displays. *Proceedings of the 8th International World Wide Web Conference*. Amsterdam. Retrieved (8 May 2008) from http://citeseer.ist.psu.edu/cache/papers/cs/3503/http:zSzzSzwww. cs.mdx.ac.ukzSzresearchzSzweb8.pdf/jones99improving.pdf.

Lee, S. & Blake, R. (1999). Rival ideas about binocular rivalry. *Vision Research*, 39, 1447–54.

Malkin, J., Li, X., & Bilmes, J. (2005). Energy and loudness for speed control in the Vocal Joystick. IEEE *Workshop on Automatic Speech Recognition and Understanding*, San Juan, Puerto Rico. Retrieved (8 May 2008) from http:// ssli.ee.washington.edu/vj/files/ASRU2005.pdf.

Martin-Emerson, R. & Wickens, C. D. (1997). Superimposition, symbology, visual attention, and the head-up display. *Human Factors*, 39, 581–601.

Mazumder, S., Drury, C. G., & Helander, M. G. (1997). Binocular rivalry as an aid in visual inspection. *Human Factors*, 39, 642–50.

Meenes, M (1999). A phenomenological description of retinal rivalry. *American Journal of Psychology*, 42, 260–69.

Minkov, Y., Perry, S., & Olan-Gilan, T. (2007). The effect of display size on performance of operational tasks with UAVs. *Proceedings of the Human Factors and Ergonomics Society 51st Annual Meeting*. Baltimore, MD (pp. 1091–5).

Muthard, E. K. & Wickens, C. D. (2004). Compensation for display enlargement in flight control and surveillance (Report No. AHFD-04-03/NASA-04-1). Champagne-Urbana: University of Illinois Human Factors Division.

Neumann, J. (2006). Effect of operator control configuration on unmanned aerial system trainability. Doctoral dissertation, University of Central Florida. *Dissertation Abstracts International*, 67, 11B, 6760.

Ni, T., Bowman, D., & Chen, J (2006). Increased display size and resolution improve task performance in information-rich environments. *Proceedings of Graphics Interface*. Quebec. (pp. 139–46).

Owens, D.A. & Wolf-Kelly, K. (1987). Near work, visual fatigue, and variations of oculomotor tonus. *Investigative Ophthalmology and Visual Science*, 28, 743–9.

Parasuraman, R. (1986). Vigilance, monitoring, and search. In K. R. Boff, L. Kaufman, & J. P. Thomas (eds), *Handbook of Perception and Human Performance*. New York: Wiley.

Pettitt, R. A., Redden, E. S. & Carstens, C. B. (2008). *Scalability of Robotic Controllers: An Evaluation of Alternatives*. ARL-TR-4457 Aberdeen Proving Ground, MD: U.S. Army Research Laboratory.

Redden, E. S., Pettitt, R. A., Carstens, C. B. & Elliott, L. R. (forthcoming). *Scalability of Robotic Displays: Display Size Investigation*. ARL-TR-4456 Aberdeen Proving Ground, MD: U.S. Army Research Laboratory.

Rempel, D., Williams, K., Anshel, J., Jaschinski, W., & Sheedy, J. (2007). The effects of visual display distance on eye accommodation, head posture, and vision and neck symptoms. *Human Factors*, 49(5), 830–38.

Renfro, M. B., Merlo, J. L., Duley, A., Gilson, R., & Hancock, P. A. (2007). A multimodal approach to unihabited vehicle operations in dynamic environments. Presented at the *Division 19/21 Midyear Conference of the American Psychological Association*. Fairfax, VA.

Ross, S. (2007). Zooming in on small displays: Microsoft research news and highlights. Retrieved (14 January 2008) from http://research.microsoft.com/displayarticle.aspx?id=1034.

Stafford, S., Hancock, P. A., Graham, J., & Merlo, J. (2007a). Equipment to meet the cognitive and physical requirements of the soldier. *Proceedings of Human Performance in Extreme Environments*. Baltimore, MD: HPEE.

Stafford, S. C., Jingjing, W., Merlo, J., & Hancock, P. A. (2007b). Soldier opinions of future military display technologies. *Proceedings of Human Performance in Extreme Environments*. Baltimore, MD: HPEE.

Stafford, S. C. (2008). Personal communication, 18 January.

Stark, J. M., Comstock, J. R., Prinzel, L. J., Burdette, D. W., & Scerbo, M. W. (2001). A preliminary examination of situation awareness and pilot performance in a synthetic vision environment. *Proceedings of the Human Factors and Ergonomics Society 45th Annual Meeting* (pp. 40–43). Santa Monica, CA: Human Factors and Ergonomics Society.

Stelzer, E. & Wickens, C. (2006). Pilots strategically compensate for display enlargements in surveillance and flight control tasks. *Human Factors*, 48, 1, 166–81.

Sterling, B. S. & Perala, C. H. (2007). *Controlling Unmanned Systems in a Simulated Counter-insurgency Environment* (ARL-TR-4145). Aberdeen Proving Ground, MD: U.S. Army Research Laboratory.

Tan, D. S., Gergle, D., Scupelli, P. G., & Pausch, R. (2003). With similar visual angles, larger displays improve spatial performance. *Proceedings of CHI 2003*. Fort Lauderdale, FL (pp. 217–24).

Wickens, C. (2005). *Attention and Aviation Display Layout: Research and Modeling* (Tech. Rep. AHFD-05-21/NASA-05-08). Urbana-Champaign: University of Illinois, Aviation Human Factors Division.

Wickens, C., Muthard, E., Alexander, A., van Olffen, P., & Podczerwinski, P. (2003). The influences of display highlighting and size and event eccentricity for aviation surveillance. *Proceedings of the 47th Annual Meeting of the Human Factors and Ergonomics Society*. Santa Monica, CA.

Wikipedia (2008). Isaac Asimov. Retrieved (18 January 2008) from http://en.wikipedia.org/wiki/Isaac_Asimov.

PART V
Cross-platform Research

Chapter 19

Lessons Learned from Human-Robotic Interactions on the Ground and in the Air

Nancy J. Cooke and Roger A. Chadwick

Arizona State University, New Mexico State University

Introduction

Unmanned or, more specifically, remotely-operated vehicles have contributed to the changing nature of warfare, and have taken center stage for civilian applications ranging from broader security and monitoring of wild fires, to search and rescue, and in surveillance of disaster locations. Technological advances now make it possible for soldiers to use (unmanned combat aerial vehicles (UCAVs) to fight the enemy in the Middle East from a trailer in Nevada (Moran & Morris, 2006). It also makes it possible for search efforts to proceed in places too difficult for humans to enter, such as the Katrina-ravaged neighborhoods or the ruins of the World Trade Center, through the use of rescue robots (Murphy, 2004a, 2004b). These robotic vehicles have taken humans out of harm's way in situations known to be "dull, dirty, or dangerous."

Although the technology has progressed, it is not without setbacks. Remotely operated air vehicles have a high crash rate that has been attributed to early fielding of the technology, sensitivity to weather, and human factors concerns (Rash et al., 2006; Williams, 2004). Similarly, ground vehicles have had a share of problems. Spatial disorientation is an inherent and consistently reported problem in the operation of ground vehicles (Chadwick & Pazuchanics, 2007; Tittle et al., 2002). Poor reliability of search robots (Carlson & Murphy, 2003, 2004) and lack of adequate vehicle situated awareness (Riley & Endsley, 2004) continue to limit their usefulness.

Despite the presumed role of human factors in the failure of these technologies, an aspect of the technology that has received minimal attention is the human-robotic interface. This is often overlooked in lieu of a focus on the hardware and a tendency to "forget about" the operators who are located elsewhere (Cooke et al., 2006b).

Absence of humans from the vehicle does not mean that the vehicles are autonomous or that there are no human factors concerns. There has been some recent impetus for considering human factors aspects of these vehicles, manifested in a series of workshops on human factors of unmanned aerial vehicles (UAVs), an edited volume that followed, and numerous conference talks and publications (Cooke et al., 2006b), and in each venue lessons have been learned.

Unfortunately there is little cross-talk between the air- and ground-vehicle camps. This may not be surprising given that the emphasis on the technology implies that these are two very different types of vehicles. On the other hand, from the perspective of human factors, the focus should not be the vehicle, but rather the human and the human-robotic interface (Cooke & Pedersen, 2009). This latter focus also suggests that there may be commonality between the two vehicles at the human-robotic interface. The purpose of this chapter is to summarize lessons learned regarding human-robotic interaction (HRI) in the context of air vehicles and ground vehicles and to highlight those lessons that occur at the intersection of the two. Our objective is to abstract the essence of human-robotic interaction from the vehicle that is being controlled.

We begin by laying out a series of lessons learned regarding human factors of UAVs and, subsequently, unmanned ground vehicles (UGVs). We then compare and contrast these two sets of lessons to identify common and discrepant lessons and, ultimately, summarize lessons learned for remotely operated vehicles more generally. Overall we view the human-robotic interface of these systems to be at least as critical as the vehicle itself.

Human Factors Lessons Learned Concerning Unmanned Aerial Vehicles

The term "unmanned" is a misnomer when it comes to UAVs and may imply to some that there should be no relevant human factors issues. On the contrary, these platforms are rife with human factors issues. Although the vehicle itself is uninhabited, there are numerous people on the ground engaged in remote operation, sensing, data interpretation, communications, maintenance, training, and the larger command-and-control function. The vehicles are not fully autonomous, but are remotely operated with various degrees of automation.

The number of humans involved, coupled with a high mishap rate—by some counts 100 times higher than that of manned aircraft (Jackson, 2003)—suggests that there may be some human factors issues. Indeed accident analyses have tied 33–43 percent of the mishaps to human factors issues (Schmidt & Parker, 1995; Seagle, 1997). The high-profile case of a Predator crash near the Arizona–Mexico border was attributed to human factors issues with the interface (NTSB, 2006). Weather is another factor that has been blamed for the high rate of UAV mishaps. However, in a complex system it is typically the case that there are a number of issues leading to a mishap and, at some level, humans are likely to figure in the mix. In this section we highlight some of the human factors lessons learned concerning UAVs (see Table 19.1).

Table 19.1   Lessons learned from human factors research on unmanned aerial vehicles

UAV Lesson 1: Piloting a UAV is not exactly like flying a plane.
UAV Lesson 2: The UAV needs to be integrated into the NAS with other unmanned planes as well as manned planes.
UAV Lesson 3: Automation is not the answer.
UAV Lesson 4: Multiple UAV control can be difficult.
UAV Lesson 5: Human intervention is critical, yet challenging.
UAV Lesson 6: Remote operation has implications for society.

UAV Lesson 1: Piloting a UAV is Not Exactly Like Flying a Plane

This is one of the main premises of this paper, and the motivation for comparing and contrasting findings from air and ground vehicles. Of course, there are some similarities between UAVs and manned aviation. The human needs to navigate and control the position and speed of the vehicle. Landing and take-off are difficult tasks for both platforms. Other similarities such as human factors issues of spatial disorientation, crew coordination, fatigue, and communications are covered in Cooke and Pedersen (2009). However, there are many differences, the biggest being that perception and control occur remotely, making UAV flight more analogous to robotic control or video game interaction. Perception occurs through the sensor displays, and there are significant delays between control and vehicle response. The visual experience of a UAV pilot is very different and likened to looking at the world through a soda straw.

The term "unmanned aerial vehicle" has also been replaced in some circles by "unmanned aerial system" to reflect the fact that the vehicle, the plane, is only part of the equipment. Equally or more important is the ground control station from which the vehicle is flown. This trailer looks very little like a plane or a cockpit. Some may argue that piloted planes are also part of the larger system including the National Airspace (NAS), but they are nonetheless distinct from UAVs. The remotely operated vehicle, unlike the airplane, is inextricably tied to those on the ground. Without ground control, it will not function.

UAV Lesson 2: The UAV Needs to be Integrated into the NAS with other Unmanned Planes as Well as Manned Planes

Although the case can be made that UAV operation differs substantially from manned aviation, nonetheless both manned and unmanned air vehicles are anticipated to operate in the same air space. This is not only true for military operations, but if the unmanned vehicles are to be used for civilian applications such as air freight transportation, agricultural surveillance, border security, and urban security then they will need to be integrated into the National Airspace. This integration has raised a host of issues, especially given the relatively high mishap rates of these vehicles (Jackson, 2003). The safety of surrounding air vehicles and those on the ground has been called into question given the remote piloting challenges discussed here. Issues pertaining to see-and-avoid technology, air frame certification, and operator training have been at the forefront, with the objective to do no harm. Standards and requirements have been suggested, but development in this area is challenging given the vast array of UAV platforms and applications (Cooke et al., 2006a; Hottman & Sortland, 2006).

UAV Lesson 3: Automation is Not the Answer

It is easy to forget about the human operator, especially when the vehicle is in Iraq and the humans operating it are miles away in Nevada. It is likewise easy to forget completely about the people "behind the curtain" and to see the vehicle as autonomous. Certainly there is much automation involved in UAV platforms. For instance, some platforms are basically on autopilot once in flight and are flown by human entry of coordinate waypoints.

Automation is also often the first answer to problems related to human factors. The assumption is that if the functions are completely automated, removing the human from the loop, then errors attributed to humans will likewise be eliminated. However, automation is not the answer. This is a lesson that has been learned more generally in human factors (Howell & Cooke, 1989; Parasuraman & Riley, 1997; Sheridan, 1987, 2002).

Automation changes the human's task to that of overseer, often removing the human from the system such that there is a loss of situation awareness when things go wrong (Endsley & Kiris, 1995). Issues of trust in automation and breakdowns in automation are also important. Mode errors can result when the human expectations do not match the automated mode. This was the case in the Predator crash near the Arizona–Mexico border, which was partially attributed to controls having different functions when used in different modes (i.e., flying the vehicle versus camera operations). With automation, the task from a human's point of view can get even more difficult and error prone. This would be the case for instance, if additional automation was developed to help move a UAV from point A to point B when the workload stems from tasks that take place at the target waypoints (Pew & Mavor, 2007).

UAV Lesson 4: Multiple UAV Control can be Difficult

Following the sentiment regarding automation described under UAV Lesson 2 and a push in military circles for lighter staffing requirements there has been demand for control of multiple UAVs by a single individual. The general idea is that if we cannot eliminate the human, then perhaps we can reduce the number of humans by having them control multiple vehicles, each with higher degrees of autonomy than currently exist.

There are a number of issues wrapped up in this objective. Although dependent on platform, among the dozens of personnel associated with a UAV are typically at least two operators. One is the Air Vehicle Operator or pilot and the other is the Payload Operator or Sensor Operator. These two or more individuals need to communicate and coordinate extensively to achieve the objectives of the mission. In fact, there have been UAV mishaps attributed to problems with teamwork (Tvaryanas et al., 2005). Thus, the starting point for single operator–multiple vehicle control is not single operator–single UAV control, but rather multiple operators–single UAV control. Interestingly, some versions of multiple UAV control have decoupled the functions of the two operators and maintain multiple sensor operators, each assigned to a single vehicle such that only the flight function of multiple vehicles is assigned to a single operator. It has become clear that the task of exercising the full functionality of a single UAV under current automation scenarios stretches the limits of human capabilities.

Another complicating factor is that workload is not constant or identical for different roles. Some operators describe workload as long periods of intense boredom followed by snippets of "extreme terror." Workload tends to be greatest when a target is reached and when replanning occurs. Based on studies (e.g., Wickens et al., 2006) and discussions with operators (Pederson et al., 2006), multiple UAV control may be possible with very high degrees of automation (but then see UAV Lesson 2) or when all vehicles are in a normal point-to-point state of flight. However, when the situation changes, when a single UAV is in trouble, or when a target is reached, multiple UAV control by a single individual could range from difficult to impossible.

Ultimately, successful multi-vehicle control hinges on reliable automation that can work in concert with the human. Having automation that is not reliable may be worse than no automation at all (Wickens et al., 2006). There are various solutions proposed to optimize automation, ranging from interfaces that adapt to operator workload (Hou & Kobierski, 2006) to algorithms for controlling teams of vehicles (Lewis et al., 2006) and strategies for delegating control (Parasuraman & Miller, 2006). The bottom line is that multiple vehicle control is complicated and a simple increase in vehicle autonomy is not the solution

UAV Lesson 5: Human Intervention is Critical, Yet Challenging

Aside from the downfalls of clumsy automation, there are a number of reasons why completely automating the human out of the UAV loop is not such a good idea. Human intervention is required in these systems to interpret sensor imagery and perform dynamic retasking and replanning when the situation changes. These functions set UAV platforms apart from satellites or surveillance cameras, and they are functions that are currently best suited for humans. Through remote operation an area can be monitored in real time for threats, and based on subtle cues in the environment monitoring can be redirected as required. Until automation is able to seamlessly carry out these functions, inclusion of the human allows the system to be adaptive and flexible.

The role of sensor operator is integral to UAV operation, yet it is often not considered part of the core UAV system. On the contrary, operators often describe UAV operation as "flying a camera." Not only is the camera functional as system payload but it is also provides the visual input for flight. Without the camera, remote UAV control would not be possible.

There are many human factors issues that are relevant to the human's interaction with camera systems. UAV operators can become spatially disoriented and have visual perception that is limited by the camera angle, and then exacerbated by being physically removed from the feedback of the vehicle. Indeed, UAV mishaps, particularly at the time of landing, have been attributed to problems of spatial disorientation (Self et al., 2006). The task of imagery analysis itself is also difficult especially when (as is the case for military applications) the target is dynamic, uncertain, and actively trying to stay hidden. Thus, although the human plays a pivotal role in UAV imagery analysis, it is a difficult task even for the human and one that warrants support through training and interface design.

UAV Lesson 6: Remote Operation has Implications for Society

There are a number of social ramifications of the application of UAVs to military and civilian problems. These types of issues are not traditionally construed as human factors, but they are human issues that have the potential to affect the operator and ultimately the system performance. In military applications, UAV operations reflect modern-day warfare which is conducted not on the battlefield as much as in control rooms or trailers in front of networked computer monitors. So-called "remote termination" or the deploying of weapons remotely raises some interesting psychological questions. How does remote control of a weapons system separated by thousands of miles from the target affect the perception of reality? Is the act of killing more like video game play? Sometimes UAV operators report feelings of guilt associated with not being in harm's way, and other times a general apathy regarding killing the enemy or destroying targets is reported. How does post-traumatic stress syndrome factor into this new battlefield? How easy is it to reintegrate into society (e.g., family life, children, civilians) when the switch between battlefield and civilian life happens several times a day? These are all very timely and interesting questions that have to our knowledge not received adequate attention from the research community.

There are other social implications for civilian applications. They range from privacy issues associated with having "spy planes" fly over our cities and suburbs to attitudes associated with suggesting that future passenger flights may be pilotless. These issues should also be considered as UAVs are integrated into the National Air Space.

Human Factors Lessons Learned Concerning Unmanned Ground Vehicles

The vision for the future includes widespread use of remote unmanned ground vehicles or UGVs in a variety of civilian and military mission roles. At the dawn of the twenty-first century, while UAVs were already making a name for themselves in extensive action over Iraq and Afghanistan, exerting a much greater influence than in their previous limited usage in the first Gulf War of the early 1990s (Biass & Braybrook, 2001), UGVs have yet to reach their envisioned potential. UGVs have been used in limited roles in urban search and rescue (USAR) operations (e.g., Casper & Murphy, 2003) and depended upon for explosive ordnance disposal (EOD) in the current Iraq conflict (e.g., Montgomery, 2005), but plans call for much more extensive roles (Evans et al., 2006; Killion, 2006). As the complexities involved in deploying productive remotely operated ground vehicles become clear, the challenges facing the realization of the vision are proving difficult to overcome.

Before discussing lessons learned from UGV research, we should examine the vision alluded to in the first paragraph of this section. Conferences relevant to military endeavors over the past few years have showcased dreams of myriad UGVs filling the battlefield with a host of vehicle types fulfilling various roles, including transport, EOD disposal, checkpoint search, reconnaissance, urban search, perimeter defense, and attack (e.g., Killion, 2006). Kurzweil (2006, p. 1) describes the near future of warfighting as "a remote-controlled, robotic, robust, size-reduced, virtual-reality paradigm." With the constant advance in robotics and computational efficiency, the realization of a variety of ground vehicles equipped to perform a multitude of tasks without risk to human lives seems very plausible, and the pursuit of these dreams a viable expenditure of time and effort.

The problem with developing useful UGVs is essentially that navigating through real-world environments and productively accomplishing tasks on the ground via robotic systems requires great powers of agility and cognition. With regard to cognition, two basic approaches are vehicle autonomy and remote operation by human beings. Artificial intelligence required for vehicle autonomy remains a disappointing technology despite exponential improvements in computing cost and density (for a discussion of exponential growth of information technologies see Kurzweil, 2006). Remote perception required for teleoperation is notoriously problematic (e.g., Tittle et al., 2002). Teleoperation of UGVs has often been likened to viewing the world through a soda-straw (Casper & Murphy, 2003; Voshell & Woods, 2005), but it is more akin to driving a car from the back seat of another vehicle with an awkward joystick while viewing the roadway through a digital camera display. The other possibility, that of autonomy via artificial intelligence, suffers from the restrictions of far lesser intelligence for processing rather limited artificial sensations, coupled with far greater rigidity in decision making and comprehension.

The choices for UGV operation (autonomy vs. remote operation) are to use an inadequate artificial mind on the vehicle or a remotely positioned human mind handicapped by greatly restricted and unnatural sensing abilities. The choice is not a dichotomous one however; combinations of human supervision over various levels of achievable autonomy probably hold the most promise for the near term. Recent advances in automatic systems for ground robots still depend heavily upon the operator for perception of objects and the environment (e.g., Zeng et al., 2007). Still, there are a limited number of valid reasons for using unmanned vehicles, and research and development to improve these systems and resolve these problems continues.

With the lessons we have learned from UGV research, perhaps a vision can be realized not in the original form, but as an evolving set of solutions which ultimately provide another avenue for technological superiority. Because full autonomy in UGVs is yet to reach the level of practical application, much work has been focused on human-robotic interaction with UGVs. Table 19.2 lists a set of human factors lessons gleaned from UGV research.

UGV Lesson 1: Remote Perception is Poor Perception

As mentioned previously, the problems of remote perception (e.g., Darken et al., 2001; Tittle et al., 2002) have been well documented. Although some solutions have been proposed, the solutions often conflict with other realistic constraints on true remote vehicle operation. For example, the use of multiple cameras and perspective folding (Voshell & Woods, 2005) has been shown to improve teleoperated navigation, but at the expense of precious communication bandwidth. These solutions have their applications because bandwidth is not an issue in tethered vehicles (with high bandwidth fiber optic or radio frequency links), but tethering is itself a major problem and constraint. In general, researchers need to search for schemes which will maximize perception of both objects and space while minimizing communication bandwidth.

Table 19.2   Lessons learned from human factors research on unmanned ground vehicles

UGV Lesson 1: Remote perception is poor perception.
UGV Lesson 2: Reducing the human to robot ratio remains challenging.
UGV Lesson 3: Spatial abilities are not created equal.
UGV Lesson 4: View integration is the key.
UGV Lesson 5: Remote possibilities are not realized.

Research on UGV search operations efficiency using miniature models and scale environments (Chadwick, 2008) reveals some of the consequences of remote perception deficiencies. In studies conducted at New Mexico State University, participants searching for victims in a simulated disaster scene made some common errors. Victims appearing in the video display were often not perceived, an issue of degraded object perception (see Pazuchanics et al., this volume). Search movements often excluded the extreme periphery: participants often failed to fully look to the left or right after entering a search area, missing targets which would otherwise have been easy to find. UGV operators in these studies often became so engrossed in navigating past obstacles that the goal of finding victims seemed to lose its priority. Finally, when target objects were successfully identified, the designation of the target's location is often in error, an issue of spatial comprehension (Chadwick & Pazuchanics, 2007).

The kinds of problems referred to in the preceding paragraph are not surprising when remote perception in a search operation is contrasted with normal human perception in natural environments. When performing an active object search people are likely to move their heads and eyes to a great extent, and have the benefit of full-fidelity vision. The next time you misplace your car keys, try searching for them using the display of your digital camera, without moving your head from side to side. Near-sighted individuals searching for their glasses can truly appreciate the difficulties of object perception under even slightly degraded visual circumstances.

UGV Lesson 2: Reducing the Human to Robot Ratio Remains Challenging

One important lesson to be learned about using remotely operated vehicles, which applies to both the air and ground variety, is that human workload is not eliminated by the creation of robotic proxies. At the beginning of our research, we were tasked with examining interface designs for one-man-to-many-ground-vehicles. This request in and of itself reflects the thinking of visionaries of the time, that one human operator could manage more than one vehicle. The capability for one intelligent person to be (at least virtually) in multiple places at one time is attractive, and would provide the very essence of force multiplication. But research suggests that this is not practically feasible.

In several studies the single operator control of multiple UGVs in search operations was studied directly (Chadwick, 2005; Rehfeld et al., 2005). In these cases there was no improvement in performance when more than one UGV was employed. The ability to control multiple UGVs productively depends on reliable autonomous vehicle operation. In a study in which operators could take advantage of rather simple (speed and heading) autonomous navigation modes (i.e., Lif et al., 2007) there was some improvement in performance when two UGVs were used, but no further improvement when three were used. It is important to note that the task in Lif et al.'s (2007) study involved waypoint navigation under simple autonomous control modes of operation. In this type of task a UGV can be sent into a productive mode of operation (heading toward the waypoint until an obstacle is encountered) and ignored while making progress toward a goal. When the only consequence of being ignored is the cessation of progress (i.e., striking an obstacle and stopping to await further instruction), efficiency can be gained with multiple vehicles. If the consequences of being ignored are more severe, such as damage to the vehicle or bystanders, then any gains are quickly eliminated. The key to success in controlling multiple vehicles is periods of autonomous operation during which the operator is not required to process contextual information from the productive UGV. This issue is worthy of some further elaboration.

In order to further examine the issue of single operator control of multiple vehicles, we designed a simulator for up to four semi-autonomous ground vehicles supervised by a single human operator. Studies using these simulators showed that the operator's ability to process actual contextual information coming in from the vehicles was severely impaired when more than one vehicle was in use (Chadwick, 2006). A detailed explanation of this conclusion follows.

If autonomous vehicles reach the level of *full* autonomy, or even near-perfect semi-autonomy, an operator could supervise many vehicles. In this scenario, however, there is by definition very little for the human operator to be responsible for. The robotic proxy, being fully autonomous in this case, knows its mission and sets out to accomplish it. The only human decision making involved in using fully autonomous vehicles would be telling the vehicles where to go, what to accomplish, and perhaps when to return home. Full autonomy is of course not realistic given the present state of technology. Given the more realistic situation involving semi-autonomous vehicles, the human operator's tasks can be broken down into monitoring status indications from the robot, responding at decision points, and detecting and correcting anomalous behavior. Our studies (Chadwick, 2006) showed that the detection of anomalous behavior in semi-autonomous vehicles might be the most difficult of these tasks.

Any semi-autonomous vehicle can, at times, take actions which are not consistent with the goals of its human supervisor. During these moments of inappropriate action, the important thing to note is that the robotic vehicle "thinks" it is doing the right thing, but in fact is doing something anomalous. The robotic vehicle in these cases will not generate an alert or request for decision from the human operator, but will continue with its autonomous behavior as if nothing is wrong because, algorithmically, nothing is detectably wrong. Only an intelligent human operator, monitoring the context of the situation, can determine that there is an anomaly and take corrective action.

Our studies showed that even when an autonomous UGV's anomalous behavior was as simple as being stuck in a repeating loop, the time to respond to such an anomaly went from just a few seconds with one robot to almost two full minutes with four robots (Chadwick, 2006). Thus, while human operators can monitor (e.g., alarms and vehicle status) and respond to cued requests for decisions from multiple robotic vehicles, their ability to detect anomalous behavior from the context of the situation as perceived via the remote imagery is drastically impaired when operating more than one vehicle. Detecting anomalous behavior requires a dedicated amount of attention, and when attention is divided amongst multiple vehicles the operator does not appear to have the dwell time or memory capacity to comprehend the nature or context of the situation. The ability for one operator to supervise multiple UGVs depends on the ability of the UGV to remain engaged in productive autonomous activity while being ignored by the operator whose attention, in being productive with multiple units, is diverted towards another vehicle.

In the real world there are often multiple humans involved with a single robot. In search applications, teams of human beings often cluster around the information received from a single robotic asset (Casper & Murphy, 2003; Tittle et al., 2005). The workload is distributed amongst the team members, with each person having a well-defined task such as vehicle operation, mapping, object (victim) identification, and structural evaluation. A common complaint or observation in these reports is that human team members with various responsibilities must (physically) compete for the robot's camera display. One simple solution is to provide external video connections for onsite distribution of robot video to multiple monitors. The lack of this capability and foresight in the design of the interface systems reportedly used in actual USAR scenarios indicate that the human robot ratio has often been underestimated.

Some interesting solutions for distributing the information from a single robot to multiple human team members have been proposed involving the oversampling of visual information using wide-angle camera lenses and digital reconstruction of multiple views which are tailored to each human team member's needs (Tittle, et al., 2005). Another idea worthy of further investigation is the incorporation of digital video playback features in UGV interfaces which allow operators to review past imagery and context leading up to the current situation. Video playback features (see Chadwick, 2006) might improve task switching situation awareness acquisition in multiple vehicle or intermittent attention scenarios.

UGV Lesson 3: Spatial Abilities are Not Created Equal

The UGV operator's spatial ability is a key cognitive component for many ground vehicle tasks because spatial comprehension via remote perception is extremely difficult (e.g., Chadwick & Gillan, 2006; Chadwick & Pazuchanics, 2007; Darken et al., 2001) and necessary for most UGV tasks. When operating a remote ground vehicle spatial skills are needed for navigation, localization of objects, and safe use of manipulators (e.g., Zeng et al., 2007). Spatial ability varies considerably between individuals (e.g., Dabbs et al., 1998). Spatial ability in UGV studies is often measured (somewhat indirectly) by target localization error in tasks in which the operator must find an object and designate its location on a map. In all of our studies on remote operation of ground vehicles we observed great differences in skill associated with the spatial aspects of the tasks (e.g., Chadwick & Gillan, 2006; Chadwick et al., 2006).

Based on many observations of participants in research tasks, we believe that UGV operators should be selected for spatial abilities. Testing paradigms should be developed that adequately define and measure the relevant abilities (which may be task dependent). Training designed specifically to enhance spatial abilities relevant to UGV operations should also be developed, but after witnessing hundreds of participants in remote vehicle operations studies, it appears that some individuals excel in this type of skill and seem to have some natural ability.

UGV Lesson 4: View Integration is the Key

UGV operational scenarios almost invariably involve perceptual sensing of the environment (e.g., target identification) and localization (of both the UGV itself and identified targets or other objects of interest) in space. Maps are inherently important in UGV operations, whether they are hand drawn by USAR team members examining a rubble pile or sophisticated informative charts prepared in advance for reconnaissance or attack missions. Integrating the vehicle-centric imagery provided by the UGV sensors with global space views, whether in the form of map displays or live aerial views, is an important and challenging cognitive task associated with UGV operations. While we are beginning to understand the cognitive process of ground and air view integration (see Chadwick & Gillan, 2006), facilitative interface features have yet to be demonstrated. In a recent unpublished study (Chadwick, 2008) we demonstrated the usefulness of live aerial views in conjunction with UGV search operations. The use of live aerial views necessitates some means of communication of aerial view direction, and simply auto-tracking the ground vehicle may be the best solution for most situations. There are many variables associated with presenting these views that remain to be studied, and the difficulty involved in matching a ground-viewed scene with an aerial view varies greatly with features of the terrain and the number and similarity of objects present. Remote vehicle search of an area populated with many similarly shaped and colored buildings, perhaps with identical landmarks at different spatial locations, could be extremely confusing for the remote operator.

Providing the UGV operator with information on vehicle position and orientation on a highly detailed map display will go a long way towards resolving these issues, but problems may still occur. In one (unpublished) study involving the teleoperation of two robotic ground vehicles we provided the operators with a map display showing the UGV's position and orientation on a rather simple map of a straight roadway. Due to the distraction of multi-tasking, operators at times became confused and drove for several minutes in the wrong direction along the road before noticing landmarks they had previously encountered. The participants in these studies often used the map to confirm their disorientation rather than verify their orientation. In other studies in which both position and orientation of the ground view camera were given, participants made a preponderance of errors in depth, a dimension for which no direct information was provided and depth perception from two-dimensional images had to be relied upon (Chadwick &

Gillan, 2006). There is no doubt that using aerial views or maps can be extremely useful to the UGV operator, but the difficulties of view integration (see Thomas et al., 1999) remain a challenge for interface designers.

UGV Lesson 5: Remote Possibilities are Not Realized

While the goal of UGV deployment is remote operation, most current scenarios are not truly remote in nature. As aptly described by Witus et al. (2006, p. 3 of abstract), "Remote teleoperation from the egocentric perspective of the on-board camera provides a limited field of view, and has been blamed for disorientation in teleoperation. Soldiers operating small robotic vehicles tend to prefer to stand behind the UGV close enough to have an overwatch perspective." The scenario described by Witus and associates involves robotic vehicle inspection at military checkpoints. It is much safer to remain truly remote in such situations inside a protected bunker, and the choice to view the vehicle directly rather than operate in a true remote fashion reflects the severity of the remote operation problem.

The more successful use of robotic vehicles to date involves teleoperation with direct viewing. Soldiers disabling explosive devices using EOD robots generally have a clear view of the vehicle or are at least in the general vicinity of the robot. Being truly remote does not necessarily relate to distance from the robot; USAR teams are often in close proximity to their vehicle, which disappears into a remote void. USAR teams must rely entirely on remote perception, and the difficulties associated with navigation, obstacle avoidance, and object perception are well documented (Casper & Murphy, 2003).

The difficulties in vehicle operation will become exacerbated when operation is truly remote. Interface systems for true remote operation will have to be more than small television monitors coupled with awkward joystick controls. Perhaps large-screen, high-resolution display technology coupled with sophisticated navigation and map displays and augmented cognition tools will be critical to successful deployment of this technology.

Synthesis of Lessons Learned

The lessons learned from air and ground vehicles are remarkably consistent. Both platforms are used for tasks that would otherwise be dull, dirty, or dangerous for the human. As a result, the vehicles are consistently working in environments that bring with them numerous constraints on perception and action. From rubble piles to enemy air space, these environments are not especially conducive to exploration. There also appears to be an underlying tension in both arenas between "true" autonomy and remote control, recognizing that even true autonomy involves human developers and technicians. Examining air and ground systems as a whole, this tension is not best characterized as a dichotomy, but rather a continuum that allows for various degrees of human intervention. Ground systems may be tethered and vehicles may be operated remotely in the line of sight (as is typical of UAV take-off and landing) or using camera vision. Cameras allow the operator to be just outside the rubble pile or thousands of miles from the target. The environment and the degree of autonomy are both critical factors for human-robotic interaction.

There are also common human factors themes. Two appear consistently for both types of vehicles and are the most pressing concerns. The first has to do with *multiple vehicle control* with the push in both camps coming from the military and the need to reduce staffing requirements. It is clear that the solution to this problem depends on adequate vehicle autonomy in which the technology can carry out critical cognitive tasks now relegated to humans. The bottom line is that our autonomous capabilities are not sufficient yet and the stakes are high (e.g., crash of a UAV into a suburban neighborhood). Curiously, for both types of vehicles, human factors researchers and most operators recognize the irony of the reality which in many cases requires multiple humans to operate a single vehicle.

The second area of human factors attention concerns the *degraded visual environment* that comes with the use of cameras to perceive the world. Although action is also degraded due to time delays, the degraded perception seems to present the predominant human-robotic interaction problems. Disorientation is common and for the ground vehicle, more than the air vehicle, it is not even clear how the vehicle is oriented much less where the vehicle is located in space. It is also important to recognize that although navigation is challenging in this degraded environment, the camera in both platforms is used not only to navigate, but also to perform the surveillance task. The constrained visual environment coupled with the restricted camera views present serious camera constraints to the operator. Solutions to this problem will determine the ultimate value of the vehicle.

In terms of human-robotic interaction there are more similarities between the two types of vehicles than differences. Based on the lessons learned that are described in this chapter we can identify a few areas of concern voiced in one domain, but not the other. However, this does not indicate that the concerns are not equally relevant for the other domain. For instance, the importance of spatial abilities for operation of ground vehicles is also relevant to air vehicles. In this case the analogy to manned aviation, and the concomitant importance of spatial abilities, may result in this issue being taken for granted in the UAV world. Similarly, there is no reason why the social implications that are highlighted in regard to UAVs are not equally important for UGVs that also enable remote warfare and threaten individual privacy. In addition, considering these systems more broadly, they are perceptual extensions of the human, and the camera functionality extends beyond navigation to the essence of the targeting or surveillance task itself. Visual perceptual issues are rampant and of central importance for both types of vehicle.

The single vehicle difference that can be discerned from these lessons learned has to do with the issues surrounding the integration of UAVs into the National Airspace. Until we start placing UGVs on our nations' highways amidst manned ground vehicles, there is not a comparable concern at the moment.

Conclusion

In summary, unmanned aerial and ground vehicles are robots first and vehicles second. The operation of the unmanned vehicles differs significantly from their manned counterparts. The vehicles are perceptual and motor extensions of their human operators. Air and ground vehicle human factors considerations (i.e., human-robotic interaction) overlap almost completely (see Table 19.3). At center stage are human-robotic interaction issues of degree of autonomy and degraded perceptual and motor environments. By addressing these human-robotic interaction issues we will be able to maximize return on these remotely operated systems.

Table 19.3   Lessons learned from human factors research on unmanned vehicles

UV Lesson 1: Unmanned aerial and ground vehicles are robots first and vehicles second.
UV Lesson 2: Remote perception is poor perception.
UV Lesson 3: Spatial abilities are not created equal.
UV Lesson 4: Remote control poses challenges for humans.
UV Lesson 5: Automation alters the human's task; it does not eliminate it.

References

Biass, E. H. & Braybrook, R. (2001). The UAV as sensor platform-from Pioneer to Global Hawk. *Armada International*, 5, 2–11.

Carlson, J. & Murphy, R. R. (2003). Reliability analysis of mobile robots. *IEEE Conference on Robotics and Automation*, 1, 274–81.

Carlson, J. & Murphy, R. R. (2004). How UGVs physically fail in the field. *IEEE Transactions on Robotics*, 21, 423–37.

Casper, J. & Murphy, R. R. (2003). Human-robotic interactions during the robot-assisted urban search and rescue response at the World Trade Center. *IEEE Transactions on Systems, Man, and Cybernetics, Part B: Cybernetics*, 33, 367–85.

Chadwick, R. A. (2005). The impacts of multiple robots and display views: An urban search and rescue simulation. *Proceedings of the Human Factors and Ergonomics Society 49th Annual Meeting*. Santa Monica, CA: Human Factors and Ergonomics Society, 387–91.

Chadwick, R. A. (2006). Operating multiple semi-autonomous robots: Monitoring, responding, detecting. *Proceedings of the Human Factors and Ergonomics Society 50th Annual Meeting*. Santa Monica, CA: Human Factors and Ergonomics Society, 329–33.

Chadwick, R. A. (2008). Costs and benefits of aerial views in UGV search operations (unpublished manuscript).

Chadwick, R. A. & Gillan, D. J. (2006). Strategies for the interpretative integration of ground and aerial views in UGV operations. 25th Army Science Conference poster session (November).

Chadwick, R. A. & Pazuchanics, S. (2007). Spatial disorientation in unmanned ground vehicle operations: Target localization errors. *Proceedings of the Human Factors and Ergonomics Society 51st Annual Meeting*. Santa Monica, CA: Human Factors and Ergonomics Society, 161–5.

Chadwick, R. A., Gillan, D. J., & Pazuchanics, S. L. (2006). What the robot's camera tells the operator's brain. In N. J. Cooke, H. Pringle, H. K. Pedersen, and O. Conner (eds), *Advances in Human Performance and Cognitive Engineering Research: Human Factors of Remotely Piloted Vehicles* (pp. 373–84). Amsterdam: Elsevier.

Cooke, N. J. & Pedersen, H. K. (2009). Human factors of unmanned aerial vehicles. In Wise, J. A., Hopkin, V. D., & Garland, D. J. (eds), *Handbook of Aviation Human Factors* (2nd edn). Boca Raton, FL: CRC Press.

Cooke, N. J., Gesell, L.E., Hartman, J., Pack, W., Pedersen, H. K., & Skinner, M. (2006a). *Human Factors in Unmanned System Training*. Technical Report for NASA sponsored Unmanned Aerial Vehicles Alliance, Research and Curriculum (UAV-ARC) Development Partnership Project.

Cooke, N. J., Pringle, H., Pedersen, H. K., & Connor, O. (eds) (2006b), *Human Factors of Remotely Piloted Vehicles: Advances in Human Performance and Cognitive Engineering Research*, Amsterdam: Elsevier.

Dabbs, J. M., Chang, E-L., Jr., Strong, R. A., & Milun, R. (1998). Spatial ability, navigation strategy, and geographic knowledge among men and women. *Evoution and Human Behavior*, 19, 89–98.

Darken, R., Kempster, K., & Peterson, B. (2001). Effects of streaming video quality of service on spatial comprehension in a reconnaissance task. *Proceedings of I/ITSEC*, Orlando, FL.

Endsley, M. & Kiris, E. (1995). The out-of-the-loop performance problem and level of control in automation. *Human Factors*, 37, 381–94.

Evans, A. W., III, Hoeft, R. M., Jentsch, F., Rehfeld, S. A. & Curtis, M. T. (2006). Exploring human-robot interactions: Emerging methodologies and environments. In N. J. Cooke, H. Pringle, H. K. Pedersen, & O. Connor, (eds), *Human Factors of Remotely Piloted Vehicles: Advances in Human Performance and Cognitive Engineering Research* (pp. 345–58). Amsterdam: Elsevier.

Hottman, S. B. & Sortland, K. (2006). UAV operators, other airspace users, and regulators: Critical components of an uninhabited system. In N. J. Cooke, H. Pringle, H. K. Pedersen, & O. Connor (eds), *Human Factors of Remotely Piloted Vehicles: Advances in Human Performance and Cognitive Engineering Research* (pp. 71–88). Amsterdam: Elsevier.

Hou, M. & Kobierski, R. D. (2006). Operational analysis and performance modeling for the control of multiple uninhabited aerial vehicles from an airborne platform. In N. J. Cooke, H. L. Pringle, H. K. Pedersen, & O. Connor (eds), *Human Factors of Remotely Piloted Vehicles: Advances in Human Performance and Cognitive Engineering Research* (pp. 267–82). Amsterdam: Elsevier.

Howell, W. C. & Cooke, N. J. (1989). Training the human information processor: A look at cognitive models. In I. L. Goldstein and Associates (eds), *Training and Development in Organizations* (pp. 121–82). New York: Jossey Bass.

Jackson, P. (ed.) (2003). *Jane's all the World's Aircraft 2003–2004*. Alexandria, VA: Jane's Information Group.

Killion, T. H. (2006). Technology for future warfighting. 25th Army Science Conference, 27 November. Orlando, FL.

Kurzweil, R. (2006). Warfighting in the early 21st century: A remote-controlled, robotic, robust, size-reduced, virtual reality paradigm. 25th Army Science Conference, 27 November. Orlando, FL.

Lewis, M., Polvichai, J., Sycara, K., & Scerri, P. (2006). Scaling-up human control for large UAV teams. In N. J. Cooke, H. L. Pringle, H. K. Pedersen, & O. Connor (eds), *Human Factors of Remotely Piloted Vehicles.* Volume in *Advances in Human Performance and Cognitive Engineering Research* Series, pp. 237–50. Amsterdam: Elsevier.

Lif, P., Hedström, J., & Svenmarck, P. (2007). Operating multiple semi-autonomous UGVs: Navigation, strategies, and instantaneous performance. In D. Harris (ed.), *Engineering Psychology and Cognitive Ergonomics* (pp. 731–40). Berlin: Springer-Verlag.

Montgomery, J. (2005). EOD robots performing tech wonders in Iraq. *Army News Service*, 10 January, from http://www.globalsecurity.org/military/library/news/2005/01/mil-050110-arnews02.htm.

Moran, T. & Morris, D. (2006). Predators fight the war from Vegas. ABS Nightline Broadcast, 2 May 2006.

Murphy, R. (2004a). Human-robot interaction in rescue robotics. *IEEE Transactions on Systems, Man and Cybernetics, Part C: Applications and Reviews*, 34, 138–53.

Murphy, R. (2004b). Rescue robotics for homeland security. *Communications of the ACM*, 27, 66–9.

NTSB (National Transport Safety Board) (2006). Report on crash of U.S. Border Patrol Predator-B UAV, 25 April. NTSB Identification: CHI06MA121.

Parasuraman, R. & Miller, C. (2006). Delegation interfaces for human supervision of multiple unmanned vehicles: Theory, experiments, and practical applications. In N. J. Cooke, H. L. Pringle, H. K. Pedersen, & O. Connor (eds), *Human Factors of Remotely Piloted Vehicles.* Volume in *Advances in Human Performance and Cognitive Engineering Research* Series, pp. 251–66. Amsterdam: Elsevier.

Parasuraman, R. & Riley, V. (1997). Humans and automation: Use, misuse, disuse, abuse. *Human Factors*, 39, 230–53.

Pazuchanics, S. L., Chadwick, R. A., Sapp, M. V., & Gillan, D. J. (2010). Robots in Space and Time: The role of object, motion and spatial perception in the control and monitoring of UGVs (pp. 83–102, this volume).

Pederson, H. K., Cooke, N. J., Pringle, H. L., & Connor, O. (2006). UAV human factors: operator perspectives. In N. J. Cooke, H. L. Pringle, H. K. Pedersen, & O. Connor (eds), *Human Factors of Remotely Piloted Vehicles.* Volume in *Advances in Human Performance and Cognitive Engineering Research* Series, pp. 21–33. Amsterdam: Elsevier.

Pew, R. W. & Mavor, A. S. (eds) (2007). *Human-System Integration in the System Development Process: A New Look*. Washington, DC: National Academies Press.

Rash, C. E., LeDuc, P. A., & Manning, S. D. (2006). Human factors in U.S. military unmanned aerial vehicle accidents. In N. J. Cooke, H. L. Pringle, H. K. Pedersen, & O. Connor (eds), *Human Factors of Remotely Piloted Vehicles*. Volume in *Advances in Human Performance and Cognitive Engineering Research* Series, pp. 117–31, Amsterdam: Elsevier.

Rehfeld, S. A., Jentsch, F. G., Curtis, M., and Fincannon, T. (2005). Collaborative teamwork with unmanned ground vehicles in military missions. *Proceedings of the 11th Annual Human-Computer Interaction International Conference*. Las Vegas, NV, August.

Riley, J. M. & Endsley, M. R. (2004). The hunt for situation awareness: Human-robot interaction in search and rescue. *Proceedings of the Human Factors and Ergonomics Society 48th Annual Meeting*. Santa Monica, CA: Human Factors and Ergonomics Society, 693–7.

Schmidt, J. & Parker, R. (1995). Development of a UAV mishap factors database. *Proceedings of the 1995 Association for Autonomous Vehicle Systems International Symposium*, 310–15.

Seagle, Jr., J. D. (1997). *Unmanned Aerial Vehicle Mishaps: A Human Factors Approach*. Unpublished master's thesis. Norfolk Embry-Riddle, VA: Aeronautical University.

Self, B. P., Ercoline, W.R., Olson., W. A., & Tvaryanas, A. P. (2006). Spatial disorientation in uninhabited aerial vehicles. In N. J. Cooke, H. L. Pringle, H. K. Pedersen, & O. Connor (eds), *Human Factors of Remotely Piloted Vehicles.* Volume in *Advances in Human Performance and Cognitive Engineering Research* Series, pp. 133–46. Amsterdam: Elsevier.

Sheridan, T. (1987). Supervisory control. In G. Salvendy (ed.), *Handbook of Human Factors*, pp. 1244–68. New York: Wiley.

Sheridan, T. (2002). *Humans and Automation*. Santa Monica and NY: Human Factors and Ergonomics Society & Wiley.

Thomas, L. C., Wickens, C. D., & Merlo, J. (1999). Immersion and battlefield visualization: Frame of reference effects on navigation tasks and cognitive tunneling. Aviation Research Laboratory Institute of Aviation Technical Report ARL-99-3/FED-LAB-99-2.

Tittle, J. S., Roesler, A., & Woods, D. D. (2002). The remote perception problem. *Proceedings of the Human Factors and Ergonomics Society 46th Annual Meeting* (pp. 260–64). Santa Monica, CA: Human Factors and Ergonomics Society.

Tittle, J., Elm, W., & Potter, S. (2005). Functional requirements for effective decision making in human robot teams: Lessons learned from operational settings. *Proceedings of the Human Factors and Ergonomics Society 49th Annual Meeting* (pp. 452–6). Santa Monica, CA: Human Factors and Ergonomics Society.

Tvaryanas, A. P., Thompson, B. T., & Constable, S. H. (2005). U.S. Military UAV mishaps: Assessment of the role of human factors using HFACS. Paper presented at the CERI Second Annual Human Factors of UAVs Workshop, 25–26 May. Mesa, AZ.

Voshell, M. & Woods, D. (2005). Overcoming the keyhole in human-robot coordination: Simulation and evaluation. In *Proceedings of the Human Factors and Ergonomics Society 49th Annual Meeting* (pp. 442–446). Santa Monica, CA: Human Factors and Ergonomics Society.

Wickens, C. D., Dixon, S. R., & Ambinder, M. S. (2006). Workload and automation reliability in unmanned air vehicles. In N. J. Cooke, H. L. Pringle, H. K. Pedersen, & O. Connor (eds), *Human Factors of Remotely Piloted Vehicles.* Volume in *Advances in Human Performance and Cognitive Engineering Research* Series, pp. 209–22. Amsterdam: Elsevier.

Williams, K. W. (2004*). A Summary of Unmanned Aircraft Accident/incident Data: Human Factors Implications.* Technical Report DOT/FAA/AM-04/24. U.S. Department of Transportation, Federal Aviation Administration, Office of Aerospace Medicine, Washington, DC.

Witus, G., Smuda, W. J., & Gerhart, G. (2006). Experiments in augmented teleoperation for soldier-robot checkpoint inspection systems. 25th Army Science Conference Poster Session, 27–30 November. Orlando, FL.

Zeng, J. J., Yang, R. Q., Zhang, W, J., Weng, X. H., & Qian, J. (2007). Research on semi-automatic fetching for an EOD robot. *International Journal of Advanced Robotic Systems*, 4, 247–52.

Chapter 20

On Maximizing Fan-Out: Towards Controlling Multiple Unmanned Vehicles

Michael A. Goodrich

Computer Science Department, Brigham Young University

Understanding the models and metrics that govern the number of unmanned vehicles (UVs) that a single human can manage will promote advances in both autonomy and interface design. This chapter reviews and extends literature on such models and metrics, organized around the conceptual notion of *fan-out*, defined as the number of UVs that a single human can manage. From the literature, there are two different types of fan-out models, which we refer to as *pigeon-hole* and *scheduling* models. The two types of model differ in who waits for whom; can the UV wait for the human to give it attention or must the human always be available to service the UV? Developing these models allows us to identify which technologies are most likely to help maximize mission effectiveness by maximizing fan-out. The most important technologies center on the notion of *delegation*, which influences autonomy, interface design, and team organization.

Introduction

Accomplishing a dangerous and complicated mission, while minimizing risks and costs, is a goal in many military situations. One potential way to accomplish a goal with minimum risk is to employ one or more unmanned vehicles (UVs) to provide information and/or munitions deployment without exposing a human to harm. For example, many Intelligence, Search, and Reconnaissance (ISR) missions can benefit from using unmanned aerial vehicles (UAVs) and unmanned ground vehicles (UGVs). Similarly, Suppression of Enemy Air Defenses (SEAD) missions can benefit from using one or more UVs.

Unfortunately, managing even a single UV can require many humans (Cummings et al., 2007; Murphy et al., 2006). Thus, although UVs can reduce risk, they can potentially increase the cost of a mission both by requiring significant manpower during the mission and by requiring considerable pre-mission training (Hunn, 2006). It is desirable to develop UV technology that allows a single human (or small group of humans) to manage multiple vehicles. Thus, a goal of developing UV technologies is to *create an organization of humans, software agents, and UVs such that mission effectiveness is maximized at a minimum cost.*

Although there are many factors that can significantly constrain the efficiency of such an organization (such as the length of the chain of command), a key factor is *span of control*. In this context, the term "span of control" refers to the maximum number of agents that a single human can effectively manage (Urwick, 1956). In the literature on human-robot interaction (HRI), span of control has been associated with a metric known as *fan-out* (Olsen & Wood, 2004; Crandall et al., 2005). Roughly speaking, fan-out is the maximum number of UVs that can be managed by a single human at some acceptable level of performance.

Maximizing fan-out has often been treated as a goal of designing autonomy and interfaces. Although, high fan-out appears to be correlated with mission effectiveness (Olsen et al., 2004), we must be cautious about naively focusing on maximizing this metric for at least two reasons. First, some have questioned the myopic focus on maximizing the number of vehicles that a single human can manage (Hancock et al., 2007); it may be better to have multiple people managing multiple assets with a high average ratio of UVs to people (Lewis et al., 2006). Second, high fan-out can be achieved artificially by, for example, having UVs make glacially slow progress toward a goal so that their neglect time is high enough to allow interactions with a large number of UVs. Clearly, maximizing fan-out is only desirable if it is associated with some notion of mission effectiveness.

Unfortunately, maximizing fan-out is not easy. As identified by Cummings et al. (2007), mission performance requirements push toward high fan-out while human factors considerations push toward low fan-out. Simply put, it is easy to overwhelm the cognitive resources of a single human if he or she is required to manage too many UVs.

The objective of this chapter is to develop a theoretical and behavioral framework that describes the key factors that affect fan-out. We will argue that maximizing fan-out requires delegation. We identify different types of delegation that are being explored in the literature, and discuss key research questions that arise as a result of different types of delegation. Throughout the remainder of this chapter, we use the term *delegate agent* or simply *delegate* to indicate the human, UV, or software agent to whom a commander delegates responsibility. We refer to the set of situations within the delegate's capabilities and within the tactical scope as the delegate's *region of support* (ROS).[1]

---

1   The use of the terms *strategic*, *tactical*, and *operational* is meant to invoke notions of level of detail and planning horizon. These or similar terms have been used both in the human factors literature, such as Rasmussen's knowledge-based, rule-based, and skill-based levels of behavior (Rasmussen, 1983), and in the intelligent control literature (see, for example, Saridis, 1989; Albus, 1991). Strategic elements of the response address longer time frames than tactical and operational elements.

Elements of Fan-Out

This section introduces two of the key factors that affect fan-out, namely *neglect time* and *interaction time* (Crandall et al., 2005; Olsen & Goodrich, 2003).

Neglect Time

Eventually, every interesting dynamic situation will evolve to something outside a delegate's region of support unless either the delegate is completely autonomous (which means that the commander is not necessary) or the delegate is capable of stopping and waiting for the commander when necessary. Barring these conditions, it is necessary for the commander to interact occasionally with the delegate either (a) to perform operational control; (b) to instruct or inform the delegate so that it "re-centers" in its region of support; or (c) to adapt tactical scope by changing instructions or tactical parameters. The amount of time that the commander can spend attending to something other than the delegate is denoted the *neglect time* of that delegate. The purpose of delegation is to create sufficient neglect time to permit the commander to manage other operational tasks within some mission context. The number of other tasks that can be performed depends on the neglect time; small neglect time, for example, means that there is little time to attend to other things.

There are a number of factors that determine the duration of the neglect interval, including the following:

a. Situation Complexity. Tactics will need to shift more often in complex situations, and the operational skill of the delegate will more often reach the limits of his/her expertise.
b. Delegate Capability. An unskilled delegate will not only have a smaller region of support, but also may be less trustworthy and reliable.
c. Operational Tempo. If the operational tempo is high, then the situation evolves more quickly.
d. Communications Bandwidth. Bandwidth constrains not only a commander's situation awareness, but also his or her ability to give high-fidelity instructions to a delegate. Thus, the quality of information exchanged between them may not support periods of trusted neglect.
e. Undo. If the commander can "undo" actions, then it is possible temporarily to tolerate a delegate operating outside of its region of support.
f. Performance Standards. Low performance standards imply a larger region of support.

Interaction Time

Precisely how many "other things" can be done during the neglect interval clearly depends on how long it takes to do those other things. Thus the second significant factor that determines fan-out is the amount of time required to interact with the delegate. This period of time is denoted the *interaction time* of the delegate.

There are a number of factors that determine interaction time, including the following:

a. Shared Situation Awareness. A mutually aware commander and delegate have more efficient interactions. For example, a commander familiar with operational practices and a delegate familiar with tactical considerations can potentially have low interaction times.

b. Situation Complexity. A complex situation is typically harder to respond to than a simple one, especially if the situation demands a shift in tactics or strategies.

c. Time Elapsed. If the time since the last interaction is very short then switching contexts may be easy, but a lengthy elapsed time might force the commander to spend considerable time reacquiring situation awareness.

d. Intervening Task Complexity. A complex task performed during a neglect interval can consume short term memory and impose executive control overhead (Baddeley, 1986; Goodrich, 2004) that makes reacquiring situation awareness difficult. Moreover, similarity with an intervening task can cause confusion.

Modeling Fan-out

Given the concepts of neglect time and interaction time as a foundation, we are now in a position to evaluate models of fan-out. Results from papers in these areas make predictions and observations that are relevant to a commander's span of control. Although terminology varies across authors, we will use the term *fan-out* to mean the maximum number of delegates that a commander can effectively control.

The remainder of this section reviews models of fan-out. Although we touch briefly on heuristic models, the majority of this section describes two model types: pigeon-hole and scheduling models. The primary difference between these two model types is whether the delegate can wait for the human or must receive immediate attention when situations evolve outside the region of support.

A Heuristic Model

An early discussion of span of control was published by Urwick (1956). He argues that an executive can effectively manage no more than six[2] people over whom

2   Urwick's limit of six may recall Miller's classic article on the limits of working memory (Miller, 1956). Although working memory is certainly a limiting factor on how many delegates a single human can manage, the temporal scope and task complexity

he or she has direct management responsibilities. Although this number was obtained from qualitative observations, Urwick cites its existence in a number of organizations. He argues that supposed exceptions to the fan-out limit of six are, in practice, exceptions in name only. Urwick notes that effective organizations that seem to violate this heuristic limit may allow information to come to a manager from more than six people, but the manager only delegates directly to no more than six of these people.

One relevant reason for Urwick's heuristic limit is that the management process often requires the manager to gather and interpret information about ambiguous situations from multiple interdependent sources. Each source provides information that varies in accuracy and reliability. Thus, working with subordinates to construct accurate situation awareness takes considerable mental effort and time, which limits the number of subordinates that the commander can oversee. Additionally, developing a response for the organization and coordinating efforts of subordinates can be very complex and time consuming.

Cummings cites two important exceptions to Urwick's rule: the maximum number of airplanes that a single air traffic controller can manage and the maximum number of tactical Tomahawk missiles that a single operator can manage (Cummings & Guerlain, 2003). These exceptions are important for two reasons. First, they indicate that it is possible for fan-out to exceed Urwick's rule but, second, only if there is an appropriately high level of trustworthy autonomy. In the case of air traffic control, two facts make it possible to exceed Urwick's limit: (a) each airplane has a highly capable pilot on board and (b) the controller's primary task is to sequence and deconflict aircraft; the resulting interactions between controller and pilot are at a high level of interaction. Managing more than six tactical Tomahawks is possible because the Tomahawks can loiter, unsupervised, for a considerable period of time prior to being commanded to strike a target. This allows the commander sufficient flexibility in planning when and how each missile will receive his or her attention.

Although Urwick's heuristic limit is useful, the presence of exceptions to the rule suggests that it would be useful to have more systematic models of fan-out that lend themselves to more quantitative predictions. This leads us to the next section.

Pigeon-Hole Models

The most simple systematic model for fan-out can be referred to as a "pigeon-hole" model. In this model, each period of neglect represents a hole into which the commander fits as many interactions with other delegate agents as possible. A pigeon-hold model assumes that it is unacceptable for a delegate to wait for the

---

associated with managing multiple delegates causes other cognitive factors (such as information processing limitations, limits on planning horizons, etc.) to contribute to Urwick's limit.

commander's interaction, but that it is desirable for the commander to manage as many delegates as possible. This assumption may apply when, for example, the UV would be destroyed, would violate rules of engagement, or would produce unacceptable levels of collateral damage if the commander does not interact in a timely manner.

The most simple pigeon-hole model is Olsen's original fan-out model (Olsen & Wood, 2004; Crandall et al., 2005):

$$FO = 1+NT/IT$$

This equation means that the number of delegates that can be managed by a single commander (fan-out, denoted by FO) is determined by finding the number of interaction times (IT) that fit into a single neglect time (NT). The ratio NT/IT gives this number, and indicates how many other delegates can be managed while one delegate is being neglected. The fan-out is this ratio plus one, since the ratio does not count the delegate currently being serviced. Since this fan-out model assumes that the NT and IT values are equal and constant, we call this model *homogenous*.

The homogeneous pigeon-hole model is very useful as a conceptual tool for determining the tradeoffs between autonomy and interfaces (Crandall et al., 2005). However, it makes two very limiting assumptions. First, it assumes that there is no variability in the system: all delegates are homogeneous; neglect and interaction times do not vary over time; and neglect and interaction times do not vary across delegates or situations. Second, it assumes that there is no management overhead.

It is useful to emphasize that the defining attribute of pigeon-hole models is that no delegate ever has to wait to receive attention from the commander. This attribute means that the commander can wait for the delegates, but not vice versa.

Pigeon-hole models with variable NT and IT   When delegates are homogeneous, the average value for both NT and IT is constant across all delegates. Suppose that both NT and IT vary by some value in the enumerable[3] sets Δ^{NT} and Δ^{IT}. Variability in both NT and IT can occur when world complexity and operational tempo change, which will almost always happen in realistic missions.

In pigeon-hole models, fan-out must consider the worst case so that no delegate ever has to wait for the commander's attention. The worst case occurs when neglect time is smallest and interaction time is largest. Thus, under variability fan-out becomes:

---

3   In theory, since NT and IT are real-valued, the set of possible variabilities should be drawn from real-valued intervals. Such intervals are not enumerable, which implies that the maximum and minimum operators should be replaced by the supremum and infinum operators. This technical issue largely disappears if we assume that the intervals are closed, though it is still technically incorrect to enumerate the variabilities.

$$FO = 1 + \frac{\min_{\delta \in \Delta NT}(NT + \delta)}{\max_{\delta' \in \Delta IT}(IT + \delta')}$$

The effect of variability on fan-out has important implications if the restrictions of the pigeon-hole model are applied. Specifically, the commander may spend a substantial portion of his or her time waiting for delegates if the actual variability differs dramatically from the worst-case variability. To help illustrate, consider the following example.

Example
Suppose that the nominal values for *NT* and *IT* are 12 and 3 units of time, respectively. Then nominal fan-out is 5=1 + 12/3. Suppose that *NT* can vary up to plus or minus 3 units of time, and that *IT* can vary up to plus or minus 2 units of time. Then worst case fan-out is given by

$$
\begin{aligned}
FO &= 1 + \frac{\min_{\delta \in \Delta NT}(NT + \delta)}{\max_{\delta' \in \Delta IT}(IT + \delta')} \\
&= 1 + \frac{12 - 3}{3 + 2} \\
&= 2.8
\end{aligned}
$$

This implies that the worst-case fan-out is two delegates rather than the nominal five delegates, meaning that during an interval of neglect the operator interacts with only one other delegate.

This can be very costly to the operator, especially since it is possible that the actual values of NT and IT are such that best-case fan-out occurs. To illustrate, suppose that, by chance, variabilities work out in favor of allowing the maximum number of delegates:

$$
\begin{aligned}
FO_{bestcase} &= 1 + \frac{\max_{\delta \in \Delta NT}(NT + \delta)}{\min_{\delta' \in \Delta IT}(IT + \delta')} \\
&= 1 + \frac{12 + 3}{3 - 2} \\
&= 16
\end{aligned}
$$

Thus, there is an eightfold difference in fan-out between best case and worst case. If the system is designed for worst-case fan-out and best-case *NT* and *IT* obtain, then the commander will spend only 1 unit of time interacting with one delegate while he or she neglects the other delegate for up to 15 units of time. This means that 14 out of every 15 units of time are wasted.

As this example illustrates, the potential cost can be high if (a) the commander cannot afford to let a delegate leave its region of support or (b) the delegate cannot detect and then wait for interaction if it is about to leave its region of support. Importantly, the impact of variability grows larger as more agents are considered.

In other words, fan-out is most negatively impacted by variability when a lot of delegates are possible, such as when interaction is efficient, thereby producing low interaction times. This large negative impact of variability on the pigeon-hole model leads us to explore scheduling models. Prior to doing so, it is useful to discuss the impact of heterogeneity on pigeon-hole models.

Pigeon-hole models with heterogenous delegates   Beyond the effects caused by variations from nominal *NT* and *IT* values, it is useful to evaluate fan-out when there is variability in *NT* and *IT* *between delegates*. To this end, let NT_i and IT_i denote the neglect and interaction times, respectively, for delegate *i*. We refer to this as the case of *heterogeneous delegates*. When delegates are heterogeneous, the make-up of the set of delegates that will be serviced during a neglect interval has a strong effect on fan-out. If we restrict attention to a set of delegates that require little interaction then fan-out will be high, but if we restrict attention to a set of delegates that require a lot of interaction then fan-out will be low. This complicates the definition of fan-out for heterogeneous delegates.

Suppose that we are given a finite population of possible delegates, denoted by $A=\{a_j\}$. We can determine whether any subset of these delegates, $S \subseteq A$, is feasible. The first feasibility condition is that it must be possible to service all other delegates during a neglect interval (Goodrich et al., 2005). This condition can be written as:

$$\min_{\delta \in \Delta_i^{NT}} (NT_i + \delta) \geq \sum_{j \neq i} (\max_{\delta' \in \Delta_j^{IT}} (IT_j + \delta'))$$

where Δ_i^{NT} and Δ_j^{IT} denote the enumerated variability in neglect time experienced by delegate *i* and in interaction time experienced by delegate *j*, respectively.

The second feasibility condition is that the delegates must be capable of accomplishing the mission. Assuming that we are free to choose the subset, the fan-out of a team of heterogeneous delegates is the cardinality of the "best" feasible set, where "best" can be defined by any external performance criterion, including that which maximizes fan-out. Thus:

$$FO = |arg\ max_{S\ is\ feasible}\ U(S)|$$

where *U(S)* denotes the utility of some subset of delegates, and where |*arg max U()*| denotes the cardinality of the set that maximizes *U*.

There are a number of different possible ways to construct *U(S)* that match different tactical or strategic goals; we consider three examples. First, in some ISR missions, the goal is to cover as much ground as possible at the highest level of detail. A standard mathematical way of representing such a goal (Russell & Norvig, 1995) is to assume that *U(S)* is a weighted sum of each delegate in the set *S*, *U(S)* $= \Sigma_{a \in S} w_a U(a)$, where w_a is the importance of delegate *a* and *U(a)* is the individual contribution to the mission from delegate *a*. The summation grows quickly as any delegate succeeds, which implies that as more delegates make contributions the perceived usefulness of the team grows as more delegates are added. Second, in some SEAD missions, the quality of the performance of the team is limited by the contribution of the worst-performing delegate. A mathematical way of representing such a goal is to assume that $U(S)$ is the weighted product of each delegate in the set S, $U(S) = \Pi_{a \in S}\, w_a U(a)$. The product only grows if all delegates have high utility; low utility makes the entire product small. Third, if the goal is simply to maximize the number of delegates involved, then $U(S) = |S|$ where the cardinality of the set S is denoted by $|S|$.

Although it is easy to show that variability again becomes a severe problem for heterogeneous models, we wish to focus on a different lesson from pigeon-hole models of heterogeneous agents. Specifically, heterogeneity implies the need for an explicit performance metric. The presence of an explicit performance metric is an important aspect of the class of models discussed in the following section.

Scheduling Models

Pigeon-hole models assume that performance is unacceptable if a delegate ever has to wait for the commander. By contrast, allowing UVs to wait in a "degraded state" (Cummings et al., 2007) for the commander can dramatically increase fan-out by allowing the commander's attention to be appropriately scheduled. We refer to such problems as *scheduling problems* and note that researchers are (rediscovering) the usefulness of scheduling algorithms in problems where a single human must manage multiple unmanned vehicles (Lewis et al., 2006; Mau & Dolan, 2006; Cummings & Guerlain, 2007).

To obtain an estimate of span-of-control under scheduling theory, it is necessary to distinguish between neglected delegates that are functioning well and those that are waiting in a degraded state. Fan-out is a function of the number of neglected delegates that are performing well; we cannot include degraded delegates waiting for attention in the fan-out measure because this can be arbitrarily high if some non-productive delegates can be neglected for unrealistically and arbitrarily long periods of time.

Using scheduling algorithms has the potential to improve performance because (a) such models explicitly use a performance criterion that should be maximized (or, equivalently, a cost to be minimized) while (b) they maximize fan-out by maximizing human utilization even if an individual UV may have to wait for a period of time. Even with the ability for delegates to wait for the commander, we will see that it is still sometimes desirable to have the commander operate at less than peak workload all the time.

Simple scheduling rules: queuing theory Queuing theory is a subset of scheduling theory that deals with how one or more servers can best respond to a series of tasks that arrive at variable times (called the arrival rate) and take variable amounts of times to complete (called the service time) (Stallings, 2000). Queuing theory (a) assumes that the jobs to be scheduled can wait for some period of time before receiving attention; (b) models the arrival rate and service time of jobs; (c) invokes a performance criterion; and (d) identifies a scheduling rule that (approximately) maximizes this performance criterion. It is useful to describe queuing theory models using the terms of fan-out. The probabilistic model of arrival times is encoded in a model of neglect times; jobs in the queue are analogous to delegates that are able to function for some period of time in a degraded state; and the model of service times is encoded as the model of interaction times.

Cummings et al. have recently applied queuing theory to the problem of a single commander managing multiple UVs. In their paper, they contrasted a queuing theory model with pigeon-hole models (Cummings & Guerlain 2007). They conclude that, when UVs are allowed to "wait in a degraded state," queuing theory models are better predictors than pigeon-hole models of a human's capability in managing multiple UVs.

A general rule of thumb, obtained from specific probabilistic models (e.g., Poisson arrival rate and exponential service time) and scheduling rules (e.g., first-in, first-out), is that the commander can work at no higher than a 60–70 per cent utilization (Stallings, 2000). Simply put, the variability in neglect times means that more than one delegate will sometimes need interaction from the commander at approximately the same time. To keep these delegates from waiting too long for interaction under peak conditions, the commander has to operate at no more than 60–70 per cent of the best-case workload.

Making it possible for delegates to wait in a queue produces both higher average and higher worst-case fan-out. Although 60–70 percent seems like poor use of the commander's time, it is significantly better than planning for worst-case performance without allowing waits; in the pigeon-hole example from the previous section, scheduling for worst case meant that the commander could be utilized in as little as one in 15 seconds, or around 7 percent utilization.

Consistent with observations from queuing theory, Lewis et al. (2006) observe that adding people who can support the commander is often effective for dealing with queuing bottlenecks (see also Chapter 21 in this book). Lewis et al.'s implementation of this observation uses the metaphor of a "call room" where multiple people wait to help whatever delegate needs attention next. Each person in the call center is capable of helping a delegate quickly resume operations when it needs attention. Although this simple metaphor has great potential for increasing fan-out, one important factor limits its application in managing multiple UVs. Specifically, each person in the call room may be required to coordinate tactics subject to a strategic response. Coordinating with other commanders can begin to consume significant portions of available neglect time or cause large downtimes (Shattuck & Woods, 2000).

Sophisticated scheduling rules   Under some circumstances, it is possible to do better than the 60–70 per cent queuing theory limit by applying specific scheduling rules. In this section, we review a few scheduling algorithms, identify the performance criterion that they try to maximize, and explain the assumptions under which they achieve optimal results. In scheduling theory, a key performance measure relates to how much time the delegate spends in an unproductive state. Such a state is called "downtime" in scheduling theory. In the context of this chapter, downtime is the sum of (a) time that the delegate waits for attention from the commander and (b) the interaction time: $DT = WT + IT$. Scheduling algorithms are typically designed to either minimize the *average downtime* across all delegates or to minimize the *worst-case downtime* experienced by a single delegate.

In terms of mission performance, minimizing average downtime assumes that the mission depends on having as many delegates as possible operating at a time. Thus, when "more active delegates" means "higher mission effectiveness" then algorithms that minimize average downtime should be chosen. By contrast, minimizing worst-case downtime assumes that the mission depends on having each delegate functioning. Thus, if mission effectiveness requires each delegate to accomplish a task rather than having the maximum number of active delegates, then algorithms that emphasize minimum worst-case downtime should be used.

The precise algorithm for optimizing the average or worst-case downtime depends on a number of problem characteristics. First, the problem is much easier to solve and the optimal schedule is much easier to implement if we assume that all delegates arrive simultaneously rather than sequentially in some stochastic manner. Second, the problem is much easier to solve if there is no due date that determines when a task must be completed.

Scheduling without due dates   The two most famous algorithms for minimizing average downtime and maximum downtime are, respectively, the shortest processing time (SPT) algorithm and the first-in-first-out (FIFO) algorithm. The SPT algorithm "schedules tasks by non-decreasing processing times" (Mau & Dolan, 2006). Under the assumption that all tasks arrive simultaneously, SPT minimizes the mean waiting time and the maximum waiting time. The FIFO algorithm assumes stochastic arrival times and schedules tasks "in the order they arrive" (Mau & Dolan, 2006). It minimizes the worst-case time that the task waits in the queue plus the time spent in interaction. Unfortunately, some short tasks can be stuck in the queue waiting for longer tasks to complete, which means that the algorithm can produce poor average performance.

Interestingly, the longest processing time (LPT) algorithm has been observed in experiments with humans. LPT schedules tasks by non-increasing processing times. Although this algorithm is not considered efficient and can maximize the worst-case downtime if all tasks arrive simultaneously, LPT can give a psychological satisfaction that may lead some people to erroneously adopt it for scheduling tasks (Moray et al., 1991).

Scheduling with due dates   In a survey paper, Dessouky et al. (1995) review the scheduling literature, identify a classification scheme for scheduling algorithms, and show how the terminology used in the scheduling literature corresponds to the terminology used in human factors. The survey makes it clear that to use a scheduling algorithm in the presence of due dates, it is essential that a performance criterion be identified.

The use of a due date/time as an element of a human scheduling problem was introduced by Tulga and Sheridan (1980) and formed the basis for an experiment by Moray et al. to determine when humans can use optimal scheduling algorithms in supervisory control (Moray et al., 1991). Perhaps the most important observation from Moray et al.'s work is that the algorithm for optimally scheduling tasks in the presence of due dates may be so complex that a human supervisor has difficulty implementing it.

Mau and Dolan noted that static scheduling algorithms are inappropriate when tasks arrive in the service queue at variable times (2006). They propose a non-preemptive algorithm that reschedules tasks as they arrive in the queue; the algorithm is run whenever the commander completes the interaction with one delegate and is available to interact with the next. They present experimental evidence that indicates that the algorithm produces lower average downtime than either the SPT or FIFO algorithm, and conclude that fan-out is increased using their algorithm. Mau and Dolan assume "[the commander] will take the advice of the [scheduled] supervisory control list to provide [him or her] with an efficient order in which to attend to the robot tasks" (2006). Unfortunately, this means that the commander loses the ability to apply his or her expertise to maximizing mission effectiveness by selecting which delegate to service next. Additionally, as with any scheduling algorithm, the amount of interaction time must be known prior to creating a schedule.

Toward Maximizing Fan-Out

The ability to neglect a delegate for a sufficiently long period of time makes it possible for a commander to manage multiple delegates. This means that the commander assumes management responsibilities such as sharing attention between delegates, sequencing periods of interaction, and efficiently transitioning between interactions with delegates. In the words of Drucker, this presents significant challenges: "To manage a business in the age of 'automation' is simply not possible on the basis of 'intuition,' 'hunch,' or even of experience. It requires systematic knowledge. Above all, it requires of [commanders] a real ability for decision-making as a logical and strict process following strict laws and requiring a great deal of systematic study and analysis" (Drucker, 1955, p. 159).

Although "logical and strict," the simple pigeon-hole approach of counting the number of interactions that fit into a neglect interval is far too sensitive to variability. Assuming that the UV can wait for interaction, scheduling models are much more resilient, but introduce a set of new challenges identified in the previous pages. Specifically, managing multiple delegates imposes some overhead as the commander must monitor several delegates, identify problems, choose which delegate needs attention most, etc. Management becomes a meta-task that must be serviced just as regularly and effectively as the task of interacting with multiple delegates. This meta-task can be very demanding (Drucker, 1955; Moray et al., 1991) and performing it will consume a portion of the available neglect interval. Table 20.1 summarizes information relevant to the management meta-task for SPT, FIFO, LPT, Mau and Dolan's algorithm (2006), and algorithms in Dessouky, et al. (1995) and Sheridan (1992).

Compounding this problem, workload and time pressure can further impede the commander's performance. This can be understood by noting that even a small misstep in interacting with a delegate agent (or even selecting which delegate agent requires interaction) can consume enough of the available neglect time to throw off a schedule. Furthermore, stress can contribute to the frequency and severity of missteps made by the commander and thus make more missteps likely as time pressure builds. For example, Moray et al. showed that the psychological effects of having to perform scheduling under time pressures cause the performance of a human scheduler to degrade (1991).

There is evidence in human factors studies that the management meta-task can be a major influence on fan-out. For example, Cummings et al. (2007) identify an influence on fan-out that they describe as the amount of time that a delegate must wait because the commander is unaware that it needs assistance. This type of "wait time due to situation awareness" indicates that the commander does not have sufficient situation awareness of the management meta-task to appropriately respond to a delegate in a timely manner. As another example, Moray et al. (1991) identify situations where the time pressure on a commander is intense enough that he or she can no longer follow an optimal scheduling rule. The commanders in Moray's study adapted by using management decision rules that were suboptimal (though highly effective in some situations). Crandall and Cummings (2007) collect sources of management overhead into a metric class that they call "attention

Table 20.1   Types of scheduling algorithms, key assumptions, and performance criteria

Schedule algorithm	Arrival type	Performance criteria	Due date	Complex?
SPT (Dessouky et al., 1995)	Simultaneous	Average DT	No	No
	Simultaneous	Worst-case DT		
FIFO (Mau & Dolan, 2006; Sheridan, 1992)	Stochastic	Average DT	Yes	Yes
	Stochastic	Worst-case DT		
LPT	Stochastic			
		Worst-case DT	No	No
		Average DT	No	Yes
		Psychological benefit?	N/A	No

allocation efficiency." Similarly, Wang et al. (2008; see also Chapter 21 in this volume) define a "coordination demand" metric that depends on how much time two interdependent tasks overlap in the way that they are controlled.

The purpose of this section is to identify technologies that can be used to maximize fan-out by providing support for the management meta-task.

Adjustable Autonomy

Adjustable autonomy means that the commander has the ability to alter the region of support. Historically, adjustable autonomy switches between two extremes: supervisory control and teleoperation (Sheridan, 1992). More recently, adjustable autonomy has replaced these two extremes with more *levels of autonomy* between the two extremes (Sheridan & Verplank, 1978; Kaber & Endsley, 2004; Parasuraman et al., 2000).

One form of adjustable autonomy is to decide who performs which tasks, such as delegating UAV control to a computer algorithm but having the human being responsible for interpreting video. Another form of adjustable autonomy is to allow the commander to specify how much authority a delegate has over particular functions. For example, in so-called "shared control" a human may teleoperate a UGV through a joystick, but the UGV can either fine-tune these commands based on real-time feedback from sensors (Fong et al., 1999; Crandall & Goodrich, 2002) or veto some commands altogether, and thus "safeguard" a path to prevent collisions (Fong et al., 2001). A final form of adjustable autonomy is to allow the commander to alter specific control parameters. An easily understood example of such adjustment is to allow the commander to specify "no-fly zones" for a UAV or to restrict the set of GPS points that a UGV can enter. Other ways of adjusting autonomy include interactive learning, learning from demonstration (Saunders et al., 2007), and programming by reward (see, for example, Nicolescu & Matarić, 2001).

Adjustable autonomy can help a commander manage multiple UVs by allowing him or her to decide how to allocate attention to various UVs. Adjusting autonomy changes the neglect and interaction times of delegates and thus allows a commander flexibility in scheduling interactions. As noted previously, giving the commander the opportunity to adjust schedules can impose a significant management burden for all but very simple schedules.

Adaptive Autonomy

Since requiring the human to explicitly adjust autonomy adds to the management meta-task, managing autonomy can be a limitation on maximizing fan-out. Adaptive autonomy seeks to address these limitations by granting a UV some authority to adjust its own autonomy. Like adjustable autonomy, adaptive autonomy allows for changes in the delegate's region of support. Unlike adjustable autonomy, in which only the commander has authority to change the region of support, the defining feature of adaptive autonomy is that the delegate can change its own region of support.

An important question to ask about adaptive autonomy is "If adaptive autonomy allows a delegate to shift its region of support, how is adaptive autonomy different than just giving the delegate a larger region of support?" There are two answers to this question. The first answer is that there is a qualitative shift in the behavior of the delegate between the region of support before the change and after. This qualitative difference means that the neglect and interaction properties change. For example, the most simple form of adaptive autonomy is when a UV simply stops moving because it is no longer confident that it can succeed. A stopped UV can perhaps be neglected for a significant period of time, but interacting with this UV may take a considerable amount of time to get it "back on task." Since types of autonomy are often organized into discrete levels, adaptive autonomy typically involves a discrete switch in the region of support and, consequently, in both neglect and interaction times.

The second answer to the question above is that the change in region of support may be triggered not by something intrinsic in the environment, but rather by something that has changed with the commander. The automation trigger is the event or state that causes the delegate to adapt (Hancock, 2007). If the commander is occupied with a time-consuming task with a large queue of waiting processes, a delegate may change to a less effective but more neglect-tolerant form of autonomy to allow the commander to better cope with his or her workload; in the example, the automation trigger is the workload of the commander.

Adaptive automation has the potential to improve fan-out not only by changing neglect and interaction times, but also by decreasing the amount of time spent by the commander servicing the management meta-task. Unfortunately, although performance can improve, there is at least some evidence that when a computer system "controls the option to automate" there can be an increase in "subjective perception of ... demand" and hence operator workload (Hancock, 2007). Additionally, although research has started to explore appropriate automation triggers such as using physiological cues (Di Nocera et al., 2003; Rani et al., 2003), there is not enough known about appropriate triggers to make adaptive autonomy reliable and trustworthy over a wide range of scenarios. Finally, adaptive autonomy risks leaving the operator "out of the loop," meaning that there is a risk of not allowing the human to contribute in the way that most benefits the system. This means that interaction times can grow dramatically when an operator tries to resume control of a task that has adapted its region of support.

Mixed Initiative Interaction

One approach to avoiding leaving the operator out of the loop but still benefiting from adaptive automation is to design a system that includes both adaptive and adjustable autonomy. Although the phrase *mixed initiative interaction* is used in many ways (Fong et al., 2003; Hearst, 1999; Kortencamp et al., 1997), we use this phrase to mean a mixture of adaptive and adjustable autonomy. The word "initiative" in this context indicates the authority to change the region of support for a delegate. Thus, "mixed initiative" means that the commander and the delegate each have authority to initiate a change in the level or focus of autonomy.

Naturally, mixed initiative interaction has some of the same challenges as adaptive and adjustable interaction, including potential mode confusion and too little understanding of automation triggers. However, there is some evidence that mixed initiative interaction can improve mission effectiveness (Kaber & Endsley, 2004). In prior work, we showed that a mixed initiative interaction between a human and a team of interdependent UVs had performance advantages over purely adjustable or purely adaptive automation, largely because it was easier for the human to use more vehicles in accomplishing the mission (Goodrich et al., 2007).

Delegation to Teams

Miller et al. (2002) observed that almost every task or mission that must be performed actually consists of a combination of subtasks. Indeed, human factors analysis techniques like goal-directed task analyses (Endsley et al., 2003) and cognitive work analyses (Vicenti, 1999) are specifically designed to uncover the structure of and information flow between these subtasks. This observation leads to *delegation interfaces* where a "play is called" that specifies a level of autonomy for each subtask rather than specifying a monolithic level of autonomy for the entire task (Miller et al., 2002). Typical delegation interfaces use parameterized plays that are associated with a desirable qualitative behavior, such as "patrolling a perimeter" or "rendezvousing at a point."

Because parameterized plays naturally extend to coordinating multiple UVs, most of the reported plays are designed for coordinating delegates. As this applies to fan-out, the ability to rely on interdependencies between delegates to produce robustness decreases the number of one-to-one interactions between a commander and a group of delegates (Goodrich et al., 2007). Consequently, delegation interfaces and plays can increase overall fan-out. However, managing plays and clusters of delegates does not completely solve the fan-out problem. Instead, it shifts the responsibility from one of designing efficient one-to-one interactions to one of designing efficient one-to-many interactions. This invokes Drucker's caution about the need to make sure that commanders are trained to systematically manage subordinate delegates (1954), and also invokes command and control issues such as the length of the chain of command, effectiveness of communicating commander intent, control technologies to support command, and amount of interdependence of delegates (Wang et al., 2008).

Temporal Displays and Scheduling Support

Scheduling models indicate that the order in which delegates receive attention is important. One way to increase fan-out is to delegate all or a portion of the scheduling responsibility to an artificial agent. This *scheduling delegate* would be responsible for assisting the commander in scheduling interactions so that a given performance standard is met. Like any of the UVs, the scheduling agent could have varying levels of autonomy. For example, near one extreme the scheduling agent could make suggestions as to which other delegate needs attention next, and at the other extreme the scheduling agent could select a particular delegate and insist that the commander interact with that delegate. Scheduling agents are important because, as noted in Moray et al. (1991), some optimal scheduling algorithms are too complex for a human to manage in a reasonable interaction time.

In Tulga and Sheridan's seminal paper on human scheduling (1980), they introduce a visual task that graphically represents the scheduling problem. Cummings has studied the benefits of using a graphical representation of the scheduling problem using so-called "temporal displays" (Cummings & Guerlain, 2003). Interestingly, the displays developed by Cummings are qualitatively similar to the task introduced by Tulga and Sheridan. The results by Cummings suggest that temporal displays can help a commander better manage the scheduling meta-task, especially when supported by scheduling agents that operate at various levels of autonomy. Although the results are promising, they also suggest that some high levels of autonomy can introduce new problems, such as overreliance on artificial scheduling support and the need to properly balance how automation is invoked.

Goodrich et al. (2007) explored a variation of the temporal display that allowed a human to explore how different scheduling, cost, or parameter selection choices affected the consequences of team behavior. These results suggest that augmenting temporal displays to allow a human to explore how scheduling decisions affect mission effectiveness may reduce some of the negative effects uncovered by Cummings.

Conclusions

For both pigeon-hole and scheduling models of fan-out, the variability of interaction and neglect times substantially limits fan-out: in pigeon-hole models variability means that the commander must plan for the worst case scenario; and in scheduling models variability means that lessons from queuing theory must apply and therefore that the commander cannot operate at peak workload.

In addition to variability, heterogeneity of delegates affects fan-out. Since managing multiple delegates imposes a management meta-task, all fan-out problems involve managing heterogeneous agents. Defining fan-out for heterogeneous agents requires the definition of a performance criterion, but optimizing this criterion in scheduling models can be complicated enough that it adds to the management meta-task.

Given the inherent heterogeneity and variability of fan-out problems, it is desirable to endow delegates with adjustable and adaptive autonomy. Such mixed-initiative interaction promotes enough flexibility to allow the commander to use scheduling models. Even with these models, results from queuing theory and from human factors analysis suggest that it is probably useful to include scheduling support and other humans to maximize fan-out.

Major Points

- Olsen's fan-out metric assumes that a UV cannot wait for human input and leads to pigeon-hole models of task scheduling.
- Variability in interaction and neglect times severely limits fan-out under such circumstances.
- Scheduling theory can increase fan-out, but requires that a UV is capable of waiting for human input.
- Variability in interaction and neglect times still limits fan-out under scheduling, but not as bad as pigeon-hole models.
- Scheduling interactions imposes potentially prohibitive management overhead.
- Emerging technologies, including adaptive and adjustable autonomy, can facilitate higher fan-out.

References

Albus, J. S. (1991). Outline for a theory of intelligence. *IEEE Transactions on Systems, Man and Cybernetics*, 21(3), 473–509.

Baddeley, A. (1986). *Working Mmemory*. Oxford: Oxford University Press.

Crandall, J. W. & Cummings, M. L. (2007). Developing performance metrics for the supervisory control of multiple robots. In *Proceedings of the ACM/IEEE International Conference on Human-Robot Interaction*.

Crandall, J. W. & Goodrich, M. A. (2002). Characterizing efficiency of human robot interaction: A case study of shared-control teleoperation. In *Proceedings of the 2002 IEEE/RSJ International Conference on Intelligent Robots and Systems*. Lausanne, Switzerland.

Crandall, J. W., Goodrich, M. A., Olsen, Jr., D. R., & Nielsen, C. W. (2005). Validating human-robot interaction schemes in multi-tasking environments. *IEEE Transactions on Systems, Man, and Cybernetics, Part A: Systems and Humans*, 35(4), 438–49.

Cummings, M. L. & Guerlain, S. (2003). The tactical tomahawk conundrum: Designing decision support systems for revolutionary domains. In *IEEE International Conference on Systems, Man, and Cybernetics* (pp. 1583–8).

Cummings, M. L. & Guerlain, S. (2007). Developing operator capacity estimates for supervisory control of autonomous vehicles. *Human Factors*, 49(1), 1–15.

Cummings, M. L., Nehme, C. E., Crandall, J., & Mitchell, P. (2007). Predicting operator capacity for supervisory control of multiple UAVs). In J. S. Chahl, L. C. Jain, A. Mizutani, & M. Sato-Ilic (eds), *Innovations in intelligent machines 1* (pp. 11–37). Berlin/Heidelberg: Springer-Verlag.

Dessouky, M. I., Moray, N., & Kijowski, B. (1995). Taxonomy of scheduling systems as a basis for the study of strategic behavior. *Human Factors*, 37(3), 443–72.

Di Nocera, F., Lorenz, B., Tattersal, A., & Parasuraman, R. (2003). New possibilities for adaptive automation and work design. In G. R. J. Hockey, A. W. K. Gaillard, & O. Burov (eds), *Operational Functional State* (pp. 363–72). Washington, DC: IOS Press.

Drucker, P. F. (1954). *The practice of management*. New York: Harper Collins.

Drucker, P. F. (1955). The management horizon. *The Journal of Business*, 28(3), 155–64.

Endsley, M., Bolte, B., & Jones, D. (2003). *Designing for Situation Awareness: An Approach to User-centered Design*. London/New York: Taylor & Francis.

Fong, T., Thorpe, C., & Baur, C. (1999, March). Collaborative control: A robot-centric model for vehicle teleoperation. In *AAAI 1999 Spring Symposium: Agents with Adjustable Autonomy*. Stanford, CA.

Fong, T., Thorpe, C., & Baur, C. (2001, August). A safeguarded teleoperation controller. In *IEEE International Conference on Advanced Robotics (ICAR)*. Budapest, Hungary.

Fong, T., Thorpe, C., & Baur, C. (2003). *Collaboration, Dialogue, and Human-robot Interaction.* Berlin: Springer.

Goodrich, M. A. (2004, October). Using models of cognition in HRI evaluation and design. In *Proceedings of the AAAI 2004 Fall Symposium Series: The Intersection of Cognitive Science and Robotics: From Interfaces to Intelligence*. Arlington, VA.

Goodrich, M. A., McLain, T. W., Anderson, J. D., Sun, J., & Crandall, J. W. (2007). Managing autonomy in robot teams: Observations from four experiments. In *Proceedings of the ACM/IEEE International Conference on Human-Robot Interaction*.

Goodrich, M. A., Quigley, M. R., & Cosenzo, K. A. (2005). Task switching and multi-robot teams. In *Proceedings of the 2005 NRL Workshop on Multi-Robot Systems*.

Hancock, P. A. (2007). On the process of automation transition in multitask human-machine systems. *IEEE Transactions on Systems, Man, and Cybernetics, Part A: Systems and Humans*, 37(4), 586–98.

Hancock, P. A., Mouloua, M., Gilson, R., Szalma, J., & Oron-Gilad, T. (2007). Provocation: Is the UAV control ratio the right question? *Ergonomics in Design*, 15(1), 7.

Hearst, M. A. (1999). Mixed-initiative interaction: Trends and controversies. *IEEE Intelligent Systems*, 14–23.

Hunn, B. P. (2006, January). *Unmanned Aerial System, New System Manning Prediction* (Tech. Rep. ARL-TR-3702). Army Research Laboratory.

Kaber, D. B. & Endsley, M. R. (2004,). The effects of level of automation and adaptive automation on human performance, situation awareness and workload in a dynamic control task. *Theoretical Issues in Ergonomics Science*, 5(2), 113–53.

Kortenkamp, D., Bonasso, P., Ryan, D., & Schreckenghost, D. (1997). Traded control with autonomous robots as mixed initiative interaction. In *AAAI Symposium on Mixed Initiative Interaction*. Stanford, CA.

Lewis, M., Wang, J., & Scerri, P. (2006). Teamwork coordination for realistically complex multirobot systems. In *NATO Symposium on Human Factors of Uninhabited Military Vehicles as Force Multipliers*.

Mau, S. & Dolan, J. (2006, October). Scheduling to minimize downtime in human-multirobot supervisory control. In *Workshop on Planning and Scheduling for Space*. Pittsburgh, PA.

Miller, C. A., Funk, H. B., Dorneich, M., & Whitlow, S. D. (2002, November). A playbook interface for mixed initiative control of multiple unmanned vehicle teams. In *Proceedings of the 21st Digital Avionics Systems Conference* (vol. 2, pp. 7E4-1–7E4-13).

Miller, G. A. (1956). The magical number seven, plus or minus two: Some limits on our capacity for processing information. *The Psychological Review*, 63, 81–97.

Moray, N., Dessouky, M. J., Kijowski, B. A., & Adapathya, R. (1991). Strategic behavior, workload, and performance in task scheduling. *Human Factors*, 33(6), 607–29.

Murphy, R., Stover, S., Pratt, K., & Griffin, C. (2006, October). *Cooperative Damage Inspection with Unmanned Surface Vehicle and Micro Unmanned Aerial Vehicle at Hurricane Wilma*. IROS 2006 video session.

Nicolescu, M. N., & Matarić, M. J. (2001). Learning and interacting in human-robot domains. *IEEE Transactions on Systems, Man and Cybernetics, Part A: Systems and Humans*, 5(5), 419–30.

Olsen, D. R. & Goodrich, M. A. (2003). Metrics for evaluating human-robot interactions. In *Proceedings of PerMIS 2003*.

Olsen, Jr., D. R. & Wood, S. B. (2004). Fan-out: Measuring human control of multiple robots. In *Proceedings of Human Factors in Computing Systems*.

Olsen, Jr., D. R., Wood, S. B., & Turner, J. (2004). Metrics for human driving of multiple robots. In *Proceedings of the 2004 IEEE International Conference on Robotics and Automation* (vol. 3, pp. 2315–20).

Parasuraman, R., Sheridan, T. B., & Wickens, C. D. (2000, May). A model for types and levels of human interaction with automation. *IEEE Transactions on Systems, Man, and Cybernetics, Part A: Systems and Humans*, 30(3), 286–97.

Rani, P., Sarkar, N., Smith, C. A., & Adams, J. A. (2003). Anxiety detection for implicit human-machine interaction. In *Proceedings of the IEEE International Conference on Systems, Man, and Cybernetics*.

Rasmussen, J. (1983). Skills, rules, and knowledge; signals, signs, and symbols, and other distinctions in human performance models. *IEEE Transactions on Systems, Man, and Cybernetics*, 13(3), 257–66.

Russell, S. & Norvig, P. (1995). *Artificial Intelligence: A Modern Approach*. Englewood Cliffs, NJ: Prentice Hall.

Saridis, G. N.(1989). Analytic formulation of the principle of increasing precision with decreasing intelligence for intelligent machines. *Automatica*, 25(3), 461–7.

Saunders, J., Nahaniv, C. L., Dautenhahn, K., & Alissandrakis, A.(2007). Self-imitation and environmental scaffolding for robot teaching. *International Journal of Advanced Robotics Systems*, 4(1), 109–24.

Shattuck, L. G. & Woods, D. D. (2000). Communication of intent in military command and control systems. In C. McCann & R. Pigeau (eds), *The Human in Command: Exploring the Modern Military Experience* (pp. 279–92). New York: Kluwer Academic/Plenum Publishers.

Sheridan, T. B. (1992). *Telerobotics, Automation, and Human Supervisory Control*. Cambridge, MA: MIT Press.

Sheridan, T. B. & Verplank, W. L. (1978). *Human and Computer Control of Undersea Teleoperators* (Technical Report). Cambridge, MA: MIT Man-Machine Systems Laboratory.

Stallings, S. (2000). *Queuing Analysis*. A tutorial on W. Stallings's Computer Science Student Resource Site (http://williamstallings.com/StudentSupport.html).

Tulga, M. K. & Sheridan, T. B.(1980). Dynamic decisions and work load in multitask supervisory control. *IEEE Transactions on Systems, Man, and Cybernetics*, 10(5), 217–32.

Urwick, L. F. (1956). The manager's span of control. *Harvard Business Review*, May–June, 39–47.

Vicenti, K. (1999). *Cognitive Work Analysis: Toward Safe, Productive and Healthy Computer-based Work*. Mahwah, NJ: Lawrence Erlbaum Associates.

Wang, J., Wang, H., & Lewis, M. (2008). Assessing cooperation in human control of heterogeneous robots. *Proceedings of the 3rd ACM/IEEE Conference on Human-Robot Interaction* (pp. 9–16). Amsterdam.

Chapter 21

Coordination and Automation for Controlling Robot Teams

Michael Lewis and Jijun Wang

University of Pittsburgh/Quantum Leap Innovations, Inc.

Introduction

Effective military use of unmanned vehicles (UVs) will depend on coordination of large teams. Automating an aircraft squadron, for example, would not only require automating the piloting of planes, but also leadership functions, support roles, and incidental forms of cooperation such as battle damage assessment (BDA) or escort functions. To illustrate anticipated scales, one scenario proposed by RAND (Vick et al., 2000) attack transporter-erector-launchers (TELs), using low-cost autonomous attack system (LOCAAS), calls for the use of up to 1,000 uninhabited aerial vehicles (UAVs) in a massed attack. In this chapter we develop three key ideas for making human control of large UV teams possible:

1. a conjecture that the difficulty of an operator's task can be predicted by command complexity, a metric that assigns greatest difficulty to coordination;
2. a measure for coordination demand based on extending existing neglect tolerance metrics from multirobot control;
3. existing multiagent coordination algorithms that could provide a mechanism for making large UV teams controllable.

Command Complexity

Because some functions such as selection of targets or authorization to attack may doctrinally require human input, evaluating the operator's span of control as the number of controlled entities scales is critical for designing feasible human-automation control systems. Current human span of control limitations are severe. Miller (2004), for example, showed that under expected target densities a controller who is required to authorize weapon release for a target identified by an unmanned combat aerial vehicle (UCAV) could control no more than 13 UAVs even in the absence of other tasks. A similar breakpoint of 12 was found by Cummings and Guerlain (2004) for retargeting Tomahawk missiles. Recent Air

Force Research Laboratory (AFRL) studies (Ruff et al., 2002) target an even more modest 4 UAVs/operator. Similar numbers (3–9) (Olsen and Wood, 2004) have been found for unmanned ground vehicles (UGVs). To extend these numbers to large-scale teams will require breaking new ground in redefining the role of the operator and taking advantage of new forms of automation. Roles traditionally reserved for a human commander, such as verifying each UCAV-found target prior to weapon release (Lewis et al., 2006; Miller, 2004), may become unfeasible in rapidly evolving missions as the size of teams increases unless novel control architectures can be developed.

To extend operator span of control to large teams we must consider how control difficulty for different control tasks grows with increases in team size. Computational complexity theory (Papadimitriou, 1994) from computer science offers one possible approach. Complexity theory bounds the increase in the number of computations required by an algorithm as the number of data elements grows. An algorithm that increments each of n data elements by one, for example, would have a complexity of order n, written $O(n)$ because the number of computations is linear in the number of elements. Borrowing concepts and notation from computational complexity theory authorization for weapon release after operator verification of each UAV-detected target can be considered $O(n)$ because demand increases linearly with the number of UAVs to be serviced. Another form of control, such as designation of an attack region by drawing a box on a graphical user interface (GUI), being independent of the number of UAVs, would be $O(1)$. Practical applications are likely to require some mixture of control regimes. In our prior work with wide area search munitions (Lewis et al., 2006), for example, the operator specified search and jettison areas and ingress and egress routes, $O(1)$, but was also required to authorize attacks and allowed to command UAVs directly, both tasks of $O(n)$ complexity. Examined from this perspective the most complex tasks faced in controlling large teams are likely to be those that involve choosing and coordinating subgroups of UVs. Simply choosing a subteam to perform a particular task (the iterated role assignment problem), for example, has been shown to be $O(mn)$ (Gerkey and Matarić, 2004). Taking command complexity into consideration could provide a useful guide for designing human-automation control regimes if the operator workload could be shown to be related to it. If our conjecture were correct an effective control architecture for large teams should either eliminate, automate, or distribute human tasks that are $O(n)$ or higher.

Coordination Demand

Despite the apparent analogy between command complexity and the workload imposed by a command task there is no guarantee that human operators will experience difficulty in precisely the same way. The performance of human-robot teams is complex and multifaceted, reflecting the capabilities of the robots, the operator(s), and the quality of their interactions. Recent efforts to define common metrics for human-robot interaction (Steinfeld et al., 2006) have favored sets of metric classes to measure the effectiveness of the system's constituents and their interactions as well as the system's overall performance. In this chapter we present new measures of the demand coordination places on operators of multirobot systems (MRS) and three experiments evaluating our approach and the usefulness of these measures.

Controlling multiple robots substantially increases the complexity of the operator's task because attention must constantly be shifted among robots in order to maintain situation awareness (SA) and exert control. In the simplest case an operator controls multiple independent robots interacting with each as needed. A search task in which each robot searches its own region would be of this category, although minimal coordination might be required to avoid overlaps and prevent gaps in coverage. Control performance at such tasks can be characterized by the average demand of each robot on human attention (Crandall et al., 2005). Under these conditions increasing robot autonomy should allow robots to be neglected for longer periods of time, making it possible for a single operator to control more robots.

For more strongly cooperative tasks and larger teams, individual autonomy alone is unlikely to suffice. The round-robin control strategy used for controlling individual robots would force an operator to plan and predict actions needed for multiple joint activities, and be highly susceptible to errors in prediction, synchronization, or execution. Estimating the cost of this coordination, however, proves a difficult problem. Established methods of estimating MRS control difficulty, neglect tolerance, and fan-out (see Goodrich, Chapter 20 this volume) are predicated on the independence of robots and tasks. In neglect tolerance, the period following the end of human intervention but preceding a decline in performance below a threshold is considered time during which the operator is free to perform other tasks. If the operator services other robots over this period, the measure provides an estimate of the number of robots that might be controlled. Fan-out can also be measured empirically (Olsen and Wood 2004), working from the opposite direction, by adding robots and measuring performance until a plateau without further improvement is reached. Both approaches presume that operating an additional robot imposes an additive demand and that the threshold for acceptable performance remains unchanged as robots are added. These measures are particularly attractive because they are based on readily observable aspects of behavior: the time an operator is engaged controlling the robot, interaction time (IT), and the time an operator is not engaged in controlling the robot, neglect time (NT).

Measuring Coordination Demand

To separate coordination demand (CD) from the demands of interacting with independent robots we have extended Crandall et al.'s (2005) neglect tolerance model by introducing the notion of occupied time (OT), as illustrated in Figure 21.1.

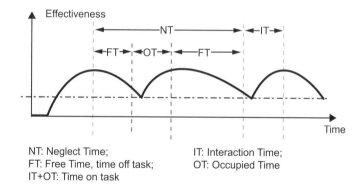

NT: Neglect Time;   IT: Interaction Time;
FT: Free Time, time off task;   OT: Occupied Time
IT+OT: Time on task

Figure 21.1   Extended neglect tolerance model for cooperative robot control

The neglect tolerance model describes an operator's interaction with multiple robots as a sequence of control episodes in which an operator interacts with a robot for period IT, raising its performance above some upper threshold, after which the robot is neglected for the period NT until its performance deteriorates below a lower threshold when the operator must again interact with it. To accommodate dependent tasks we introduce occupied time (OT) to describe the time spent controlling other robots in order to synchronize their actions with those of the target robot. The episode depicted in Figure 21.1 starts just after the first robot is serviced. The ensuing free time (FT) preceding the interaction with a second dependent robot—the OT for robot-1 (that would contribute to IT for robot-2), and the FT following interaction with robot-2 but preceding the next interaction with robot-1—together constitute the neglect time for robot-1. Coordination demand, CD, is then defined as:

$$CD = 1 - \frac{\sum FT}{NT} = \frac{\sum OT}{NT} \qquad (1)$$

where CD for a robot is the ratio between the time required to control cooperating robots and the time still available after controlling the target robot, i.e. the portion of a robot's free time that must be devoted to controlling cooperating robots. Note that OT_n associated with $robot_n$ is less than or equal to NT_n because OT_n covers only that portion of NT_n needed for synchronization.

Most MRS research has investigated homogeneous robot teams where additional robots provide redundant (independent) capabilities. Differences in capabilities such as mobility or payload, however, may lead to more advantageous opportunities for cooperation among heterogeneous robots. These differences among robots in roles and other characteristics affecting IT, NT, and OT introduce additional complexity to assessing CD. Where tight cooperation is required, as in box-pushing, task requirements dictate both the choice of robots and the interdependence of their actions. In the more general case requirements for cooperation can be relaxed, allowing the operator to choose the subteams of robots to be operated in a cooperative manner as well as the next robot to be operated. This general case of heterogeneous robots cooperating as needed characterizes the types of field applications our research is intended to support. To accommodate this more general case, the neglect tolerance model must be further extended to measure coordination between different robot types and for particular patterns of activity.

The resulting expression for generalized CD (Wang 2007) measures the way in which the team's capabilities or resources are combined to accomplish the task without reference to the operation or neglect of particular robots. So, for example, it would not distinguish between a situation in which one robot of type X was never operated while another was used frequently, from a situation in which both robots of type X were used more evenly. The incorporation of action patterns further extends the generality of the approach to accommodate patterns of cooperation that occur in episodes such as dependencies between loading and transporting robots. When an empty transporter arrives, its brief IT would lead to extended OT as the loaders do their work. When it leaves the dependency would be reversed.

Automating Coordination

If command complexity is most strongly affected by coordination demands and these demands impose a corresponding load on operator attention, automating coordination will likely be needed to bring task demands within the capabilities of a human commander. Our research focuses on human control over 4–200 cooperating UVs. Although various control strategies might be effective at the lowest end of this continuum (4–8) most of the team sizes will depend upon automating high-complexity tasks involving subteam formation and cooperation. These high-complexity tasks are required for the temporary formation of subteams to perform cooperative actions such as flush and ambush, simultaneous attack, or battle damage assessment (BDA).

"Commandability" is a closely related issue. While large UV teams may require substantial autonomy, a human is needed somewhere in the loop if for nothing else than to set goals for the team to accomplish. In some coordination architectures even commanding particular actions may be difficult. Roth, Hanson et al. (2004), for example, describe an "intent matrix" interface for specifying the relative value of target classes to a model-based architecture. Particular targets could only be selected indirectly by changing class values and computing new plans until one including the desired target(s) was found. Simple $O(1)$ forms of control that set goals for the entire team such as designating sets of potential targets or search zones are easy to accomplish within any architecture; but $O(n)$ commands such as directing a UAV through a waypoint or detailed commands to a subteam may be difficult or impossible in some architectures. Teamwork architectures such as Machinetta or other approaches based on plan libraries such as "Play Book" (see Parasuraman/Miller chapter) are particularly amenable to human control because their explicit representation of plans, roles, and information offer a variety of levers for influencing behavior. Human intentions can be expressed through goals, waypoints, explicit human roles in team plans, or instantiation of new plans.

Teamwork and Scalability

The Teamcore/Machinetta approach (Tambe, 1997) is based on *team-oriented plans* (TOPs) which describe joint activities in terms of the individual *roles* to be performed and any constraints between those roles. TOPs are instantiated dynamically from TOP templates at runtime when preconditions associated with the templates are filled. Typically, a large team will be simultaneously executing many TOPs. A team of UCAVs, for example, might execute a variety of attack TOPs. When a UCAV identifies a target in an open area it might instantiate a simple attack TOP and send out a request to fill second attacker and BDA roles. After the roles are filled, two UCAVs attack the target and the third follows to record the damage. Another UCAV spotting a convoy of trucks near cover might instantiate a more complex simultaneous attack plan requiring filling multiple attacker roles in order that they might attack together to catch the convoy in the open. Constraints between these roles will specify interactions such as required execution ordering and whether one role can be performed if another is not currently being performed.

This level of cooperation, however, is no easier for machines than for people. If team members must maintain accurate models of one another the problem becomes NP hard making the coordination of more than ~20 UVs computationally intractable. In recent work (Xu et al., 2005), however, we have found a heuristic solution that allows scaling teamwork algorithms to very large teams. Our algorithms make use of the *small world* property of large networks. Each member of the network maintains communications with a small number of associates. Connections between associates are used to move information around the network. While executing a plan, members filling its roles maintain accurate models of one another, but when the plan terminates they revert to exchanging messages only with their permanent associates. While this scheme can no longer guarantee optimal coordination, roles are filled, plans are deconflicted, and other operations are performed correctly with high probability. If conflicting plans to attack the same target were instantiated by different UCAVs, for example, the conflict would likely be discovered when some common acquaintance hears about both plans. The acquaintance would then instantiate a plan for resolving the conflict and execute that plan, leading to the termination of one of the competing attacks. In this way communications and modeling effort expended on the larger team can be kept low, while subteams executing TOPs are able to maintain accurate models and full communications among themselves. Our work additionally shows that even small biases toward choosing the appropriate acquaintance to pass information to can lead to substantial increases in efficiency. If a request seeking to fill a role in an attack on a convoy, for example, were passed on to an acquaintance who had previously mentioned a convoy, it might be more likely to reach someone to fill the role than if it had been passed on randomly.

The current Machinetta implementation incorporating these innovations makes feasible automating coordination for both the numbers of UVs and complexity of plans likely to be needed for military operations. Together our three key ideas provide the premise for the reported research: consideration of command complexity suggests that coordinating teams of UVs will impose the greatest workload on human operators, our CD measures attempt to assess this demand, while the Machinetta implementation demonstrates the feasibility of automating these coordination tasks.

Environment and Task

Unified System for Automation and Robot Simulation (USARSim)

The reported experiments were performed using the USARSim robotic simulation with 2–6 simulated UGVs performing Urban Search and Rescue, USAR (experiments 1 and 3) or box-pushing (experiment 2) tasks. USARSim is a high-fidelity simulation of USAR robots and environments developed as a research tool for the study of Human Robot Interaction (HRI) and multirobot coordination. USARSim supports HRI in ways lower-fidelity simulations cannot by accurately rendering user interface elements (particularly camera video); accurately representing robot automation and behavior; and accurately representing the remote environment that links the operator's awareness with the robot's behaviors. USARSim can be downloaded from http://.sourceforge.net/projects/usarsim and serves as the basis for the Virtual Robots Competition of the RoboCup Rescue League. The current version of USARSim includes detailed replicas of National Institute of Standards and Technology (NIST) USAR arenas, standard reference environments for testing robots, as well as large-scale indoor and outdoor hypothetical disaster scenarios. USARSim complements these maps with high-fidelity models of commercial (Pioneer P2DX, P2AT, iRobot's ATRV Jr., Foster-Miller's Talon, and Telerob's Telemax) and a variety of experimental robots, including a snake (Soyu from Tohoku University), air (Air-robot helicopter), and a large Ackerman-steered UGV (Hummer) and sensor models for many common robotic sensing packages. Validation studies showing agreement for a variety of feature extraction techniques between USARSim images and camera video are reported in Carpin et al. (2006a). Close agreement in detection of walls and associated Hough transforms for a simulated Hokuyo laser range finder are described in Carpin et al. (2005). Validation studies showing close agreement in behavior between USARSim models and the real robots being modeled are reported in Carpin et al. (2006b), Lewis et al. (2005), Pepper et al. (2007), Taylor et al. (2007), and Zaratti et al. (2006).

MrCS: The Multirobot Control System   A multirobot control system (MrCS) was developed to conduct these experiments by scaling to allow control of different numbers of robots, reconfiguring to accommodate different human-robot interfaces, and facilitating the testing of different control algorithms. The MrCS was developed around the Machinetta proxy-based, multiagent framework to provide tight integration with automated coordination.

The level of autonomy (LOA) strongly impacts the effectiveness of multirobot control. Parasuraman et al. (2005) have shown that autonomy can decrease workload but degrade SA and performance. Conversely, human involvement can increase performance and SA, but at the cost of increased workload. When the task or situation is sufficiently complex, human involvement may even have a negative effect on the overall performance (Schurr et al., 2005). Many factors, such as the task, the environment, and robot capability, can influence the optimal level of autonomy. Adaptive autonomy (AA) allows the level of autonomy to be changed to match the situation. For instance, studies of delegation interfaces (Parasuraman et al., 2005) showed that manual changes to the level of autonomy can improve performance and SA with only slightly increased workload. The desirability of AA is further supported by Fong (2001), who documents the need for control at different LOAs to enable the human to assist the robot in such tasks as getting unentangled from obstacles or designating a "wall" of grass as traversable terrain. In keeping with these observations, our MrCS follows an inclusive LOA policy allowing all levels of control up to a designated ceiling. So, for example, a system that allowed control through waypoints must also allow teleoperation.

The user interface of the MrCS is shown in Figure 21.2. The interface is reconfigurable to allow the user to resize the components or change the layout. Shown in the figure is a configuration that we used in the RoboCup 2006 competition, in which a single operator controls six robots. On the upper and center portions of the left-hand side are the robot list and team map panels, which show the operator an overview of the team. The destination of each of robot is displayed on the map to help the user keep track of current plans. Using this display, the operator is also able to control regional priorities by drawing rectangles on the map. On the center and lower portions of the right-hand side are the camera view and mission control panels, which allow the operator to maintain situation awareness of an individual robot and to edit its exploration plan. On the mission panel, the map and all nearby robots and their destinations are represented to provide partial team awareness so that the operator can switch between contexts while moving control from one robot to another. The lower portion of the left-hand side is a teleoperation panel that allows the operator to teleoperate a robot.

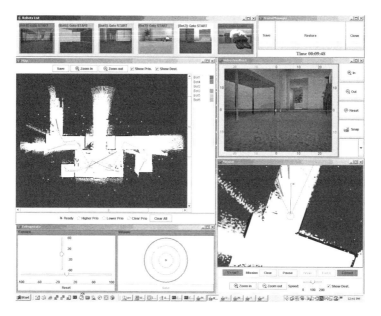

Figure 21.2 MrCS user interface

Experiments

One approach to investigating coordination demand is to design experiments that allow comparisons between "equivalent" conditions with and without coordination demands. The first experiment and one comparison within the third experiment follow this approach. The first experiment compares search performance between a team of autonomously coordinating robots, manually (waypoint) controlled robots, and mixed-initiative teams with autonomously coordinated robots that accepted operator inputs. The impact of coordination demand was observable through the difference in performance between the manually controlled teams and the mixed-initiative ones. The fully automated teams provided a control ensuring that the benefits in the mixed-initiative condition were not due solely to the superior performance of the automation.

While experiment 1 examines coordination demand indirectly by comparing performance between conditions in which it was filled either manually or through automation, experiments 2 and 3 attempt to manipulate and measure coordination demand directly. In experiment 2 robots perform a box-pushing task in which CD is varied by control mode and robot heterogeneity. By making the actions of each robot entirely dependent on the other, this choice of task eliminates the problem of distinguishing between interactions intended to control a target robot and those needed to coordinate with another. The third experiment attempts to manipulate coordination demand in a loosely coordinated task by varying the proximity needed to perform a joint task in two conditions, and by automating coordination within subteams in the third. Because robots must cooperate in pairs and interaction for control needs to be distinguished from interaction for coordination for this task, CD is computed between robot types (generalized CD) rather than directly between robots (equation 1), as done in experiment 2.

All three experiments used paid participants from the University of Pittsburgh and lasted approximately one and a half hours. All used repeated measures designs and followed a standard sequence, starting with collection of demographic data. Standard instructions for the experiment were presented followed by a 10-minute training session during which the participant was allowed to practice using the MrCS. Participants then began their first trial, followed by a second with a short break in between. Experiments 2 and 3 included a third trial with break. At the conclusion of the experiment participants completed a questionnaire.

Experiment 1

Participants were asked to control three P2DX robots simulated in USARsim to search for victims in a damaged building. Each robot was equipped with a pan-tilt camera with 45 degrees field of view (FOV) and a front laser scanner with 180-degree FOV and resolution of 1 degree. When a victim was identified, the participant marked its location on NIST Reference Test Arena, Yellow Arena (Jacoff et al. 2001). Two similar testing arenas were built using the same elements with different layouts. In each arena, 14 victims were evenly distributed in the world. We added mirrors, blinds, curtains, semitransparent boards, and wire grid to add difficulty in situation perception. Bricks, pipes, a ramp, chairs, and other debris were put in the arena to challenge mobility and SA in robot control.

Presentation of mixed-initiative and manual conditions was counterbalanced. Under mixed initiative, the robots analyzed their laser range data to find possible exploration paths. They cooperated with one another to choose execution paths that avoided duplicating efforts. While the robots autonomously explored the world, the operator was free to intervene with any individual robot by issuing new waypoints, teleoperating, or panning/tilting its camera. The robot returned back to auto mode once the operator's command was completed or stopped. While under manual control robots could not autonomously generate paths and there was no cooperation among robots. The operator controlled a robot by giving it a series of waypoints, directly teleoperating it, or panning/tilting its camera. As a control for the effects of autonomy on performance we conducted "full autonomy" testing as well. Because MrCS does not support victim recognition, based on our observation of the participants' victim identification behaviors we defined detection to have occurred for victims that appeared on camera for at least 2 seconds and occupied at least 1/9 of the thumbnail view. Because of the high fidelity of the simulation, and the randomness of paths picked through the cooperation algorithms, robots explored different regions on every test. Additional variations in performance occurred due to mishaps such as a robot getting stuck in a corner or bumping into an obstacle, causing its camera to point to the ceiling so no victims could be found. Sixteen trials were conducted in each area to collect data comparable to that obtained from human participants.

Result

All 14 participants found at least 5 of a possible 14 (36 percent) victims in each of the arenas. These data indicate that participants exploring less than 90 percent of the area consistently discovered 5 to 8 victims, while those covering greater than 90 percent discovered between half (7) and all (14) of the victims. Within-participant comparisons found wider regions were explored in mixed-initiative mode, t(13) = 3.50, p < .004, as well as a marginal advantage for mixed-initiative mode, t(13) = 1.85, p = .088, in number of victims found. Comparing with "full autonomy," under mixed-initiative conditions two-tailed t-tests found no difference (p = 0.58) in the explored regions.

No difference was found between area explored in autonomous or mixed-initiative searches; however, autonomously coordinating robots explored significantly more regions, t(44) = 4.27, p < .001, than under the manual control condition (see Figure 21.3). Participants found more victims under both mixed-initiative and manual control conditions than under full autonomy with t(44) = 6.66, p < .001, and t(44) = 4.14, p < .001, respectively (see Figure 21.3) The median number of victims found under full autonomy was five. Comparing the mixed-initiative with the manual control, most participants (79 percent) rated team autonomy as providing either significant or minor help.

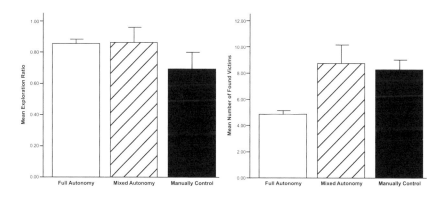

Figure 21.3   Regions explored and victims found under full-autonomy, mixed-autonomy, and manual-control modes

Human Interactions

Participants intervened to control the robots by switching focus to an individual robot and then issuing commands. Measuring the distribution of attention among robots as the standard deviation of the total time spent with each robot, no difference (p = .232) was found between mixed-initiative and manual-control modes. However, we found that under mixed initiative, the same participant switched robots significantly more often than under manual mode (p = .027).

Across participants, the frequency of shifting control among robots explained a significant proportion of the variance in number of victims found for both mixed initiative, $R^2 = .54$, $F (1, 11) = 12.98$, $p = .004$, and manual, $R^2 = .37$, $F (1, 11) = 6.37$, $p < .03$, modes (Figure 21.4).

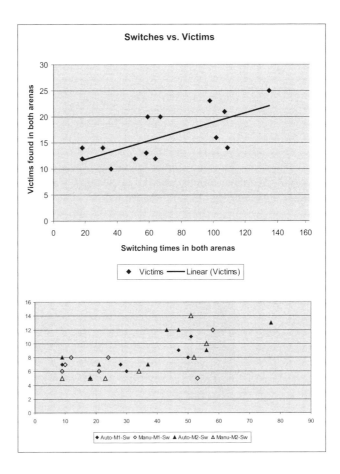

Figure 21.4   Victims found vs. switches under mixed-autonomy and manual modes

In this experiment, cooperation was limited to deconfliction of plans so that robots did not re-explore the same regions or interfere with one another. The experiment found that even this limited degree of autonomous cooperation helped in the control of multiple robots. The results showed that cooperative autonomy among robots helped the operators explore more areas and find more victims. The fully autonomous control condition demonstrates that this improvement was not due solely to autonomous task performance as found in Schurr et al. (2005), but rather resulted from mixed-initiative cooperation with the robotic team.

Experiment 2

Equation (1) defines CD as the ratio between occupied time (OT), the period over which the operator is actively controlling a robot to synchronize with others, and FT+OT, the time during which he is not actively controlling the robot to perform the primary task. This measure varies between 0 for no demand and 1 for maximum demand. When an operator teleoperates two robots one at a time to push a box forward, he must continuously interact with one of the robots because neglecting both would immediately stop the box. Because the task allows no free time (FT) we expect CD to be 1. However, when the user is able to issue waypoints to both robots, the operator may have FT before she must coordinate these robots again because the robots can be instructed to move simultaneously. In this case CD should be less than 1. Intermediate levels of CD should be found in comparing control of homogeneous robots with heterogeneous robots. Higher CD should be found in the heterogeneous group since the unbalanced pushes from the robots would require more frequent coordination. In experiment 2, we measured CDs under these three conditions.

The controlled robots were either two Pioneer P2AT robots (shown in Figure 21.5) or one Pioneer P2AT and one less capable three-wheeled Pioneer P2DX robot. When a robot pushes the box, both the box and robot's orientation and speed will change. Furthermore, because of irregularities in initial conditions and accuracy of the physical simulation the robot and box are unlikely to move precisely as the operator expects. In addition, delays in receiving sensor data and executing commands were modeled, presenting participants with a problem very similar to coordinating physical robots.

We introduced a simple matching task as a secondary task to allow us to estimate the FT available to the operator. Participants were asked to perform this secondary task as possible when they were not occupied controlling a robot. Every operator action and periodic time-stamped samples of the box's moving speed were recorded for computing CD.

A within-subject design was used to control for individual differences in operators' control skills and ability to use the interface. To avoid having abnormal control behavior, such as a robot bypassing the box, bias the CD comparison, we added safeguards to the control system to stop the robot when it tilted the box.

Figure 21.5   Box-pushing (Experiment 2) and explore/inspect (Experiment 3) tasks

The operator controlled the robots using the MrCS modified to present two equally sized camera views. Each robot was immediately controllable without selection from mirrored teleoperation controls or by adding waypoints to the map.

Participants performed three testing sessions in counterbalanced order. In two of the sessions, the participants controlled two P2AT robots using teleoperation alone or a mixture of teleoperation and waypoint control. In the third session, the participants were asked to control heterogeneous robots (one P2AT and one P2DX) using a mixture of teleoperation and waypoint control. The participants were allowed eight minutes to push the box to the destination in each session.

As we expected no free time was recorded for robots in the teleoperation condition and the longest free times were found in controlling homogeneous robots with waypoints. The box speed along the hallway reflects the interaction effectiveness (IE) of the control mode. The IE curves in this picture show the delay effect and the frequent bumping that occurred in controlling heterogeneous robots revealing the poorest cooperation performance.

None of the 14 participants was able to perform the secondary task while teleoperating the robots. Hence, we uniformly CD=1 for both robots under this condition. Within-participants comparison found that under waypoint control the team attention demand in heterogeneous robots is significantly higher than the demand in controlling homogeneous robots, $t(13) = 2.213$, $p = .045$. No significant differences were found between the homogeneous P2AT robots in terms of the individual cooperation demand; however, when a participant controlled a P2AT robot, lower CD was required in the homogeneous condition than in the heterogeneous condition, $t(13) = -2.365$, $p = .034$. The CD required in controlling the P2DX under the heterogeneous condition is marginally higher than the CD required in controlling homogenous P2ATs, $t(13) = -1.868$, $p = .084$. Surprisingly, no significant difference was found in CDs between controlling P2AT and P2DX robots under the heterogeneous condition ($p = 0.79$).

This study demonstrates that, as a generic HRI metric, CD is able to account for the various factors that affect HRI and could be used in HRI evaluation and analysis. Although only 14 participants were involved in this experiment, using measured CDs we were able to quickly identify three aberrant robot control modes, affirming the usefulness of the measure.

Experiment 3

To test the usefulness of the generalized CD measurement for a weakly cooperative MRS, we conducted an experiment assessing generalized CD using a task of complexity suitable to exercise heterogeneous robot control. In the experiment participants were asked to control explorer robots equipped with a laser rangefinder but no camera, and inspector robots with only cameras. Finding and marking a victim required using the inspector's camera to find a victim to be marked on the map generated by the explorer. The capability of the robots and the cooperation autonomy level were used to vary the coordination demand of the task.

Three simulated Pioneer P2AT robots (explorers) and three Zergs (Balakirsky et al. 2007) (scouts), a small experimental robot were used. Each P2AT was equipped with a front laser scanner with 180-degree FOV and resolution of 1 degree. The Zerg was mounted with a pan-tilt camera with 45-degree FOV. The robots were capable of localization and able to communicate with other robots and control station. The P2AT served as an explorer to build the map, while the Zerg could be used as an inspector to find victims using its camera. To accomplish the task the participant must coordinate these two robot types to ensure that when an inspector robot finds a victim, it is within a region mapped by an explorer robot so the position can be marked.

Three conditions were designed to vary the coordination demand on the operator. Under condition 1, the explorer had a 20-meter detection range, allowing inspector robots considerable latitude in their search. Under condition 2, scanner range was reduced to 5 meters, requiring closer proximity to keep the inspector within mapped areas. Under condition 3, explorer and inspector robots were paired as subteams in which the explorer robot with a sensor range of 5 meters followed its inspector robot to map areas being searched. We hypothesized that generalized CDs for explorer and inspector robots would be more evenly distributed under condition 2 (short-range sensor) because explorers would need to move more frequently in response to inspectors' searches than in condition 1, in which generalized CD should be more asymmetric with explorers exerting greater demand on inspectors. We also hypothesized that lower generalized CD would lead to higher team performance. Three equivalent damaged buildings were constructed from the same elements using different layouts. Each environment was a mazelike building with obstacles such as chairs, desks, cabinets, and bricks, and 10 evenly distributed victims. A fourth environment was constructed for training. Figure 21.5 shows the simulated robots and environment.

A within subjects design with counterbalanced presentation was used to compare the cooperative performance across the three conditions. The same control interface allowing participants to control robots through waypoints, or teleoperation was used in all conditions.

Overall performance was measured by the number of victims found, the areas explored, and the participants' self-assessments. To examine cooperative behavior in finer detail, generalized CDs were computed from logged data for each type of robot under the three conditions. We compared the measured generalized CDs between condition 1 (20-meter sensing range) and condition 2 (5-meter sensing range), as well as condition 2 and condition 3 (subteam). To further analyze the cooperation behaviors, we evaluated the total attention demand in robot control and control action pattern as well.

A paired t-test shows that in condition 1 (20-meter range scanner) participants explored more regions, $t(16) = 3.097$, $p = 0.007$, and also found more victims, $t(16) = 3.364$, $p = 0.004$, than under condition 2 (short-range scanner). In condition 3 (automated subteam) participants found marginally more victims, $t(16) = 1.944$, $p = 0.07$, than in condition 2 (controlled cooperation), but no difference was found for the extent of regions explored.

Most participants (74 percent) thought it was easier to coordinate inspectors with explorers with the long-range scanner; 12 of the 19 participants (63 percent) rated auto-cooperation between inspector and explorer (the subteam condition) as improving their performance, with only two participants (11 percent) judging team autonomy to make things worse.

From log files, generalized CDs were computed for each type of robot according to the extended form of equation (1). A typical (IT,FT) distribution under condition 1 (20-meter sensing range) in the experiment yielded a calculated generalized CD for the explorer of 0.185 and a generalized CD for the inspector of 0.06. The low generalized CDs reflect that, in trying to control six robots, the participant ignored some robots while attending to others. The generalized CD for explorers is roughly twice the generalized CD for inspectors. After the participant controlled an explorer, he needed to control an inspector multiple times, or multiple inspectors, since the explorer has a long detection range and large FOV. In contrast, after controlling an inspector, the participant needed less effort to coordinate the explorer.

We predicted that when the explorer has a longer detection range, operators would need to control the inspectors more frequently to cover the mapped area. Therefore a longer detection range should lead to higher generalized CD for explorers. This was confirmed by a two tailed t-test that found higher coordination demand, $t(18) = 2.476$, $p = 0.023$, when participants controlled explorers with large (20-meter) sensing range.

As auxiliary parameters, we also evaluated the total attention demand, i.e. the occupation rate of total interaction time in the whole control period, and the action pattern, the ratio of control times between inspector and explorer. A paired t-test shows that under long sensing conditions, participants interacted with robots more times than under short sensing, which implies that more robot interactions occurred. The mean action patterns under long- and short-range scanner conditions are 2.31 and 1.9 respectively. This means that with 20- and

5-meter scanning ranges, participants controlled inspectors 2.31 and 1.9 times respectively after an explorer interaction. Within participant comparisons show that the ratio is significantly larger under a long sensing condition than under the short-range scanner condition, $t(18) = 2.193$, $p = 0.042$. Generalized CD again proved sensitive to aberrant control behavior, easily discriminating a trial in which a robot became entangled with a desk and two others in which participants failed to control two of their robots.

Conclusions

The reported research demonstrates potential advantages of automating co-ordination among UVs (experiments 1 and 3), and has developed CD measures of coordination demand to assess how much of an operator's control efforts have been expended on coordination. This research was motivated by a theory of command complexity that predicts that control efforts for realistically complex team tasks will come to be dominated by coordination demands as team size increases. Algorithms developed with our collaborators (Xu et al., 2005) now offer the possibility of automating the coordination of even very large teams. Tested in the reported experiments, use of these coordination algorithms significantly improved performance even for small teams (2–6) and relatively weak coordination demands where they should be least effective.

The superiority of mixed-initiative control in the first experiment was far from a foregone conclusion, since earlier studies with comparable numbers of individually autonomous robots (Crandall et al., 2005; Nielsen et al., 2003; Squire et al., 2003; Trouvain et al., 2003) found poorer performance for higher levels of autonomy at similar tasks. We believe that differences between navigation and search tasks may help explain these results. In navigation, moment-to-moment control must reside with either the robot or the human. When control is ceded to the robot the human's workload is reduced, but task performance declines due to loss of human perceptual and decision-making capabilities. Search, by contrast, can be partitioned into navigation and perceptual subtasks allowing the human and robot to share task responsibilities improving performance. This explanation suggests that increases in task complexity should widen the performance gap between cooperative and individually autonomous systems. Automated coordination was similarly effective in the third experiment, where it both increased the number of victims found and significantly reduced coordination demand on the operator.

The most interesting finding from the first experiment involved the relation between performance and switching of attention among the robots. In both the manual and mixed-initiative conditions participants divided their attention approximately equally among the robots but in the mixed-initiative mode they switched among robots more rapidly. Psychologists have found task switching to impose cognitive costs, and switching costs have previously been reported (Kirlik, 1993; Schurr et al., 2005) for multirobot control. Higher switching costs might be expected to degrade performance; however in this study more rapid switching was associated with improved performance in both manual and mixed initiative conditions. We believe that the map component at the bottom of the display helped mitigate losses in awareness when switching between robots, and that more rapid sampling of the regions covered by moving robots gave more detailed information about areas being explored. The frequency of this sampling among robots was strongly correlated with the number of victims found. This effect, however, cannot be attributed to a change from a control to a monitoring task because the time devoted to control was approximately equal in the two conditions. We believe instead that searching for victims in a building can be divided into a series of subtasks involving things such as moving a robot from one point to another, and/or turning a robot from one direction to another with or without panning or tilting the camera. To effectively finish the searching task, we must interact with these subtasks within their neglect time that is proportional to the speed of movement. When we control multiple robots and every robot is moving, there are many subtasks whose neglect time is usually short. Missing a subtask means we failed to observe a region that might contain a victim. So switching robot control more often gives us more opportunity to find and finish subtasks, and therefore helps us find more victims.

Results for measuring CD are more equivocal. For the entirely dependent box-pushing task we show that measured CD=1, as predicted under teleoperation conditions. Measured CD is also higher for heterogeneous robots than homogeneous pairs, as predicted. In measuring weak coordination in experiment 3 results partially followed predictions. We found greater CD for explorers in the 20-meter condition (lots of searching with the scout before needing to move the explorer), but did not find the predicted symmetric difference for scouts. Whether this reflects an actual similarity in demands and provides an argument for focused metrics that measure constituents of the human-robot system directly, or is merely an artifact of measurements limited to interaction times, is unclear. We anticipated a correlation between found victims and the measured CDs. However, we did not find the expected relationship in this experiment. From observation of participants during the experiment we believe that high-level strategies, such as choosing areas to be searched and path planning, have significant impact on the overall performance. The participants had few problems in learning to jointly control explorers and inspectors, but they needed time to figure out effective strategies for performing the task. Because CD measures control behaviors not strategies, these effects were not captured. A stronger case for the utility of the CD measures can be made for the identification of aberrant control behaviors. In both experiment 2 and 3 abnormal CD values readily identified trials in which abnormal control occurred, something not necessarily revealed by performance scores.

Extrapolating these results to military applications requires increases in scale and complexity. The current experiments were intended to test whether humans could benefit from interacting with autonomously coordinating robots (they can) and whether we could measure the demand coordination being placed on the operator (perhaps).

Control over subteams of cooperating UVs poses difficulties in both acquiring SA and in forming and conveying intent. Trying to make sense out of what many perhaps overlapping cooperating subteams are doing at any one time is a cognitively challenging task. Even in far simpler environments with relatively few robots, researchers (Schurr et al., 2005) have found it difficult to convey cooperative behavior unambiguously to an observer. As the number of vehicles increases, use of color, highlighted targets, indicated paths, and other common techniques are unlikely to continue to allow an observer to separate out cooperating cliques and infer their intentions. Because monitoring and intervening to resolve unexpected events is a primary task reserved for humans in highly automated systems, an inability to comprehend the situation without relying on system-issued alerts would make this role particularly difficult.

These problems are mirrored on the action side. Of the available levers for controlling UV teams (altering goals, setting waypoints, response to alerts, plan instantiation, and parameter tuning), most make it difficult to connect intended behaviors with the necessary control actions. Answering questions about proper techniques for displaying large cooperating systems and guiding response where control may be indirect is going to require proper experimentation with human participants. In this process, determining those tasks humans cannot perform will be as helpful for designers as finding those they can.

Team control at other levels of aggregation should prove easier. O(1) control through team goals such as designating search areas or targets are simple and direct by avoiding the complexities of subteams and their plans. Under circumstances in which O(n) tasks could be made independent of other tasks, assigning additional operators could satisfy demands for human control without undermining team coordination. In a mission such as our search task that depends on human identification of victims, multiple operators could be used to monitor video from larger teams of coordinating robots or to intervene with robots observed to be in trouble. At the next level of complexity, for UCAVs with ATR that require human verification of targeting prior to weapons release, a team of operators could be kept standing by in the command center. As authorization requests came in they could be routed to an available operator, who would examine targeting data and clear or abort the attack. Because operators would not be tied to any particular UCAV or subteam they would be available to respond to fluctuations in demand as might occur, for example, when a subgroup of UCAVs overflew an enemy emplacement. In this Call Center approach operators would respond to independent requests from a general population rather than controlling preassigned subteams. A variety of other control architectures incorporating automated coordination are imaginable.

The goal of our research is to rationalize the design of these architectures by identifying those tasks that must be automated, finding ways to convey intent through that automation, and partitioning human tasks to support the system.

Lessons Learned

- Operator activities needed to coordinate cooperating robots occupy a substantial portion of the operator's time at tasks for tasks of varying levels of dependency.
- Coordination demand (CD) estimated from interaction durations can provide diagnostic measures of performance.
- Performance at search tasks is positively related to the frequency of shifts in attention among robots.
- Measuring CD at realistic loosely structured or phased tasks is difficult and will require further research to develop accurate measures.

References

Balakirsky, S., Carpin, S., Kleiner, A., Lewis, M., Visser, A., Wang, J., and Zipara, V. (2007). Towards hetereogeneous robot teams for disaster mitigation: Results and performance metrics from RoboCup rescue. *Journal of Field Robotics*, 24(11–12), 943–67.

Carpin, S., Lewis, M., Wang, J., Balakirsky, S., Scrapper, C. (2006b). Bridging the gap between simulation and reality in urban search and rescue. Robocup 2006: Robot Soccer World Cup X. Springer, Lecture Notes in Artificial Intelligence.

Carpin, S., Stoyanov, T., Nevatia, Y., Lewis, M., and Wang, J. (2006a). Quantitative assessments of USARSim accuracy. *Proceedings of PerMIS 2006*.

Carpin, S., Wang, J., Lewis, M., Birk, A., and Jacoff, A. (2005). High fidelity tools for rescue robotics: Results and perspectives, Robocup 2005 Symposium.

Crandall, J., Goodrich, M., Olsen, D. and Nielsen, C. (2005). Validating human-robot interaction schemes in multitasking environments. *IEEE Transactions on Systems, Man, and Cybernetics: Part A*, 35(4):438–49.

Cummings, M. and Guerlain, S. (2004). An interactive decision support tool for real-time in-flight replanning of autonomous vehicles. AIAA Unmanned Unlimited Systems, Technologies, and Operations.

Fong, T. (2001). Collaborative control: A robot-centric model for vehicle teleoperation. PhD thesis, The Robotics Institute, Carnegie Mellon University.

Gerkey, B. and Matarić, M. (2004). A formal framework for the study of task allocation in multi-robot systems. *International Journal of Robotics Research*, 23(9), 939–54.

Jacoff, A., Messina, E. and Evans, J. (2001). Experiences in deploying test arenas for autonomous mobile robots. In *Proceedings of the 2001 Performance Metrics for Intelligent Systems (PerMIS) Workshop*, Mexico City, September.

Kirlik, A. (1993). Modeling strategic behavior in human automation interaction: Why an "aid" can (and should) go unused. *Human Factors*, 35, 221–42.

Lewis, M., Hughes, S., Wang, J., Koes, M., and Carpin, S. (2005). Validating USARsim for use in HRI research. *Proceedings of the 49th Annual Meeting of the Human Factors and Ergonomics Society*, Orlando, FL, 457–61.

Lewis, M., Polvichai, J., Sycara, K., and Scerri, P. (2006). Scaling-up human control for large scale systems, In N. J. Cooke, H. Pringle, H. Pedersen, and O. Connor (eds), *Human Factors of Remotely Operated Vehicles*, New York: Elsevier, 237–50.

Miller, C. (2004). Modeling human workload limitations on multiple UAV control. *Proceedings of the 47th Annual Meeting of the Human Factors and Ergonomics Society*, 526–7.

Nielsen, C., Goodrich, M., and Crandall, J. (2003). Experiments in human-robot teams. In *Proceedings of the 2002 NRL Workshop on Multi-Robot Systems*, October.

Olsen, D. and Wood, S. (2004). Fan-out: Measuring human control of multiple robots. *Proceedings of the 2004 Conference on Human Factors in Computing Systems (CHI 2004)*, 24–29 April, Vienna, Austria, ACM, 231–8.

Papadimitriou, C. (1994). *Computational Complexity*. Reading, MA: Addison Wesley.

Parasuraman, R., Galster, S., Squire, P., Furukawa, H. and Miller, C. (2005). A flexible delegation-type interface enhances system performance in human supervision of multiple robots: Empirical studies with roboflag. *IEEE Transactions on Systems, Man, and Cybernetics: Part A*, 35(4):481–93, July

Pepper, C., Balakirsky, S., and Scrapper, C. (2007). Robot simulation physics validation. *Proceedings of PerMIS '07*.

Roth, E., Hanson, M., Hopkins, C., Mancuso, V., and Zacharias, G. (2004). Human in the loop evaluation of a mixed-initiative system for planning and control of multiple UAV teams. *Proceedings of the 48th Annual Meeting of the Human Factors and Ergonomics Society*, 280–84.

Ruff, H., Narayanan, S., and Draper, M. (2002). Human interaction with levels of automation and decision-aid fidelity in the supervisory control of multiple simulated unmanned air vehicles. *Presence*, 11(4), 335–51.

Schurr, N., Marecki, J., Tambe, M., Scerri, P., Kasinadhuni, N., and Lewis, J. (2005). The future of disaster response: Humans working with multiagent teams using DEFACTO. In *AAAI Spring Symposium on AI Technologies for Homeland Security*.

Steinfeld, A., Fong, T., Kaber, D., Lewis, M., Scholtz, J., Schultz, A., and Goodrich, M. (2006). Common metrics for human-robot interaction. 2006 Human-Robot Interaction Conference, ACM, March.

Squire, P., Trafton, G., and Parasuraman, R. (2003). Human control of multiple unmanned vehicles: Effects of interface type on execution and task switching times. In Proceedings of the 2006 Human-Robot Interaction Conference, Salt Lake City, UT, 26–32.

Tambe, M. (1997). Towards flexible teamwork. *Journal of Artificial Intelligence Research*, 7, 83–124.

Taylor, B., Balakirsky, S., Messina, E. and Quinn, R. (2007). Design and validation of a Whegs robot in USARSim. *Proceedings of PerMIS '07.*

Trouvain, B., Schlick, C., and Mevert, M. (2003). Comparison of a map- vs. camera-based user interface in a multi-robot navigation task. In *Proceedings of the 2003 International Conference on Robotics and Automation*, 3224–31.

Vick, A., Moore, R., Pirnie, B., and Stillion, J. (2001). Aerospace operations against elusive ground targets, Santa Monica, CA: RAND.

Wang, J. (2007). Human control of cooperating robots. PhD dissertation, University of Pittsburgh, http://etd.library.pitt.edu/ETD/available/etd-01072008-135804/unrestricted/Wang_EDC_2007-final2.pdf (accessed 22 July 2008).

Xu, Y., Scerri, P., Yu, B., Okamoto, S., Lewis, M., and Sycara, K. (2005). An integrated token-based algorithm for scalable coordination. *Fourth International Joint Conference on Autonomous Agents and Multiagent Systems (AAMAS '05)*, 407–14.

Zaratti, M. Fratarcangeli, M., and Iocchi, L. (2006). A 3D simulator of multiple legged robots based on USARSim. Robocup 2006: Robot Soccer World Cup X. Berlin: Springer, LNAI.

Chapter 22

Model World: Military HRI Research Conducted Using a Scale MOUT Facility

Florian Jenstch, A. William Evans III and Scott Ososky
University of Central Florida

The Team Performance Laboratory (TPL) at the University of Central Florida (UCF) has a reputation for finding low-cost alternatives to high-priced simulations for use in aviation, security, and military applications. This chapter will address the latter of these areas, military applications, and discuss how researchers at TPL created the Scale Military Operations in Urban Terrain (MOUT) environment, what has been studied using this facility and the results of these investigations, as well as future direction for this type of research. In addition, a small but relevant experiment used to validate laboratory research with field study and its impact on (human-robot interaction (HRI) research as a whole will be discussed. Finally, the third major section of this chapter will discuss lessons learned, in conducting many studies, using the Scale MOUT environment.

HRI Research at TPL

Scientists at TPL have developed a research program which emphasizes the use of researchers from diverse backgrounds—including human factors psychology, modeling and simulation, computer science, and cognitive psychology, and with varying skill sets—to achieve common goals through diverse and interactive means. In 2004, this research was begun with the creation of a research facility (Evans et al., 2005) specifically designed to engage participants in a realistic environment, in which they controlled unmanned systems with simulated levels of automation that could be quickly adapted to fit the needs of the researchers and their goals. This Scale MOUT facility, an example of which is seen in Figures 22.1 and 22.2, has expanded in capability over the years and now includes the use of several unmanned ground vehicles (UGVs) coupled with at least one unmanned aerial vehicle (UAV) and various other dynamic elements, such as improvised exploding devices (IEDs), to increase the realism and scope of research being conducted. In addition, operator control areas have been improved to include multiple stations within a secure command station environment (Figure 22.3) and a recently added "virtual foxhole" control environment designed to simulate a forward observer station (Figure 22.4).

Figure 22.1   Ground-level view of Scale MOUT facility at UCF

Figure 22.2   An aerial view of the TPL Scale MOUT environment

Figure 22.3   Unmanned system control station in the C4I command station

**Figure 22.4   Cramped quarters inside the virtual foxhole forward observer
simulation**

Using this environment as a platform for a number of studies, researchers at
TPL have been able to investigate numerous HRI issues, including human–robot
ratio (Rehfeld et al., 2005), shared control effects, image sharing, communication
analysis (Hoeft et al., 2005), spatial awareness (Fincannon et al., 2008a), target
identification (Keebler et al., 2007; Fincannon, Curtis, and Jentsch, 2006), and
distributed teammate coordination (Fincannon et al., 2008b), to name a few. Data
gathered in these areas of research have helped to advance the overall development
of HRI and guided each progressive step taken toward that goal at TPL.

Using an approach emphasizing four major areas of research—1) Team
Collaboration and Performance; 2) Route Planning and Robot Control; 3)
Target Detection and Classification; and 4) Position Localization and Situation
Awareness—researchers at TPL have begun to dissect the issues involved with
creating more efficient human-robot teams. In addition to these four core areas
research at TPL has progressed through the three levels of interdependence
introduced by Saavedra et al. (1993). Using pooled, sequential, and reciprical
interdependencies to grow in scope and complexity, this multi-year effort has
yielded many significant results associated with each of the four core areas
mentioned above. The following sections provide a more detailed explanation of
the research and results, and supply an outlook for what the future may hold at
TPL within the realm of military robotics and HRI.

Early Studies

As the use of unmanned assets became a more realistic and achievable goal,
leaders within the U.S. military envisioned having a single operator control a large
number of unmanned vehicles, thus greatly multiplying the effective force of the
military. However, this initial belief that a single operator would have the ability

to control, or even supervise, a large number of robotic assets would be challenged by the early research conducted in the field.

The first study in this line of research, done by TPL, involved an investigation into the appropriate ratio of operators and robots to produce the most effective teams. Using the Scale MOUT facility described earlier, teams of one or two operators controlled one or two UGVs, depending on the condition, during a simulated reconnaissance task. Analyzing the performance data of this study showed that not only were single operators controlling multiple UGVs the worst performing teams, finding far less than 25 percent of the available targets, but the addition of a second operator to teams only controlling a single UGV increased performance by more than 170 percent (Rehfeld et al., 2005). This substantial increase reveals an improvement of performance greater than the simple doubling of operators, emphasizing the effectiveness of multiple operators.

A further investigation into the way UGVs were utilized in this study showed that single operators had a great deal of difficulty using two UGVs in an independent manner. In fact, the most often used strategy was deploying the UGVs in a 'leapfrog' type maneuver that took little advantage of the fact that the multiple UGVs could cover a larger area, more quickly, by taking independent routes. In all, this initial study provided information in several of our four major research areas, including an early look at team collaboration, performance, route planning, and target detection.

From here, we began to investigate the individual components on a more detailed level. Fincannon et al. (2006) explored target identification in a study designed to investigate the effects of both training (familiarity) of targets and the conspicuity of targets on identification. As expected, training improved target identification significantly; but, interestingly, the conspicuity component, that is the amount a target is obscured by terrain, affected both familiar and unfamiliar targets equally, making them significantly more difficult to be identified. This is an important result given that many modern opposing forces thrive on their ability to hide within the terrain, be it a forest, or jungle, or urban landscape, and we must find a way to overcome this advantage.

From a situational awareness (SA) perspective, TPL sought to investigate how various levels of automation would affect an operator's wayfinding ability and capacity to maintain SA. Rehfeld (2006) found that operators utilizing medium levels of automation (no tele-operation, but directing a UGV via waypoints) showed the highest levels of localization and SA performance and experienced only moderate levels of workload as a result, when compared to low (complete tele-operation) and high (operator performing solely as supervisor) levels of automation.

Hoeft et al. (2005) investigated the role of shared mental models in team collaboration. Results of this study showed that teams of operators who believed they had a mismatch in their mental models communicated more explicitly with one another, while teams believing they had similar mental models performed more implicitly, trying to anticipate one another's needs. The interesting part of this finding is that these results were based upon operators' perceptions of sharedness and not the actual levels by which their mental models overlapped. This finding indicates an importance for operators to feel that their teammates share their goals and ideas in order to maintain high levels of team coordination and collaboration.

Recent Research

A closer look into team collaboration took place with a study designed to allow operators to share information, images, and control of one another's robotic assets. Fincannon et al. (2008a) conducted a study involving the use of dissimilar robotic assets (in this case a UGV and UAV) to complete a reconnaissance task. This study involved three different collaboration types: information sharing via chat communication only; information sharing with chat and image sharing; and information sharing including chat, image, and assets control sharing. Preliminary results from this study show that the teams with the highest performance levels come from the second condition, chat communication with image sharing only, indicating that transferring control of an asset may be too workload intensive to be feasible at this time. In fact further investigation showed that those operators going from UAV to UGV control had a major degradation in performance and large increases in perceived workload. This could be due to the increased visual complexity associated with a UGV's "work environment," but may simply be that shifting from one control viewpoint to another is just too much for an operator to deal with.

Current Research

Current research focuses on expanding the team collaboration concept by using two- vs. three-person teams that coordinate using different modalities (i.e. instant messaging vs. audio). The third member on teams served as the information officer, often referred to as a Robotech, filling a leadership role, keeping the team on track, and recording information about targets acquired and identified. Preliminary analyses indicated that the addition of the Robotech does not increase target identification performance, but instead increases performance associated with making decisions from the targets that were identified during a mission. This means that the Robotech's true benefit can be seen downstream in the process, aiding in decision making and resource allocation based on information gathered. Furthermore, usage of an audio modality for communication appeared to increase the ability to locate targets that change location during a search task. Being able to quickly elaborate on movement and the ability to control unnecessary communication were key components in this task, as opposed to more cumbersome communication techniques such as chat communication.

Future Directions

In moving forward, researchers at TPL hope to expand the task types being performed to include more than just reconnaissance missions. Military uses for robotic assets will include a wide range of mission types, from logistical solutions to march-to-contact assault missions, and it is important that our research at TPL reflects all of these possible mission tasks in our research. In addition, future research will explore greater and varying degrees of automation and computer assistance provided by robotic systems, and will be able to be simulated utilizing the Scale MOUT facility's flexible capabilities. One planned example of this is the coordinated effort of robotic assets to provide target identification estimations, based on data gathered from teams of robots, in the simulated environment. This research will not only allow us to continue investigation in the four primary areas described earlier, but also to expand into areas such as trust-in-automation and information data filtering.

Validating Lab Research with Field Study

One criticism that has often plagued laboratory and academic research is a belief that it does not possess a strong enough relationship to actual field performance in "real-world" tasks. When discussing military research, this criticism can often become even more resolute, with some believing that university students could never truly represent soldiers and the skills they possess. However, as the following section will discuss, there are many things that can be learned from one another when discussing the differences and similarities of lab and field studies. In fact, maintaining a strong bond between the two disciplines could prove crucial to the timely success of the U.S. military's HRI research objectives.

In July of 2008, one of the authors of this chapter had the opportunity to get out from behind their desk and experience Army technology field experimentation first hand. Traveling to Ft. Dix, New Jersey, they were able to not only observe new technologies, as those technologies were being readied for soldiers, but were also given the chance to shadow a platoon sergeant on several simulated missions. This allowed the researcher to better understand the dynamics of chain-of-command and how these new technologies will truly affect soldiers of tomorrow.

While some may envision field research simply as a more realistic version of laboratory studies, in most instances there are several key areas of difference between the two techniques, and this particular experiment held true to that model. The following section will outline the experience at Ft. Dix and compare it with the line of research being conducted at TPL mentioned earlier in this chapter. Provided are some observations about the field study and what gaps may exist between field and laboratory research in the human-robot interaction field.

In addition, this section will discuss on a more direct level the usefulness of field research to validate laboratory studies, and in turn how this validation can help support and enhance work in the field.

One of the purposes of this research was to counter the criticism that lab research has often had a lack of connection with real-world activities. That is, lab studies are rarely validated with field study counterparts. This is an area where the void between lab and field research can grow unnecessarily and should be viewed with a concerned eye.

During the visit to Ft. Dix, data were able to be gathered which got at the heart of one of the most frequent criticisms of military lab research conducted with university students: college students are not similar to soldiers. To investigate this issue soldiers participating in the field study were asked to complete a series of spatial ability assessments identical to those given to the university student participants at the lab. The purpose of this was to validate the student participant pool as being similar to the soldier population on this task-relevant skill.

Scores on the Vandenberg Test of Mental Rotation (MR) and the Guilford Zimmerman Test of Spatial Orientation (SO) were compared for military personnel from Ft. Dix and undergraduates from UCF. For the military personnel from Ft. Dix, scores for the MR test ranged from 8 to 14 (Mean = 11.750, standard deviation, SD, = 2.872), and scores for the SO test ranged from -3 to 37.5 (Mean = 26.375, SD = 19.616). For the UCF students, scores for the MR test ranged from -2 to 20 (Mean = 9.452, SD = 4.878), and scores for the SO test ranged from -2 to 41 (Mean = 18.004, SD = 10.269). While the military personnel had higher mean scores for each test, these scores were either within or under the range of scores from UCF students, and a 1-tailed t-test hypothesizing higher scores by the military personnel was insignificant for both MR, $t(64) = .928$, p>.05, and SO, $t(64) = 1.490$, p>.05. This can be seen illustrated in Figure 22.5.

As seen in these results, the student participant pool is, in fact, quite similar to the soldier population. Even with an admittedly small sample size on the part of the soldiers, the insignificant statistical results provide an excellent platform on which to base this assumption. Due to the soldiers' wider range of scores, it is probable that with an increased sample size, mean scores would only become closer between the two groups. This is good news for both lab and field researchers. In the lab this means that we can explore some of the underlying principles of HRI with confidence that results will translate into usable information for the military. From a field researcher perspective, this allows the freedom to utilize laboratory studies as leverage for ongoing experiments, and means that focus can be put on using actual technologies rather than being concerned with background principles of HRI (i.e. screen resolution, team processes, communication strategies, etc.).

Overall, this result bridges an important gap between the lab and the field. Realizing that participant populations are similar in terms of individual differences allows for some generalization in results, which may not be as clear without some hard evidence to support such an idea.

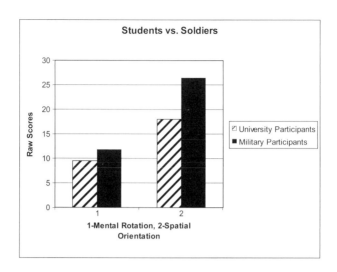

Figure 22.5   Results of soldier and student spatial ability assessments

Lessons Learned from Lab and Field Studies

The authors, and the lab in which they work, have conducted a number of studies within the subject area of human-robot teams. This work has focused primarily on the use of unmanned vehicles in a military setting, however much of this work has practical applications in other domains. As discussed earlier, previous investigations have sought to determine an optimal operator to asset ratio; examine performance issues that arise given different levels of reliability or autonomy; and explore different techniques for improving position localization when operating remote systems. More recent efforts have sought to improve team performance given different communication mediums, as well as the type of information that can be exchanged between team members.

These laboratory investigations have all been conducted using the Team Performance Lab's Scale MOUT facility (Ososky et al., 2007). The facility is built around scale model representations of a prototypical Iraqi region and a prototypical eastern European city. The facility uses radio-controlled vehicles and wireless cameras to simulate the unmanned robot systems. Operators in each of two physically separate rooms interact with the vehicles, and each another, via computer-based interfaces. These two rooms are purposely contrasting in their layouts: the "C4I station" room is designed as a secure, remote operations outpost, while the virtual foxhole is intended to simulate a dismounted forward control position.

Through our experiences in conducting studies, managing technology, and working with participants at the Scale MOUT facility we have been able to compile a list of lessons learned that we believe will be useful to those conducting similar research. These lessons, and by extension this document, also serve a dual purpose of information dissemination, in an effort to make laboratory research more congruent with field research and vice versa. The following will address those lessons in each of four primary areas, beginning with issues of fidelity, moving through participant training and technology management, and concluding with the simulation of intelligence.

Environment Fidelity

One of the main reasons why the Scale MOUT environment was physically constructed, as opposed to computer generated, is that it provides a higher sense of consequence within operators, when running into buildings or people. For example, if an operator drives into the side of a building with an unmanned ground vehicle, the building will sustain damage. The operator also risks damaging the vehicle itself, or even rendering the vehicle inoperable for the remainder of the mission. Similar to a field study, operators are motivated to exercise restraint and caution. This might not always be the case when they operate in a virtual simulation, where participants (typically undergraduate students) realize that there is little consequence to running over virtual people or smashing into a building. Physics of virtual simulations will also vary by implementation; conversely, the physical properties of a live implementation, such as the Scale MOUT facility, will consistently be the same and similar to those experienced in the field.

There are drawbacks, however, to the use of a live, scale simulation. Many aspects of the Scale MOUT facility are static, which may have a negative impact on the fidelity of the simulation. A challenge of a live simulation of this nature is that an implementation of dynamic entities is extremely resource intensive. A full-scale live simulation would require a number of actors to portray friendly and enemy forces, as well as a civilian presence. Vehicles would also have to be maintained and operated to complete the experience. In the specific case of a scale simulation, these needs would have to be met either by a complex mechanical system constructed under the environment, or a team of experimenters who can manipulate those dynamic elements. Even when the resources are available, there is an added challenge of making sure the scenario is replicated consistently across all trials, otherwise the validity of the study will be threatened.

The scenario of needing a large number of resources (i.e. confederate operators) is similar to some of the challenges found in field research, where large numbers of personnel are required to provide a meaningful mission environment. For field research this is common practice and necessary for the realism of the exercise. However, this is not as acceptable in laboratory research, as lab studies are generally intended to include large numbers of participants, operating in a consistently similar environment over a large period of time. Due to this, we continue to investigate new techniques which could help us create an even more realistic research environment, while using fewer resources and remaining highly reliable.

Training Participants

In the laboratory, we are faced with the challenge of training participants to perform tasks with unmanned systems for which those in the field have been extensively trained. Although the performance goals are often simple (navigation or reconnaissance), successful completion of many of these tasks requires a diverse set of skills. These include various intellectual and psychomotor skills, for instance: basic computer operation, joystick control, mental rotation, position localization, leadership and communication, cursory knowledge of military operations, vehicle identification, and friend/foe discrimination. Like many university laboratory studies, our pool of participants typically consists of undergraduate students, whose enthusiasm and skill set vary greatly.

To overcome this challenge we have designed a solution that includes self-paced computer-based training, practice missions with feedback, and reference material for use throughout the study. Although these efforts have been successful in avoiding floor effects in the data, there is still confusion among some participants who do not fully understand all aspects of either the task or the actions required to complete the task. For example, first-time exposure to a significant amount of military terminology (FRAGO, phase lines, no-fly zones, etc.) may distract from participants' ability to retain more relevant portions of task-related information provided in the training. However, training is not always the best solution for acquiring certain skills. For example, it might be better to provide a job aid for operating a joystick (cheat sheet or labels), rather than requiring the user to remember its functions. This will allow them to focus on learning other skills necessary for successful mission execution, such as route planning or navigation. In summary, great care should be given to the design of training and the extent to which training is the best solution for enabling certain skills within participants.

Managing Technology

Although our current efforts use scale model simulation to represent our operating environment, we use a significant amount of technology to facilitate communication between participants, control the vehicles within the environment, and provide streaming video from the vehicles within the environment to the operators. The use of these systems has been highly successful, but we remain aware of those instances when reconfiguration or upgrades are necessary. However, not all shortcomings of technology necessarily have a negative impact on the overall experience of the simulation, as we have discovered through the following example.

Most of the video sources, originating from the unmanned vehicle's (UV) cameras, are delivered in an analog format. Even those sources that reach the participants in a digital format are generated from an analog component. In our specific implementation, wireless cameras, coupled with long lines of a/v cable, generate and distribute video throughout the facility to the dislocated operators (i.e. to the C4I room, virtual foxhole, and confederates). The result is an image that is not always clear or directly coupled with the operator's input. This may seem a detriment to the simulation, but the impact is very much to the contrary. In the field, video from UV systems are subject to interference from various sources, and may also be time-delayed due to the distance or medium over which the data must travel. Therefore, what originated as a design limitation resulted in a more realistic simulation of real-world operating conditions. In fact, many participants have been recorded as saying "It's amazing these robots are so smart," an ode to the fidelity of the facility.

As was mentioned early in the chapter, the capabilities and reliability of unmanned systems in controlled laboratory conditions can often exceed (through design or fabrication) their real-world counterparts, but this may not always be desired, depending upon the purpose of the investigation. Researchers should take great care in determining the fidelity of their equipment so as to not far exceed participants' expectations of a particular piece of equipment and thus reduce believability of the experiment.

Simulating Intelligence

The simulation of unmanned system capabilities, when implemented with discretion, is very useful when either the cost to replicate the actual system is too great, or the technology is simply not fully developed to the point of deployment. Artificial intelligence is one such facet of unmanned systems technology in which simulation is currently more practical than actual implementation. In the laboratory setting, confederate experimenters can operate unmanned vehicles "behind the curtain" in a manner that makes these systems appear to be semi- to fully autonomous. This allows for the possibility of investigating human-agent relationships, in addition to the relationships between multiple human teammates.

The danger in this instance is that experimenters must remain consistent in their behaviors, even more so when reacting to human input. There is also an added challenge in determining how best to signal or provide feedback to human operators with respect to the vehicles' status. The Scale MOUT facility uses a lighting system in combination with text to indicate whether or not the vehicle is acting autonomously, as well as if it has successfully been able to reach a waypoint or destination on a map. It is also recommended that simulated agent behaviors be fully scripted to ensure consistency among experimenters. However, it is difficult to plan a behavior and response for every possible scenario that might arise during the operation of unmanned system, and even more difficult to predict what actual intelligent agents might be capable of in as little as 20 years. This further emphasizes the need for increased communication and information exchange between laboratory and field environments.

Summary

This chapter has attempted to bring together a great deal of information related to the use of scale simulations in HRI research. Included are brief descriptions of a facility designed to simulate autonomous vehicle control; research that has been and is continuing to be conducted in that facility; a validation experiment tied to U.S. Army field research; and some lessons learned by the authors in completing each of these tasks. It is the hope of the authors that this will both aid and encourage other researchers to explore new, less conventional means to study HRI and improve the overall quality of research being done in this field.

To summarize:

- Scale environments are viable and cost-effective alternatives to virtual simulations.
- Fidelity created in scale environments can be perceived as higher than that of virtual simulations.
- Research using this scale environment has yielded results in human-to-robot ratio, situation awareness, target identification, and team communication, to name a few.
- Currently, team human-to-robot ratios of at least N+1:N are most effective, especially when the +1 is able to serve as an information organizer.
- A validation study revealed that, in terms of spatial ability, university students (typical research participants) are comparable to military personnel.
- Participants should not always be burdened with a high level of domain-specific jargon: take special care in the required knowledge and training of participants.

References

Evans, A. W. III, Hoeft, R. M., Rehfeld, S., Feldman, M., Curtis, M., Fincannon, T., Ottlinger, J. & Jentsch, F. (2005). Demonstration: Advancing robotics research through the use of a scale MOUT facility. *Proceedings of the 49th Annual Meeting of the Human Factors and Ergonomics Society*, 742–6.

Fincannon, T., Curtis, M., & Jentsch, F. (2006). Familiarity and expertise in the recognition of vehicles from and an unmanned ground vehicle. *Proceedings of the Human Factors and Ergonomics Society*.

Fincannon, T., Evans, A. W. III, Jentsch, F., & Keebler, J. (2008a). Interactive effects of backup behavior and spatial abilities in the prediction of teammate workload using multiple unmanned vehicles. *Proceedings of the Human Factors and Ergonomics Society 52nd Annual Meeting*. New York.

Fincannon, T., Evans, A. W. III, Jentsch, F., & Keebler, J. (2008b). Interactive effects of spatial ability and shared visual resources in the prediction aerial vehicle backup behavior using multiple unmanned vehicles. *5th Annual Workshop on the Human Factors of Unmanned Aerial Vehicles*. Apache Junction, AZ, 15 May.

Hoeft, R. M., Jentsch, F., & Bowers, C. (2005). The effects of interdependence on team performance in human-robot teams. *Proceedings of the 11th International Conference on Human-Computer Interaction Conference*. Las Vegas, NV, August.

Hoeft, R. M., Jentsch, F., Smith-Jentsch, K. A., & Bowers, C. (2005). Shared mental models and implicit coordination in (human-robot) teams: Convergence, accuracy, or perceptions? *Proceedings of the 49th Annual Human Factors and Ergonomic Society Conference*. Orlando, FL, October.

Keebler, J. R., Harper-Sciarini, M., Curtis, M., Schuster, D., Jentsch, F., & Bell-Carroll, M. (2007). Effects of 2- dimensional and 3-dimensional media exposure training on a tank recognition task. *Proceedings of the Human Factors and Ergonomic Society 51st Annual Meeting*. Baltimore, MD.

Ososky, S., Evans, A. W. III, Keebler, J. R., & Jentsch, F. (2007). Using scale simulation and virtual environments to study human-robot teams. In D. D. Schmorrow, D. M. Nicholson, J. M. Drexler, & L. M. Reeves (eds), *Foundations of Augmented Cognition* (4th edn, pp. 183–9). Arlington, VA: Strategic Analysis, Inc.

Rehfeld, S. (2006). Investigating the mechanisms that drive implicit coordination in teams. Unpublished dissertation. University of Central Florida, Orlando.

Rehfeld, S. A., Jentsch, F. G., Curtis, M., & Fincannon, T. (2005). Collaborative teamwork with unmanned ground vehicles in military missions. *Proceedings of the 11th Annual Human-Computer Interaction International Conference*. Las Vegas, NV, August.

Saavedra, R., Earley, P. C., & Van Dyne, L. (1993). Complex interdependence in task-performing groups. *Journal of Applied Psychology*, 78(1), 62.

PART VI
Future Directions

Chapter 23

The Future of HRI: Alternate Research Trajectories and their Influence on the Future of Unmanned Systems

A. William Evans, III and Florian G. Jentsch
University of Central Florida

The editors of this volume do not begin to pretend that this is an all-inclusive review of human-robot interaction (HRI), in the military or otherwise. This volume primarily focuses on U.S. Army related research and, though the U.S. Air Force is represented, does not take into account the efforts made by other branches of the military within the U.S. As well, civilian research is limited in its representation, and, though depicted with input from Germany, Israel, and the Netherlands, many of the advances made by foreign researchers are not discussed in this volume. It is, however, the wish of the editors that this volume will have an effect and provide some benefit for each of these communities.

What this volume does provide is an insight into where HRI research and robotics research, as a whole, are headed. When the idea of using robots and unmanned systems first began developing for military scientists, the thought was that there would be a single operator able to control and supervise a large number of unmanned systems, not only keeping the human safe, but multiplying the fighting power of that single individual. Indeed, as recently as 2007 the expectation of a single operator controlling a fleet of robot assets was reiterated as a goal over the next 25 years (Department of Defense, 2007). However, in the early stages of this process, researchers quickly came to understand that this goal of a single operator and many robots would not come easily. Many researchers in the first half of this decade found that expecting a single operator to control numerous robot agents was quite an ambitious goal indeed (Burke & Murphy, 2004; Rehfeld et al., 2005). In fact, initially the ratio seemed to be headed in the exact opposite direction, with many human operators needed to control even a single robotic asset. As we have learned more about the relationships between controllers and the controlled, we have been able to bring this ratio closer to a more manageable 1:1.

As we move forward from here it is evident that HRI research can move in two distinct trajectories: either improving the efficiency of the 1:1 ratio interactions and taking existing levels of performance higher, or expanding the role of the human operators to achieve the goal of controlling multiple unmanned assets and multiplying their military force. While not mutually exclusive, most studies fall into one of these two categories, with a few that can contribute to both trajectories. Both of these paths will contain their own challenges and surely contribute their own unique solutions; however, it is important that the end goal of each of these maintains constant at a workable solution to all of the issues that have arisen and will arise in military HRI applications. As well, working along each of these research trajectories in parallel is likely to prove essential in moving human-robot team performance forward in a substantially more efficient and timely manner than if the research were to be performed in serial fashion.

This summary will discuss how the chapters within this volume fit into each of these trajectories, and the influence that each author's contribution could have on the future of unmanned systems within the military. While it is always important to be knowledgeable about where we have been, understanding where we are focusing our future efforts will help us all to achieve our common goals more quickly and effectively.

Chapter Summaries

The following sections are outlined similarly to those found in the introductory chapter of this volume. The same four areas of military HRI research will be presented—fundamentals, unmanned aerial vehicle (UAV) research, unmanned ground vehicle (UGV) research, and cross-platform research—and trajectories pointing to the future of research in each of these areas will be discussed.

Fundamentals

The fundamentals of HRI are seeded in a number of more established areas of research, including visual perception, trust in automation, stress and workload, adaptive strategies, cognition, and team communication, to name a few. The authors contributing to this book touched on several of these areas and provided a deeper understand of how each ties into the modern needs of HRI. In addressing the cognition of HRI, Gillan et al. (Chapter 4) discussed how robots are now much more than tools to increase human capabilities physically. Modern technology has also increased our ability to process information. Robots now serve a number of roles in improving human performance. Not only do they increase our strength, mobility, and sensory abilities, but robots are now charged with helping us to filter, organize, and decipher the information that they help us to acquire. Of course, the rigid nature of machines leaves humans still in front as the most complex and *adaptive* systems. *Wired* magazine recently ran an article comparing the human brain to the 'one machine' (a combination of all the machines we may use in routine life) and determined that the human brain still has roughly twice as many connections (about 100 trillion) between various pieces of information as the 'one machine' does (Kelly, 2008). It is this vast number of connections which allow humans to be efficiently adaptive, drawing on all of their experiences to overcome any current obstacles. The key to a successful relationship between man and machine is anchored in how well we are able to understand each other and utilize each other's strengths while limiting each other's weaknesses. While robotic assets can be built to enhance our abilities and aid in the rapid completion of goals, the human component of HRI is still the most powerful tool in creating a coordinated effort for human-robot teams to succeed.

Pazuchanics et al. (Chapter 6) discussed HRI from another angle. They introduced the challenges of perception that are encountered by operators during robotic asset manipulation. Robotic assets provide operators with a limited viewpoint of the environment, one in which human operators can quickly lose understanding and fail to have an accurate perspective of their surroundings. However to be effective, operators must achieve and maintain a greater understanding of a more global space, which includes areas and objects outside the assets view. Following the research trajectory aimed toward the increase of operator performance, their line of research has successfully navigated the potential pitfalls associated with expanding the role of the human operator in mission space and improving overall effectiveness of human-robot teams. Achieving this global space perspective has been shown to be rather difficult. This difficulty, however, can be overcome by providing operators with information to enhance their understanding of the environment, such as localization and orientation cues. Other authors in this volume discuss just such a solution in tactile displays, which will be touched on later in this chapter.

These chapters, along with the others in Part II of this book, show us that to truly understand HRI relationships we must first understand both man and machine on their own. Humans, while extremely adaptive and highly perceptive, can quickly lose understanding of their surroundings when overwhelmed with stress and workload, and in turn become deprived of a full slate of environmental information. Machines, as powerful as they may be physically and computationally, are very rigid in their form, and thus can fail to adequately conform to their user's needs. Understanding these shortcomings as well as each group's strengths will enable researchers, developers, and engineers to overcome many of the obstacles of HRI that will be faced in the future.

UAV Research

The authors in this book have a wide range of backgrounds and experience working with various types of unmanned systems. UAVs represent the first class of unmanned assets to be employed by the military in regular usage, and will be the first type discussed here. Used primarily for reconnaissance, UAVs were initially deployed either from a great distance and at great altitude (i.e. global hawk) for more gestalt battlefield understanding, or from a relatively close distance and low altitude (i.e. dragon eye) for more immediate reconnaissance, at the squad level. In either case, UAVs would operate as a single entity with a specific role to play, generally completing preplanned routes, while investigating specific target locations.

Researchers are now looking at new ways to operate UAVs, allowing for their control on the squadron level and with more flexibility in mission tasks. Travelling the research trajectory aimed at multiplying forces, by allowing a single operator to control many systems, Schulte and Meitinger's work (Chapter 9) involves just such a control system. They call it *co-operative control*. In this control system the automation is more responsive and receptive to individual operator's needs, adapting the level and type of automation to fit a more cohesive teammate role assisting operators more seamlessly to achieve mission goals.

Wickens, Levinthal, and Rice's (Chapter 11) focus continues more on this advanced issue of increasing the operator to UAV ratio, such that one operator could supervise a number of unmanned systems with great effect. While still in the early stages of this advance, a move from two operators per one UAV is expected to flip to a ratio of 1:2 based on their conclusions. And while a ratio of one operator to every two robotic assets may seem low, consider that even a shift as small as this would *quadruple* effective forces, and the importance of such research becomes instantly clear.

From the brief discussion above about UAV research, it is evident that aerial vehicles benefit from both their environment and a relatively longer research history, allowing for current studies involving UAVs to focus on expanding forces. With limited physical obstacles for operators to be concerned with in UAV airspace, workload can become considerably lower when compared to ground-based counterparts. This reduction in workload contributes to free up more cognitive resources, allowing for the desired force multiplication. Add to this aerial reconnaissance's history dating back beyond World War II (Staerck, 1998; Stanley, 1998), and UAVs are likely to lead the way when it comes to improving the operator-to-asset ratio.

UGV Research

Of the issues presented to UGV researchers, situational awareness has emerged as one of the most critical, given the generally tight quarters in which UGVs are being deployed. Moving back to a trajectory in line with research aimed at increasing performance of existing human-robot teams, Jansen and van Erp (Chapter 14) have explored increasing operator situation awareness through the use of telepresence. Specifically, Jansen and van Erp investigated the ability of operators, in supervisory roles, to step quickly back 'into the loop' and take control when utilizing unmanned systems in complex situations and environments. In structured, simple environments, regaining situation awareness can be achieved with relative ease, but as complexities increase, so does the difficulty to understand the situation. Improvements in this area could help to increase the usable missions for robot assets, helping to increase the performance of the teams overall while reducing the stresses put on the operators themselves. Indeed Jansen and van Erp found that telepresence did provide significant improvements in performance based primarily on operators' ability to still hear and observe as if they were actually present at the location, something not available with traditional teleoperation.

Another major research concern for UGVs has been the usability of operator control units (OCUs). In fact several chapters from this book have focused on this issue, which can have an overarching effect on unmanned system usability regardless of vehicle type, team ratio, or size. Haas and van Erp (Chapter 15) and Redden and Elliott (Chapter 18) investigated the usefulness of adaptive displays on OCUs for UGVs, utilizing multi-modal effects, as well as scalable applications, in an effort to increase performance. It is hoped that the creation of a better OCU will allow soldiers to more easily utilize all of the capabilities of the unmanned systems they look to employ, while reducing the overall workload imposed upon the operators. While initially these goals may serve to increase performance with existing operator–asset ratios, it is easily conceivable that an improved OCU could have effects which allow for a swing in that ratio, thus multiplying effective forces. Tactile display cues have been shown to be effective waypoint navigation tools and could help to spread the use of cognitive resources to free up more cognitive facilities for visual tasks, which seem to be the most burdened.

Cross-Platform Research

Many of the issues associated with unmanned systems are not tied to any one type of platform. Teaming, coordination, asset control, and visual performance issues are all present in the varying platforms associated with unmanned systems. Cooke and Chadwick (Chapter 19) addressed many of these issues in their lessons learned work. With a substantial background in HRI (Chadwick, 2005, 2006; Cooke et al., 2006; Chadwick & Gillan, 2006), Cooke and Chadwick discussed many of the recurring issues (i.e. limitations in visual performance) that affected their research. Visual performance, or more accurately the current limitations of visual performance among operators, has had significant effects on team performance in human-robot teams (Cooke & Chadwick, Chapter 19; Jentsch et al, Chapter 22), whose outcomes can be felt in virtually every type of unmanned vehicle. Improvements in this area of performance should yield far-reaching performance increases in all areas of human-robot team tasks.

With a more direct look at expanding the role of operators to handle the use of multiple unmanned assets, both Lewis and Wang (Chapter 21) and Goodrich (Chapter 20) have contributed chapters detailing their work at improving the operator–asset ratio beyond 1:1. Lewis and Wang found that the substantial time required of an operator to coordinate various assets should be taken into consideration when constructing the operator's overall duties within the team. With this substantial workload, it would be difficult at best for operators to become involved in other tasks while supervising all of their assets, depending on team and task complexity. Goodrich's work estimates that, to a large degree, fan-out models of coordination, allowing for increased operator-to-asset ratios, will be dependent on the autonomy of the vehicles in question. In fact, Goodrich suggests that adaptive automation, much like the co-operative control scheme discussed by Schulte and Meitinger earlier in this chapter, would prove to be a powerful tool in creating effective fleets of unmanned systems, controlled by only a few human operators.

The Future of HRI in the Military

In the above referenced materials, many successes and failures in research have been discussed, along with the strengths and weakness of various research tools. It is important to discuss findings such as this so that future researchers can avoid the mistakes of others and help the research community to better understand why they may have occurred. As a part of the scientific community it is important to share all of our findings, both good and bad, and welcome new data with open minds so that we might collaborate and coordinate our future efforts in a way that will more steadily advance all of the goals we wish to achieve.

How to integrate multiple humans and multiple robots into efficient and effective teams is a complex challenge which requires a number of complex answers to fully understand all of the dynamic issues involved. Previous chapters within this volume have given some insight into the various research areas which, when combined, can provide scientists with a greater overall understanding of the intricacies of HRI. Even though when viewed individually many of these research areas can seem rather tangential toward one another, it is important that all of the research in this domain maintains a similar trajectory, such that a gestalt understanding of HRI may be achieved.

As research continues to bring human and robot teammates closer together, the work of some such as Saavedra et al. (1993) can add to the discussion. Using their structures for team interdependency, HRI has moved from an initial pooled interdependency, with all of the information simply being blindly supplied by the unmanned systems, to the sequential relationships that exist today, where operators provide assets with information, who in turn generate responses to the inputs. Through progressive research, human-robot teams will soon reach the reciprocal interdependency that Saavedra et al. believe to be at the peak of team coordination.

The premise of maintaining a similar research path throughout the HRI domain may at first seem rudimentary and simple. After all, scientists in this area do all have the same goal of improving performance using robots; however, it is in the details that we can vary significantly from our intended route. It is precisely because of this issue that books, articles, papers, and presentations such as this volume are so important in sharing our thoughts, ideas, theories, and findings: to allow others to not only build off what we have already learned, but also to use previous research as a guide and a blueprint that can direct us into the future of HRI. This is not to say that every study in the HRI domain should be run in the same manner, but rather that it is imperative that we all understand *why* each experiment was conducted the way it was, so as to improve on our own ideas.

References

Burke, J. & Murphy, R. (2004). Human-robot interaction in USAR technical search: Two heads are better than one. *13th IEEE International Workshop on Robot and Human Interactive Comuunication*, 13, 307–12.

Chadwick, R. (2005). The impacts of multiple robots and display views: An urban search and rescue simulation. *Proceedings of the 49th Annual Meeting of the Human Factors and Ergonomics Society*. Santa Monica, CA: Human Factors and Ergonomics Society. 387–91.

Chadwick, R. (2006). Operating multiple semi-autonomous robots: Monitoring, responding, and detecting. *Proceedings of the 50th Annual Meeting of the Human Factors and Ergonomics Society*. Santa Monica, CA: Human Factors and Ergonomics Society. 329–33.

Chadwick, R. & Gillan, D. (2006). *Strategies for the Interpretive Integration of Ground and Air Views in UGV Operations*. Poster session presented at the 25th Army Science Conference, Orlando, FL.

Cooke, N. J., Pringle, H., Pedersen, H. K, & Connor, O. (eds) (2006). *Human Factors of Remotely Operated Vehicles*. Amsterdam: Elsevier.

Department of Defense (2007). *Unmanned Systems Roadmap: 2007–2032* (OMB No. 0704-0188). Washington, DC.

Kelly, K. (2008). Infoporn: Tap into the 12-million-teraflop handheld megacomputer. *Wired*, 16(7).

Rehfeld, S. A., Jentsch, F. G., Curtis, M., & Fincannon, T. (2005). Collaborative teamwork with unmanned ground vehicles in military missions. *Proceedings of the 11th Annual Human-Computer Interaction International Conference*, Las Vegas, NV, August.

Saavedra, R., Earley, P. C., & Van Dyne, L. (1993). Complex interdependence in task-performing groups. *Journal of Applied Psychology*, 78(1), 61–72.

Staerck, C. (ed.) (1998). *Allied Photo Reconnaissance of World War II*. San Diego, CA: Thunder Bay Press.

Stanley, R. M. (1998). *To Fool a Glass Eye: Camouflage versus Photoreconnaissance in World War II*. Washington, DC: Smithsonian Institution Press.

Index